Security, Strategy and Military Change in the 21st Century

This edited volume explores and analyses strategic thinking, military reform and adaptation in an era of Asian growth, European austerity and US rebalancing.

A significant shift in policy, strategy and military affairs is underway in both Asia and Europe, with the former gaining increasing prominence in the domain of global security. At the same time, the world's powers are now faced with an array of diverse challenges. The resurgence of great power politics in both Europe and Asia, along with the long-term threats of terrorism, piracy and sustained geopolitical instability, has placed great strain on militaries and security institutions operating with constrained budgets and wary public support.

The volume covers a wide range of case studies, including the transformation of China's military in the 21st century, the internal and external challenges facing India, Russia's military modernization program and the USA's reassessment of its strategic interests. In doing so, this book provides the reader with the opportunity to conceptualize how strategic thinking, military reform, operational adaptation and technological integration have interacted with the challenges outlined above. With contributions by leading scholars and practitioners from Europe and Asia, this book provides a valuable contribution to the understanding of strategic and operational thinking and adjustment across the world.

This book will be of much interest to students of military and strategic studies, security studies, defence studies, Asian politics, Russian politics, US foreign policy and IR in general.

Jo Inge Bekkevold is Head of the Center for Asian Security Studies at the Norwegian Institute for Defense Studies, Oslo.

Ian Bowers is Assistant Professor at the Center for Asian Security Studies at the Norwegian Institute for Defense Studies, Oslo. He has a PhD in War Studies from King's College, London.

Michael Raska is a Research Fellow at S. Rajaratnam School of International Studies, Singapore, and has a PhD from the National University of Singapore.

Cass Military Studies

Intelligence Activities in Ancient Rome
Trust in the gods, but verify
Rose Mary Sheldon

Clausewitz and African War
Politics and strategy in Liberia and Somalia
Isabelle Duyvesteyn

Strategy and Politics in the Middle East, 1954–60
Defending the northern tier
Michael Cohen

The Cuban Intervention in Angola, 1965–1991
From Che Guevara to Cuito Cuanavale
Edward George

Military Leadership in the British Civil Wars, 1642–1651
'The genius of this age'
Stanley Carpenter

Israel's Reprisal Policy, 1953–1956
The dynamics of military retaliation
Ze'ev Drory

Bosnia and Herzegovina in the Second World War
Enver Redzic

Leaders in War
West Point remembers the 1991 Gulf War
Edited by Frederick Kagan and Christian Kubik

Khedive Ismail's Army
John Dunn

Yugoslav Military Industry 1918–1991
Amadeo Watkins

Corporal Hitler and the Great War 1914–1918
The list regiment
John Williams

Rostóv in the Russian Civil War, 1917–1920
The key to victory
Brian Murphy

The Tet Effect, Intelligence and the Public Perception of War
Jake Blood

The US Military Profession into the 21st Century
War, peace and politics
Edited by Sam C. Sarkesian and Robert E. Connor, Jr.

Civil–Military Relations in Europe
Learning from crisis and institutional change
Edited by Hans Born, Marina Caparini, Karl Haltiner and Jürgen Kuhlmann

Strategic Culture and Ways of War
Lawrence Sondhaus

Military Unionism in the Post Cold War Era
A future reality?
Edited by Richard Bartle and Lindy Heinecken

Warriors and Politicians
U.S. civil–military relations under stress
Charles A. Stevenson

Military Honour and the Conduct of War
From Ancient Greece to Iraq
Paul Robinson

Military Industry and Regional Defense Policy
India, Iraq and Israel
Timothy D. Hoyt

Managing Defence in a Democracy
Edited by Laura R. Cleary and Teri McConville

Gender and the Military
Women in the armed forces of Western democracies
Helena Carreiras

Social Sciences and the Military
An interdisciplinary overview
Edited by Giuseppe Caforio

Cultural Diversity in the Armed Forces
An international comparison
Edited by Joseph Soeters and Jan van der Meulen

Railways and the Russo-Japanese War
Transporting war
Felix Patrikeeff and Harold Shukman

War and Media Operations
The US military and the press from Vietnam to Iraq
Thomas Rid

Ancient China on Postmodern War
Enduring ideas from the Chinese strategic tradition
Thomas Kane

Special Forces, Terrorism and Strategy
Warfare by other means
Alasdair Finlan

Imperial Defence, 1856–1956
The old world order
Greg Kennedy

Civil–Military Cooperation in Post-Conflict Operations
Emerging theory and practice
Christopher Ankersen

Military Advising and Assistance
From mercenaries to privatization, 1815–2007
Donald Stoker

Private Military and Security Companies
Ethics, policies and civil–military relations
Edited by Andrew Alexandra, Deane-Peter Baker and Marina Caparini

Military Cooperation in Multinational Peace Operations
Managing cultural diversity and crisis response
Edited by Joseph Soeters and Philippe Manigart

The Military and Domestic Politics
A concordance theory of civil–military relations
Rebecca L. Schiff

Conscription in the Napoleonic Era
A revolution in military affairs?
Edited by Donald Stoker, Frederick C. Schneid and Harold D. Blanton

Modernity, the Media and the Military
The creation of national mythologies on the Western Front 1914–1918
John F. Williams

American Soldiers in Iraq
McSoldiers or innovative professionals?
Morten Ender

Complex Peace Operations and Civil–Military Relations
Winning the peace
Robert Egnell

Strategy and the American War of Independence
A global approach
Edited by Donald Stoker, Kenneth J. Hagan and Michael T. McMaster

Managing Military Organisations
Theory and practice
Joseph Soeters, Paul C. van Fenema and Robert Beeres (eds)

Modern War and the Utility of Force
Challenges, methods and strategy
Edited by Jan Angstrom and Isabelle Duyvesteyn

Democratic Citizenship and War
Edited by Yoav Peled, Noah Lewin-Epstein and Guy Mundlak

Military Integration after Civil Wars
Multiethnic armies, identity and post-conflict reconstruction
Florence Gaub

Military Ethics and Virtues
An interdisciplinary approach for the 21st century
Peter Olsthoorn

The Counter-Insurgency Myth
The British experience of irregular warfare
Andrew Mumford

Europe, Strategy and Armed Forces
Towards military convergence
Sven Biscop and Jo Coelmont

Managing Diversity in the Military
The value of inclusion in a culture of uniformity
Edited by Daniel P. McDonald and Kizzy M. Parks

The US Military
A basic introduction
Judith Hicks Stiehm

Democratic Civil–Military Relations
Soldiering in 21st-century Europe
Edited by Sabine Mannitz

Contemporary Military Innovation
Between anticipation and adaptation
Edited by Dmitry (Dima) Adamsky and Kjell Inge Bjerga

Militarism and International Relations
Political economy, security and theory
Edited by Anna Stavrianakis and Jan Selby

Qualitative Methods in Military Studies
Research experiences and challenges
Edited by Helena Carreiras and Celso Castro

Educating America's Military
Joan Johnson-Freese

Military Health Care
From pre-deployment to post-separation
Jomana Amara and Ann M. Hendricks

Contemporary Military Culture and Strategic Studies
US and UK armed forces in the 21st century
Alastair Finlan

Understanding Military Doctrine
A multidisciplinary approach
Harald Hoiback

Military Strategy as Public Discourse
America's war in Afghanistan
Tadd Sholtis

Military Intervention, Stabilisation and Peace
The search for stability
Christian Dennys

International Military Operations in the 21st Century
Global trends and the future of intervention
Edited by Per M. Norheim-Martinsen and Tore Nyhamar

Security, Strategy and Military Change in the 21st Century
Cross-regional perspectives
Edited by Jo Inge Bekkevold, Ian Bowers and Michael Raska

Military Families and War in the 21st Century
Comparative perspectives
Edited by René Moelker, Manon Andres, Gary Bowen and Philippe Manigart

Private Security Companies during the Iraq War
Military performance and the use of deadly force
Scott Fitzsimmons

Security, Strategy and Military Change in the 21st Century
Cross-regional perspectives

Edited by Jo Inge Bekkevold, Ian Bowers and Michael Raska

LONDON AND NEW YORK

First published 2015
by Routledge
2 Park Square, Milton Park, Abingdon, Oxon OX14 4RN

and by Routledge
711 Third Avenue, New York, NY 10017

Routledge is an imprint of the Taylor & Francis Group, an informa business

© 2015 selection and editorial matter, Jo Inge Bekkevold, Ian Bowers and Michael Raska; individual chapters, the contributors

The right of the editor to be identified as the author of the editorial matter, and of the authors for their individual chapters, has been asserted in accordance with sections 77 and 78 of the Copyright, Designs and Patents Act 1988.

All rights reserved. No part of this book may be reprinted or reproduced or utilized in any form or by any electronic, mechanical, or other means, now known or hereafter invented, including photocopying and recording, or in any information storage or retrieval system, without permission in writing from the publishers.

Trademark notice: Product or corporate names may be trademarks or registered trademarks, and are used only for identification and explanation without intent to infringe.

British Library Cataloguing-in-Publication Data
A catalogue record for this book is available from the British Library

Library of Congress Cataloging-in-Publication Data
Security, strategy and military change in the 21st century : cross-regional perspectives / edited by Jo Inge Bekkevold, Ian Bowers and Michael Raska.
    pages cm. – (Cass military studies)
    Includes bibliographical references and index.
    1. National security–Case studies. 2. Armed Forces–Reorganization–Case studies. 3. Military policy–Case studies. 4. Strategy. 5. Security, International. I. Bekkevold, Jo Inge, editor of compilation. II. Bowers, Ian, editor of compilation. III. Raska, Michael, editor of compilation.
    UA10.5.S37356 2015
    355'.033–dc23   2014045470

ISBN: 978-1-138-83265-7 (hbk)
ISBN: 978-1-315-73589-4 (ebk)

Typeset in Baskerville
by Wearset Ltd, Boldon, Tyne and Wear

Contents

List of figures   xi
List of tables   xii
Notes on contributors   xiii
Acknowledgements   xvi
List of abbreviations   xvii

1 **Introduction**   1
   JO INGE BEKKEVOLD, IAN BOWERS AND
   MICHAEL RASKA

PART I
Military change in Asia   13

2 **Managing military change in China**   15
   DENNIS J. BLASKO

3 **Military change in Japan: National Defense Program Guidelines as a main tool of management**   36
   ISAO MIYAOKA

4 **Garuda rising? Indonesia's arduous process of military change**   55
   BENJAMIN SCHREER

5 **The management of military change: the case of the Singapore Armed Forces**   70
   BERNARD FOOK WENG LOO

6 **The sources of military change in India: an analysis of evolving strategies and doctrines towards Pakistan**   89
   S. KALYANARAMAN

7 The Indian Army adapting to change: the case of
counter-insurgency 115
VIVEK CHADHA

PART II
Military change in Europe 133

8 Perspectives on military change and transformation in
Europe 135
SVEN BERNHARD GAREIS

9 Managing military change in Russia 155
KATARZYNA ZYSK

10 Military change in Britain and Germany in a time of
austerity: meeting the challenge of cross-national pooling
and sharing 178
TOM DYSON

11 Austerity is the new normal: the case of Danish defence
reform 197
MIKKEL VEDBY RASMUSSEN

PART III
Military change in the United States and NATO 219

12 Military change in NATO: the CJTF concept – a case study
of military innovation in a multinational environment 221
PAAL SIGURD HILDE

13 A new wary titan: US defence policy in an era of military
change, Asian growth and European austerity 241
AUSTIN LONG

14 Conclusion: security, strategy and military change in the
21st century 266
JO INGE BEKKEVOLD, IAN BOWERS AND
MICHAEL RASKA

Index 274

Figures

3.1 Impression of the SDF   49
8.1 Decreasing force levels of important EU member states   141
8.2 Military expenditures in per cent/GDP of important EU
     countries and EU average   142
11.1 Danish defence expenditure, 1990 to 2012 (constant
     prices)   198
11.2 Ratio between total defence expenditure (constant 2010
     US$) and armed forces strength   199
11.3 Index 100 of defence expenditure in per cent of BNP vs.
     social expenditure in per cent of BNP   205
11.4 Cost of Danish capabilities (in millions of DKK)   210
11.5 Costing military models (in millions of DKK)   213

Tables

3.1	NDPG comparison table	43
3.2	NDPG comparison table	46
4.1	Indonesia defence spending, 2003 to 2013	60
5.1	Singapore defence spending, 1991 to 2010	78
8.1	Important multinational military structures in Europe	143
8.2	Military operations of the EU	145

Contributors

Jo Inge Bekkevold is Head of the Centre for Asian Security Studies at the Norwegian Institute for Defence Studies. His main research areas are Chinese foreign and security policies, China–Russia relations, naval developments in Asia, and Asia in the Arctic. He is a former career diplomat.

Dennis J. Blasko, Lieutenant Colonel, U.S. Army (Retired), served 23 years as a Military Intelligence Officer and Foreign Area Officer specializing in China. He was an army attaché in Beijing and Hong Kong from 1992 to 1996. He served in infantry units in Germany, Italy, and Korea and in Washington at the Defense Intelligence Agency, Headquarters Department of the Army (Office of Special Operations), and the National Defense University War Gaming and Simulation Center. He has written numerous articles and chapters on the Chinese military and is the author of *The Chinese Army Today: Tradition and Transformation for the 21st Century, second edition* (Routledge, 2012).

Ian Bowers is an Assistant Professor at the Norwegian Institute for Defence Studies. He received his PhD in War Studies from King's College London. His research focuses on Asian naval modernization and Northeast Asian security.

Vivek Chadha, Colonel (Retired) is a Research Fellow with the Institute for Defence Studies and Analyses IDSA in New Delhi. Prior to joining IDSA, he served in the Indian Army for 22 years. His research covers counter-insurgency, counter-terrorism and terrorist finance.

Tom Dyson is Senior Lecturer in International Relations at Royal Holloway, University of London. He was an Alexander von Humboldt Research Fellow based at the Chair of German and European Politics at Potsdam University and the Stiftung Wissenschaft und Politik in Berlin, Germany. He is the author of several books and journal articles on European defence policies, among others *Neoclassical Realism and Defence Reform in post-Cold War Europe* (Palgrave, 2010).

Sven Bernhard Gareis is Deputy Dean at George C. Marshall European Center for Security Studies. Since 2007 he has been teaching as a Professor of Political Science at the Westfaelische Wilhelms-Universitaet in Muenster/North Rhine-Westphalia. His research interests include the United Nations, German and European foreign and security policy and the foreign relations of the People's Republic of China. He is the author of *The United Nations: An Introduction* (Palgrave, 2012).

Paal S. Hilde is Associate Professor at the Norwegian Institute for Defence Studies. He received his DPhil in Politics in 2003 at the University of Oxford (St. Antony's College). His research focuses on Baltic security, NATO and the Arctic. He is the co-editor of *The Future of NATO: Regional Defense and Global Security* (University of Michigan Press, 2014).

S. Kalyanaraman is a Research Fellow with the Institute for Defence Studies and Analyses in New Delhi. His main research focus is on India's foreign and security policies.

Austin Long is an Assistant Professor at the School of International and Public Affairs and a Member of the Arnold A. Saltzman Institute of War and Peace Studies and the Harriman Institute for Russian, Eurasian, and East European Studies at Columbia University. He is also a non-resident Senior Fellow at the Foreign Policy Research Institute.

Bernard Fook Weng Loo is Associate Professor and Coordinator of the Master of Science (Strategic Studies) Degree Programme at the S. Rajaratnam School of International Studies (RSIS) in Singapore. He received his PhD from the University of Wales, Aberystwyth. He is the author of *Middle Powers and Accidental Wars: A Study in Conventional Strategic Stability* (Edwin Mellen, 2005) and the editor of *Transformation and Military Operations* (Routledge, 2009).

Isao Miyaoka is Professor in International Politics, Department of Political Science, Faculty of Law, Keio University. He received his DPhil in Politics in 1999 at the University of Oxford (St. Antony's College). He is the author of *Legitimacy in International Society: Japan's Reaction to Global Wildlife Preservation* (Palgrave Macmillan: 2004). His current research focuses on Japanese and American security policies as well as international relations theory.

Michael Raska is a Research Fellow with the Military Transformations Programme at the Rajaratnam School of International Studies, Nanyang Technology University. He received his PhD at the Lee Kuan Yew School of Public Policy at the National University of Singapore. His research interests focus on East Asian security and defense. He is the author of *Military Innovation in Small States: Creating a Reverse Asymmetry* (Routledge, 2015).

Mikkel Vedby Rasmussen is Professor with the Department of Political Science, University of Copenhagen. He is currently on leave from the University, working as Head of the Secretariat for Defence and Security Policy at the Danish Ministry of Defence and Head of the Centre for Military Studies at the same department. Among his publications are *The Risk Society at War: Terror, Technology and Strategy in the Twenty-First Century* (Cambridge University Press, 2006) and *The West, Civil Society and the Construction of Peace* (Palgrave Macmillan, 2003).

Benjamin Schreer is a Senior Analyst for Defence Strategy at the Australian Strategic Policy Institute (ASPI). Previously, he was deputy head of the Strategic and Defence Studies Centre (SDSC) at the Australian National University (ANU). He also worked as a research fellow at the Stiftung Wissenschaft und Politik (SWP) in Berlin. His current research interests include strategic trends in the Asia-Pacific region and nuclear dynamics in Asia.

Katarzyna Zysk is Associate Professor at the Norwegian Institute for Defence Studies. She also serves as a non-resident Research Fellow at the US Naval War College and is a member of the Arctic Security Initiative at the Hoover Institute, Stanford University. Her current research interests focus on international security studies, including Russian security and defence policy, military innovation, Arctic geopolitics, and security in transatlantic relations.

Acknowledgements

In early December 2013 academics from three continents met in Oslo to discuss global perspectives on military change. This engaging and often surprising conference led to the realization that military change, its drivers and influences differ across the world and was worthy of further examination. As a result this volume was conceived of as a way to distil some of the conclusions reached during the conference.

First, the editors would like to thank the Norwegian Embassy in New Delhi, India. Without their funding support this volume would not have been possible. A number of people have also provided invaluable advice and assistance. The staff at the Norwegian Institute for Defence Studies who participated in the conference and made invaluable contributions to the discussion are worthy of special mention. Krishnappa Venkatshamy, formerly of the Institute for Defence Studies and Analyses, New Delhi, provided great assistance during the formative stage of both the conference and this volume. Wrenn Yennie Lindgren was invaluable as the volume's proof-reader. Finally, Andrew Humphrys and Hannah Ferguson from Routledge deserve special mention for shepherding this volume through the publication process.

Abbreviations

3G SAF	Third Generation SAF
A2/AD	Anti-Access/Area Denial
ACE	Allied Command Europe
ACLANT	Allied Command Atlantic
ADIZ	Air Defence Identification Zone
AMS	Academy of Military Science (China)
ARRC	Allied Command Europe Rapid Reaction Corps
ASBM	Anti-Ship Ballistic Missile
ASDF	Air Self-Defence Force
ASEAN	Association of Southeast Asian Nations
BICES	Battlefield Information Collection and Exploitation System
C2	Command and Control
C3I	Command, Control, Communications and Intelligence
C4ISTAR	Command, Control, Communications, Computer, Intelligence, Surveillance, Targeting Acquisition and Reconnaissance
CBS	Corps Battle Schools
CCP	Chinese Communist Party
CENTRIX	CENTCOM Regional Intelligence Exchange System
CDU	Christian Democratic Union (Germany)
CI	Counter-insurgency
CIA	Central Intelligence Agency
CJTF	Combined Joint Task Force
CMC	Central Military Commission (China)
COSTIND	Commission of Science, Technology, and Industry for National Defence
COTS	Commercial off the Shelf
CSDF	Common Security and Foreign Policy
CSDP	Common Security and Defence Policy (EU)
CSE	Communications Security Establishment (Canada)
CSU	Christian Social Union (Germany)
DANCON	Danish Contingent
DCIS	Deployable Communication and Information System

DJS	Deployable Joint Staff
DJSE	Deployable Joint Staff Elements
DJTF	Deployable Joint Task Force
DSCO	Doctrine for Sub Conventional Operations
DSD	Defence Signals Directorate (Australia)
DTIB	Defence Technological and Industrial Base
EDA	European Defence Agency
EEZ	Exclusive Economic Zone
EME	Electronics and Mechanical Engineers
ESDI	European Security and Defence Identity
ESDP	European Security and Defence Policy
ESS	European Security Strategy
EUMC	European Union Military Committee
EUMS	European Union Military Staff
FDP	Free Democratic Party (Germany)
GAD	General Armaments Department (China)
GCHQ	Government Communication Headquarters (UK)
GCSB	Government Communications Security Bureau (New Zealand)
GLD	General Logistics Department (China)
GMTI	Ground Moving Target Indication
GPD	General Political Department (China)
GSD	General Staff Department (China)
GSDF	Ground Self-Defence Force (Japan)
ICBM	Intercontinental Ballistic Missile
IED	Improvised Explosive Device
IRBM	Intermediate Range Ballistic Missile
IP	Internet Protocol
ISAF	International Security Assistance Force
ISR	Intelligence, Surveillance and Reconnaissance
J&K	Jammu and Kashmir
JSF	Joint Strike Fighter
JSOC	Joint Special Operations Command (US)
KODAM	Military Area Commands (Indonesia)
LPD	Landing Platform Dock
LDP	Liberal Democratic Party (Japan)
LoC	Line of Control
MBT	Main Battle Tank
MC	Military Committee
MEF	Minimum Essential Force
MFP	Major Force Program
MJO	Major Joint Operations
MLF	Multinational Land Force
MND	Ministry of National Defence (China)
MNF	Mizo National Front

MPV	Mine-protected Vehicle
MR	Military Region (China)
MRBM	Medium Range Ballistic Missile
MRO	Military Representative Offices (China)
MSDF	Maritime Self-Defence Force
NAC	North Atlantic Council
NATO	North Atlantic Treaty Organisation
NCCT	Network Centric Collaborative Targeting
NCS	NATO Command Structure
NDPG	National Defense Program Guidelines
NDMC	National Defence Mobilization Committee (China)
NDU	National Defence University (China)
NPC	National People's Congress
NRF	NATO Response Force
NSA	National Security Agency
NSC	National Security Commission (China)
NSC	National Security Council (Japan)
NSHQ	NATO Special Operations Headquarters
NSS	National Security Strategy (Japan)
NUDT	National University of Defense Technology (China)
NWFP	North West Frontier Province
ODA	Official Development Assistance
OOTW	Operations Other Than War
PAP	People's Armed Police
PCG	Policy Coordination Group
PFP	Partnership for Peace
PKO	Peacekeeping Operations
PLA	People's Liberation Army
PSC	Policy and Security Committee
RAP	Readiness Action Plan
RMA	Revolution in Military Affairs
RR	Rashtriya Rifles
SACEUR	Supreme Allied Commander Europe
SACLANT	Supreme Allied Commander Atlantic
SAF	Singapore Armed Forces
SASTIND	State Administration of Science, Technology and Industry for National Defence
SDF	Self-Defence Forces (Japan)
SDSR	Strategic Defence and Security Review
SHAPE	Supreme Headquarters Allied Powers Europe
SIGINT	Signals Intelligence
SLBM	Submarine-launched Ballistic Missile
SLOC	Sea Lines of Communications
SOCOM	Special Operations Command (US)
SOF	Special Operations Forces

SPD	Social Democratic Party (Germany)
SRBM	Short Range Ballistic Missile
SSBN	Ballistic Missile Submarine
TNI	Tentara Nasional Indonesia
TNI-AD	Tentara Nasional Indonesia – Angkatan Darat (Indonesian Army)
TNI-AL	Tentara Nasional Indonesia – Angkatan Laut (Indonesian Navy)
TNI-AU	Tentara Nasional Indonesia – Angkatan Udara (Indonesian Air Force)
UAV	Unmanned Aerial Vehicles
ULFA	United Liberation Front of Assam
USAF	United States Air Force
VJTF	Very High Readiness Joint Task Force
WEU	Western European Union
WHAM	Winning Hearts and Minds

1 Introduction

Jo Inge Bekkevold, Ian Bowers and Michael Raska

As the 21st century reaches adolescence, established orders are being challenged across the world. On a global level an economic and political power shift is occurring as China's rise and Asia's overall economic strength places the region at the heart of world affairs. While the US is rebalancing to Asia in response to China's potential threat to the established order, Europe is still dealing with the aftermath of the great recession and the impact of austerity. The actions of Russia in annexing Crimea and redrawing the European map have placed great pressure on European nations and NATO to engage more closely with the region's defence. But the battle between welfare and warfare continues to be waged in capitals across the Western world.

These geopolitical and economic challenges are occurring in a global environment of rapidly changing technologies and ever-increasing interconnectivity. Transnational threats and non-traditional security challenges such as terrorism and piracy are not new phenomena but pose long-term if not existential difficulties for state actors and institutions.

What is happening is not merely a shift in geographic priorities but a change from the continental to the maritime, from the asymmetric threat of insurgency to the considerations of great power politics and an alteration in capitals across the globe in how threats are perceived and should be confronted. Seen together, the developments and challenges described here amount to a fundamental shift in international security and the post-Cold War order.

This volume aims to address these shifts through examining cross-regional military change and its management. Why military change? This phrase provides the editors, contributors and ultimately the readers with the opportunity to conceptualize how strategic thinking, military reform, operational adaptation and technological integration have interacted with the challenges outlined above.

A new security landscape in the making

As the world entered the 21st century, US military predominance was assured by a defence budget more than ten times larger than its greatest potential competitor, China. By 2013 the disparity had been closed significantly with China's military expenditure being estimated as just under one-third of that of the US.[1] While a gap remains, China's regional focus allows it to narrow its endeavours while the US is spread across the world, dealing with threats in Asia, the Middle East, Africa and Europe.

China's new military muscle, its growing assertiveness in pursuing its international goals, and its perceived disregard for the established rules-based status quo is increasing regional tension and endangering the stability of the world's economic powerhouses.[2] This dynamic has been further fuelled by a complex mix of historical animosity, economic competition and ever more fraught territorial disputes.

In 2012 military budgets in Asia surpassed those of NATO European states for the first time. The five biggest arms importers between 2008 and 2012 were all in Asia.[3] A shift has occurred where the forefront of large-scale modern military procurement has moved from Europe to Asia. Importantly, the maritime system which sustains Asia's economy has now become the focal point of security interaction; sea power has replaced land power as the dominant reflection of national strength and prowess. The potential for a naval arms race exists, although currently competitive procurement is not quite in evidence, despite naval modernization continuing apace.[4] Thus the most peaceful region since the end of the Cold War has become the crucible for security in competition in the 21st century, complete with shifting threat perceptions, developing alliances and the integration of new and more powerful military technologies.

India currently finds itself on the edge of this security dynamic, but its size, location and ambition mean that it will play an increasingly important role in Asian and global security. However, it faces an extraordinarily complex array of threats. Pakistan alone ensures that India has to prepare for the combined threat of nuclear weapons (Pakistan went nuclear in 1998), conventional warfare (The Kargil War in 1999), insurgency (Kashmir), and terrorism (the Mumbai attacks in 2008).

At the same time China's rise and growing political, economic and military strength is both a concern and an opportunity for New Delhi. China has shown flexibility in solving most of its land border disputes but not yet with India, while the Indian Ocean Region (IOR) is an area where maritime ambitions could clash with dangerous consequences. The visit of Xi Jinping to India in September 2014 highlighted the potential for both competition and cooperation in the relationship. Alleged Chinese encroachment in the disputed border region coincided with agreements for Chinese economic investment in India. Prime Minister Modi has now made serious future economic cooperation contingent upon the settling of the dispute.[5]

Furthermore, India also plays an important role in peacekeeping and contributes to international military operations such as anti-piracy and disaster relief operations. India has, since independence, pursued a non-aligned foreign policy, and even though it has been argued that India has 'crossed the Rubicon' in its foreign policy,[6] and signed a nuclear deal with the US in 2005,[7] it wants to remain an independent actor in foreign and security policy.

Asia is not alone in having to deal with seismic shifts in its security environment. In Europe a perfect storm has been washing over the region's militaries. Economic difficulties resulting in austerity in combination with a lack of an identifiable enemy or strategic purpose and ambivalent populations weary of costly foreign operations has resulted in years of negative inertia.[8]

The 2012 reduction in real defence spending in Europe was 1.63 per cent, which came on top of a 2.52 per cent decline in 2011. In 2012, real defence spending fell in 60 per cent of European states.[9] NATO still maintains its aspiration for member states to spend at least 2 per cent of their GDP on defence, a goal met by few. Initiatives like 'smart defence', and pooling and sharing within NATO and the EU, may lead to some rationalization. However, many European states seem to act according to national imperatives, and capability reductions are largely uncoordinated, with significant implications for combat capacity.[10]

It may be argued that such a reduction in defence spending makes sense for countries finding themselves in a relatively safe security environment. However, former NATO Secretary-General Anders Fogh Rasmussen stated in *Foreign Affairs* in 2011 that 'Libya is a reminder of how important it is for NATO to be ready, capable, and willing to act'.[11] But the operation over Libya revealed gaps in key capabilities for NATO Europe.[12] These gaps were made further evident by operations over Iraq and Syria in 2014 where European contributions have been small and in some cases were hindered by a lack of capability. Further, sharp reductions in European spending on defence may have implications for NATO credibility.

Russian actions in Crimea and the Ukraine are a stark reminder of the realities of geo-politics and the potential need for core European competencies to be maintained and modernized. Russia is in the midst of a military modernization programme which has been justified through increasingly nationalist and irredentist rhetoric. NATO and the West are at the centre of a Russian view of the world where the status quo is an imposed constraint. The willingness to use force to redress this perceived imbalance is a significant threat to stability in Europe; one which requires NATO to take concrete measures to combat. The 2014 NATO summit in Wales saw the creation of the Very High Readiness Joint Task Force (VJTF) as part of a package of responses to the Russian actions. However, the new NATO Secretary-General, Norwegian Jens Stoltenberg, will be faced with the challenge of split expectations between Eastern and Western European member states in maintaining a consistent response to Russia.

Before premonitions of a new Cold War take hold,[13] it must be noted that Russia remains a relatively weak country. While its energy exports give it a high degree of leverage over the gas-hungry states of Europe, its military, despite its large size and being bolstered by a substantial nuclear deterrent, lags behind its peers in terms of technology and training. Its economy, while being the ninth largest in terms of GDP in 2013, is dwarfed by the major economies of Western Europe. Russia accounts for 5 per cent of the world's total defence spending. While this is individually larger than any single European nation, when combined, the UK, France and Germany spending accounts for almost twice that amount.[14] Europe and NATO need to find a way to leverage their superior strength to prevent further Russian revisionism while operating within the confines of austerity and public suspicion.

The US is reassessing its strategic interests after over a decade of war and a focus on COIN. In the midst of overcoming its own economic problems and subsequent reductions in defence spending the world's largest military power is rebalancing towards Asia by increasing its diplomatic, economic and military assets in the region. This is, however, being tested by the continued proliferation of traditional and non-traditional threats across the globe. The actions of ISIS in Syria and Iraq and renewed tensions in Europe highlight the difficulty the US faces in balancing its global commitments.

In 2012, the US accounted for just under half of global defence spending (45.3%) and still outstrips that of the next 14 countries combined.[15] The US has been forced, however, to reduce its defence budget[16] and withdraw resources from Europe,[17] choosing instead to refocus its strategic outlook and military capabilities on Asia.[18] The level to which the US can continue to commit to such a wide variety of threats in such diverse geographic locations remains to be seen. Budgetary pressures and potential political disengagement call the future of US as a global enforcer of norms into question. Strategic shifts are occurring all around, and US engagement or otherwise will have a large impact upon the direction such shifts take.

Alongside this changing geo-political landscape and a renewed emphasis on great power politics, nations are facing an increasing array of non-traditional or new security challenges. Thus militaries are potentially being pulled towards a broader and broader spectrum of operations, many of which require international coordination and collaboration.

The proliferation of weapons of mass destruction (WMD), piracy and terrorism have been an ever-present feature in defence thinking around the globe. These threats require a broad range of responses, many of which are multilateral rather than state-centric and place further pressures on already stretched budgets. While piracy and terrorism cannot strictly be considered new threats, extant platforms and capabilities are not always the most efficient at countering them. As more state and non-state actors invest in developing cyber capabilities, the conventional use of force is increasingly intertwined with confrontations in and out of cyber space, cyber-attacks on physical systems and processes controlling critical information infrastructure, as well as various forms of cyber espionage. The increasing complexity of cyber threats means that the distinctions between civil and military domains, state and non-state actors, principal targets and weapons become gradually blurred. How militaries balance their capabilities and operations to deal with both the traditional and the non-traditional is a key concern. At the same time, international cooperation in tackling the converging 'hybrid' threat spectrum is required.

Military change and its management

This volume aims to understand how militaries and nations are adapting to the external and internal instabilities outlined above. In order to achieve this, military change and its management is defined as the ongoing process of adapting, changing and improving military capabilities to handle and manage threats and risks. The success and failure of this process is defined by an intervening set of variables such as institutional and political capacity, leadership skills, technology, and economics.

In other words, managing military change is linked to *strategic, organizational, and operational control and adaptability* – not only in detecting new sources of military innovation, but more importantly, changing military posture quickly and easily over time in response to shifts in geo-strategic environment, military technology, the realities of cost, performance, and organizational behaviour, and national priorities.[19]

How change is managed is reflected in the choices or judgements the actors in question in the case studies of this volume make and as such it is best revealed in the strategies, procurement, organization and operational means that a military employs.

We have already argued that the world is seeing significant changes in international security which, when taken together, will have major consequences for the established post-Cold War order. As a result of the changes in the security environment in Asia, Europe and the US, are we witnessing a major military change, or are the countries in question 'muddling through' in a slower evolutionary process of military change? This study proposes to examine this phenomenon by addressing both external and internal variables utilizing the following criteria:

- How do the geo-political changes and emerging new security landscape affect each country (if at all) with regard to threat perception and strategic thinking?
- How well do political, strategic elites and military institutions manage to translate threat perception and strategic thinking into doctrines, operational plans and procurement?
- What are the pathways, enablers and constraints of military change in each country, and what is the institutional capacity to manage military change?

To answer these questions it is important to calibrate our conceptions of military change. The literature on military change has traditionally portrayed the term largely in the context of the debate on what constitutes major military revolutions (MRs) and revolutions in military affairs (RMAs). According to Williamson Murray and MacGregor Knox, for example, 'military revolutions recast society and the state as well as military organizations. They alter the capacity of states to create and project military power.'[20] In other words, military revolutions reflect a disruptive change at the grand-strategic level that transcends the operational military-technological domain. In contrast, within or alongside the cataclysmic military revolutions are lesser RMAs characterized by Murray and Knox as

> periods of innovation in which armed forces develop novel concepts involving changes in doctrine, tactics, procedures, and technology ... RMAs [also] take place almost exclusively at the operational level of war. They rarely affect the strategic level, except in so far as operational success can determine the large strategic equation. RMAs always occur within the context of politics and strategy – and that context is everything.[21]

Similarly, Theo Farrell and Terry Terriff, distinguish major military change or 'change in the [organizational] goals, actual strategies, and/or structure of a military organization' and minor change or 'changes in operational means and methods (technologies and tactics) that have no implications for organizational strategy or structure'.[22] More recently, Michael Horowitz equated major military innovations as 'major changes in the conduct of warfare, relevant to leading military organizations, designed to increase the efficiency with which capabilities are converted to power'.[23] Dima Adamsky also focused on disruptive military innovation through the lens of military-technical revolutions (MTRs) or RMAs, when 'new organizational structures together with novel force deployment methods, usually but not always driven by new technologies, change the conduct of warfare'.[24]

The perennial question of what constitutes military revolutions, revolutions in military affairs and military-technical revolution has shaped a significant path in the contemporary strategic studies debate, with each term emphasizing the disruptive or 'revolutionary' character.[25] However, in a historical perspective, most military changes and innovations have arguably followed a distinctly less than revolutionary or transformational path, consisting of incremental, often near-continuous, improvements in existing ways and means of war.[26] In other words, while major, large-scale and simultaneous military innovations in military technologies, organizations and doctrines have been a rare phenomenon, military organizations have progressed through *a sustained spectrum of military innovation* ranging from a small-scale to a large-scale innovation that has shaped the conduct of warfare.

More importantly, the academic debate has focused predominantly on *what, why and when* the MRs and RMAs occur, with only limited insights on *how* military change diffuses, how it is managed, adopted and adapted over time. In our perspective, therefore, we attempt to conceptualize the term military change as a dynamic yet gradual (not necessarily revolutionary) process of policy change at three interrelated dimensions: *strategic change* focusing on changes in the global and regional security environment, and concomitant changes in defence strategy, and operational conduct; *defence management* embedded in the transformation of ideas and knowledge into new or improved products, processes, and services for military and dual-use applications; and *military innovation*, encompassing 'both product innovation and process innovation, technological, operational, and organizational innovation, whether separately or in combination to enhance the military's ability to prepare for, fight, and win wars'.[27]

In particular, policy-level changes may be defined as national-level changes and fall under the realm of grand strategy. Grand strategy is the highest level of direction in which military power and strategy is linked with political, economic, demographic, and other national resources to form a coherent direction for the employment of state power.[28] Thus an analysis of grand strategy would include the capacity of a state's civil and military leadership to adjust its threat assessment and grand strategic direction, to formulate and update national security strategies according to changes in a country's domestic and external security environment and to manage civil–military relations during this period of transformation.

Strategy bridges the policy and institutional levels. Identifying a single definition of strategy is a complex task given the diffusion of meanings that now surround the term. As Freedman notes, 'strategic discourse has now moved beyond its etymological roots in the arts of generals'.[29] In this volume, we use a classical definition of strategy in that it is the link between military means and political ends. The 2012 US Department of Defense Strategic Guidance – Sustaining U.S. Global Leadership: Priorities for the 21st Century Defense, for the first time signed by a US president, would fall under the definition of strategy with implications for military change. In this context, strategy is a tool with which to utilize military power and is dependent on choice and the conditions both internal and external which frame it.

On the operational level doctrine is the central element. Defining doctrine is also complex, as each country and organization utilizes and understands what doctrine is in diffuse ways. We see doctrine as a guide and, by extension, a tool of change for how to conduct strategy and direct armed forces at an operational level. Optimally it reflects a state's capabilities – in terms of technology, manpower and various other factors – its ethos, culture, training and ultimately the external and internal environments in which a military operates.[30]

However, to focus on strategy and doctrine in an analysis of military change has certain caveats. A doctrine in itself does not ensure institutional change, and may be 'developed as much for political as for strategic or operational reasons.'[32] This reinforces the need to understand how doctrine is reflected in operational success on the ground.

These three overarching levels of change may be observed in Japan's recent approach to its security. Japan is, in response to the challenge of China and other internal and external pressures, looking to adjust its constitution and by extension its policy-level posture while developing a new security strategy (enunciated in the NDPG 2014) and retooling its forces and adjusting its doctrine to meet both the challenges set by the new strategy and to take advantage of the opportunities provided by rapidly advancing technologies.

The structure of the volume

In looking to answer these questions this volume has selected major and medium military powers in both Asia and Europe. The eclectic mix of states, including the United States and Russia, is designed to highlight the diversity of how military change is managed on policy or grand strategy, strategy and doctrinal levels. While some states have pursued and managed military change on all three levels, others have been more restrictive in their efforts to deal with their security environment.

Part I examines military change in Asia. In a region of rising powers, US allies and sleeping giants, the chapters highlight the diversity of responses to military change. In Chapter 2, Dennis Blasko analyses the main drivers, enablers and constraints of China's rapid military modernization. He elicits the success or otherwise of ongoing military change projects and emphasizes the primacy of the political leadership in setting military modernization policy. In Chapter 3, Isao Miyaoka looks at how Japan's defence force is redefining its role in accordance with the changing regional security environment. By focusing on the National Defense Program Guidelines (NDPG) he highlights how the Japanese leadership has managed military change amidst varying levels of opposition from a reluctant populace and the seemingly intractable problems of the Japanese economy. In Chapter 4, Benjamin Schreer addresses how the sleeping strategic giant in Southeast Asia, Indonesia, is attempting to realize its renewed ambition for military change and the modernization of its doctrine, training and equipment. In Chapter 5, Bernard Loo discusses the case of the Singapore Armed Forces, the strategic context in which Singapore finds itself, and the challenges for Singapore as a small state with a conscript force embracing costly, high-technology weaponry.

Two chapters on India conclude Part I. Facing an array of threats, India makes an excellent case study for students of military change. Its operational and structural adaptations have to be assessed alongside economic growth but ever-present budgetary pressures. The extent to which India can cope with and adjust to future challenges will be a significant factor in regional and global security. In having to deal with both conventional and non-conventional threats since its founding, India is the only nation in Asia with a sustained history of combat operations. It is a unique case in that it has had to apply lessons learned from operations while continuing to deal with extant threats. This provides an important differentiation between it, Japan and China. The two chapters address the challenges of China, Pakistan and insurgency, reminding the reader of the sheer complexity of the security challenges facing India. In Chapter 6, S. Kalyanaraman provides an assessment of the changing Indian strategies towards conflict with China and Pakistan, placing the posture of Pakistan as the key determining factor in the Indian military's various approaches. In Chapter 7, Vivek Chadha examines how the Indian Army has adapted to change in the face of challenges emerging from insurgencies and terrorism. His analysis is conducted across three major drivers of change: doctrinal, organizational and operational.

Part II addresses military change in Europe. The chapters demonstrate how austerity and disinterested publics have challenged militaries to justify their existence through the construction of contributory expeditionary forces. The Russian annexation of Crimea and the instability in the Ukraine highlights the complexity of the current strategic situation and difficulties Europe faces in attempting to manage such a situation. In Chapter 8, Sven Bernhard Gareis paints the bigger picture of military change in Europe. He argues that in facing a number of security challenges European nations will have to align their military and security policies, advocating the eventual formation of a European army to meet the region's security needs.

In Chapter 9, Katarzyna Zysk examines the motives, drivers and enablers of Russian military reforms and modernization after 2008. She looks at Russia's understanding of the geo-strategic environment and the leadership's willingness to use force to alter perceived disadvantages in the status quo. In Chapter 10, Tom Dyson examines the ability of the UK and Germany to translate the imperative of closer defence cooperation through the EU's Common Security and Defence Policy (CSDP) into actual policy change. Denmark is a unique case as they have approached military change in an almost corporate fashion. In Chapter 11, Mikkel Vedby Rasmussen describes how defence cuts set the agenda for redefining the mission and force structure of the Danish armed forces. He argues that the armed forces are thus no longer a fighting force that can deliver independent, joint operations but are now structured to provide key combat capabilities to larger allies such as the UK and the United States.

Part III addresses military change in the United States and NATO. In Chapter 12, Paal Hilde enlightens us on how military change is managed within an alliance, using NATO and the Combined Joint Task Force Concept (CJFT) as his case. In 2011, after 17 years of troubled existence, the NATO Military Committee unceremoniously dropped the concept.

Hilde also looks at possible lessons for NATO's future military change efforts. In Chapter 13, Austin Long outlines the current state of US defence policy in the context of US domestic politics and economics, examines the substantial military change that has taken place in the United States during the past decade, and discusses the effect of Asian growth and European austerity on current and future US defence policy. The volume ends with some concluding observations about military change.

Notes

1 These figures (in constant 2011 US$ figures) are based on the SIPRI military expenditure database, http://milexdata.sipri.org/ (accessed 1 October 2014).
2 Ashley J. Tellis and Travis Tanner (eds.), 'China's Military Challenge', *Strategic Asia 2012–13*, The National Bureau of Asian Research, Seattle and Washington, DC; John J. Mearsheimer, *The Tragedy of Great Power Politics (updated edition)* (New York: W.W. Norton, 2014); Robert S. Ross, 'Balance of Power Politics and the Rise of China: Accommodation and Balancing in East Asia', *Security Studies* 15, 3 (2006); Robert D. Kaplan, *Asia's Cauldron: The South China Sea and the End of a Stable Pacific* (New York: Random House, 2014); Bjørn Grønning, 'Japan's Shifting Military Priorities: Counterbalancing China's Rise', *Asian Security* 10, 1, 2014.
3 'Trends in International Arms Transfers, 2012', *SIPRI Fact Sheet*, March 2013, http://books.sipri.org/files/FS/SIPRIFS1303.pdf (accessed 20 March 2014).
4 Geoffrey Till, *Asia's Naval Expansion: An Arms Race in the Making* (London: Routledge, 2012).
5 Gardiner Harris, 'India Takes Tough Stance with China on Kashmir', *New York Times*, 18 September 2014, www.nytimes.com/2014/09/19/world/asia/modi-pushes-xi-to-resolve-border-issue-in-kashmir.html?_r=0 (accessed 3 October 2014).
6 C. Raja Mohan, *Crossing the Rubicon: The Shaping of India's New Foreign Policy* (New York: Palgrave Macmillan, 2003).
7 Harsh V. Pant, *Contemporary Debates in Indian Foreign and Security Policy: India Negotiates its Rise in the International System* (New York: Palgrave Macmillan, 2008); Rajiv Sikri, *Challenge and Strategy: Rethinking India's Foreign Policy* (New Delhi: Sage, 2009).
8 As an example, German military readiness was exposed as being lacking in a series of media reports in September and October 2014. See Frank Jordans, 'Germany Unable to Meet NATO Readiness Target', Associated Press, 29 September 2014, www.abcnews.go.com/International/wireStory/general-germany-defense-spending-25829475 (accessed 7 October 2014).
9 The International Institute for Strategic Studies, *The Military Balance 2013*, Routledge, p. 6, www.iiss.org/publications/military-balance/ (accessed 15 April 2013).
10 The International Institute for Strategic Studies, *The Military Balance 2013*, Routledge, p. 6, www.iiss.org/publications/military-balance/ (accessed 15 April 2013); Terry Terriff, Frans Osinga and Theo Farrell, eds, *A Transformation Gap? American Innovations and European Military Change* (Palo Alto, CA: Stanford University Press, 2010).
11 Anders Fogh Rasmussen, 'NATO After Libya: The Atlantic Alliance in Austere Times', *Foreign Affairs*, July/August 2011, www.foreignaffairs.com/articles/67915/anders-fogh-rasmussen/nato-after-libya (accessed 23 April 2013).

12 United Kingdom House of Lords, European Union Committee, 'European Defence Capabilities: Lessons from the Past, Signposts for the Future' (2012), www.publications.parliament.uk/pa/ld201012/ldselect/ldeucom/292/29202.htm (accessed 23 April 2013).
13 Simon Shuster, 'Cold War II The West is Losing Putin's Dangerous Game', *Time Magazine*, 4 August 2014; Robert Legvold, 'Managing the New Cold War', *Foreign Affairs* 93, 4 (2014).
14 'The Share of World Military Expenditure of the 15 States with the Highest Expenditure in 2013', SIPRI, www.sipri.org/research/armaments/milex/milex-graphs-for-data-launch-2014/The-share-of-world-military-expenditure-of-the-15-states-with-the-highest-expenditure-in-2013.png (accessed 3 October 2014).
15 The International Institute for Strategic Studies, *The Military Balance 2013*, Routledge, p. 6, www.iiss.org/publications/military-balance/ (accessed 3 October 2014).
16 Karen Parrish, 'Hagel Presents Defense Budget Request to Congress', *American Forces Press Service*, 11 April 2013, www.defense.gov/news/newsarticle.aspx?id=119755 (accessed 15 April 2013).
17 Admiral James Stavridis, United States Navy Commander, United States European Command, *Testimony Delivered Before the 113th Congress*, 2013, www.eucom.mil/mission/background/posture-statement (accessed 10 April 2013).
18 'The U.S. Defence Rebalance to Asia', speech delivered by Deputy Secretary of Defense Ashton B. Carter at the Center for Strategic and International Studies, Washington, DC, 8 April 2013; *U.S. Force Posture Strategy in the Asia Pacific Region: An Independent Assessment*, Center for Strategic and International Studies, 2012; The White House, *Fact Sheet: The Fiscal Year 2014 Federal Budget and the Asia-Pacific*, www.whitehouse.gov/sites/default/files/docs/asia_pacific_rebalance_factsheet_20130412.pdf (accessed 10 October 2014).
19 Paul Davis, 'Defense Planning in an Era of Uncertainty: East Asian Issues'. In Natalie Crawford and Chung-in Moon, eds, *Emerging Threats, Force Structures, and the Role of Air Power in Korea* (Santa Monica, CA: RAND, 2000), pp. 25–47.
20 MacGregor Knox and Williamson Murray, eds, *The Dynamics of Military Revolution 1300–2050* (Cambridge: Cambridge University Press, 2001), p. 7.
21 Ibid., pp. 12, 180.
22 Theo Farrell and Terry Terriff, eds, *The Sources of Military Change: Culture, Politics, Technology* (London: Lynne Rienner Publishers, 2002), p. 5.
23 Michael Horowitz, *The Diffusion of Military Power: Causes and Consequences for International Politics* (Princeton, NJ: Princeton University Press, 2010), p. 22.
24 Dima Adamsky, *The Culture of Military Innovation: The Impact of Cultural Factors on the Revolution in Military Affairs in Russia, the US, and Israel* (Palo Alto, CA: Stanford University Press, 2010), p. 1.
25 Colin Gray, *Strategy for Chaos: Revolutions in Military Affairs and the Evidence of History* (London: Frank Cass, 2002).
26 Andrew Ross, 'On Military Innovation: Toward an Analytical Framework', *IGCC Policy Brief* no. 1 (2010): 14–17.
27 Tai Ming Cheung, Thomas Mahnken and Andrew Ross, 'Frameworks for Analyzing Chinese Defense and Military Innovation'. In Tai Ming Cheung, ed., *Forging China's Military Might: A New Framework for Assessing Innovation* (Baltimore, MD: The Johns Hopkins University Press, 2013), p. 18.
28 This use of the term 'grand strategy' is derived from a number of works. See: Hal Brands, *What Good is Grand Strategy? Power and Purpose in America's Statecraft from Harry Truman to George W. Bush* (Ithaca, NY: Cornell University Press, 2014); Richard N. Rosecrance and Arthur A. Stein, *The Domestic Bases of Grand Strategy* (Ithaca, NY: Cornell University Press, 1993); Barry R. Posen and Andrew

L. Ross, 'Competing Visions for U.S. Grand Strategy', *International Security* 21, 3 (winter 1996/1997), pp. 5–53.
29 Lawrence Freedman, 'Introduction', *The Adelphi Papers*, 45:379, 5-10, 9.
30 Lawrence Freedman, *Strategy: A History* (New York: Oxford University Press, 2013); John Stone, *Military Strategy: The Politics and Techniques of War* (London: Continuum, 2011); Hew Strachan, 'The Lost Meaning of Strategy'. In Thomas Mahnken and Joeseph A. Maiolo, eds, *Strategic Studies: A Reader* (London: Routledge, 2008), pp. 421–436; Barry Posen, *The Sources of Military Doctrine: France, Britain and Germany between the World Wars* (Ithaca, NY: Cornell University Press, 1984).
31 Harald Høiback, *Understanding Military Doctrine: A Multi-disciplinary Approach* (London: Taylor & Francis, 2013), pp. 22–23.
32 Farrell and Terriff, 2002, p. 5.

Part I
Military change in Asia

2 Managing military change in China

Dennis J. Blasko

Since the beginning of the reform and opening period in December 1978, military modernization in China has been included as one element of China's larger national development strategy. Military modernization was listed last among the 'Four Modernizations', following agriculture, industry, and science and technology. Although China faces multiple security threats and challenges, its civilian and military leaders believe the country faces no imminent, existential external threat. Therefore military budgets need not be excessive and are growing in coordination with economic development. China sees military modernization as a long-term, multi-faceted programme with 2049 set as the target year for completion. By that date, China seeks to build a smaller, more technologically advanced force commensurate with China's international standing. This chapter examines how political, economic, international, social and, of course, military dimensions are guiding military change in China. First, we need to take a closer look at China's threat perception, and how this influences China's defence posture.

China's threat perception and defence posture

The subordination of military modernization to the primary national objective of economic development has been the most fundamental guiding principle for managing military change in China. This principle was reinforced by the demise of the Soviet Union and continues to this day. Beijing's strategic priorities were justified by the strategic assessment announced in 1985 that China no longer faced the threat of an 'early, major, and nuclear war', but rather it was more likely to fight 'local, limited war' in the future. Without an imminent existential external threat, Chinese leaders could focus first on improving the nation's domestic conditions and not feel compelled to invest vast resources in defence. China's strategy of great economic reform with minimal political reform but considerable loosening of social restrictions contrasted with the path taken by the Soviet Union. In further contrast to the Soviet Union, which exhausted its military in Afghanistan, China has not fought a major military campaign since its brief war with Vietnam in 1979.

In the late 1990s, the pace of military modernization increased as China's economy grew and the threat of Taiwan independence heightened. The central government decided it could provide more funding to the military without jeopardizing economic growth. US arms sales and support to Taiwan and the mistaken bombing of the Chinese Embassy in Belgrade in May 1999 added impetus to the speed of modernization. Nonetheless, military modernization remained a multi-decade, multi-generational project in line with China's national development strategy. In 2006, the Chinese government announced a 'three-step development strategy' for military modernization, with milestones in 2010 and 2020 and a completion date of 2049, which parallels the programme to achieve economic modernization.[1]

Currently, the Party leadership defines 'terrorist, separatist and extremist forces' as 'on the rise' and threatening China's sovereignty and territorial integrity.[2] It also sees 'hostile Western forces' as implementing a strategy of 'Westernizing and dividing' China, seeking to 'contain' its national development.[3] The Chinese government has not defined the United States as a strategic enemy, but it alludes to it as a factor that has complicated China's security environment. Japan also has not been identified as an enemy, but rather as 'making trouble'.

For at least 15 years, however, the People's Liberation Army (PLA) has considered the military capabilities of the US (and its allies) as the 'gold standard' for advanced military operations. At the operational and tactical levels, the PLA's doctrine seeks to devise methods to counter US capabilities and evolving operational concepts. The current political threat assessment has not altered Beijing's overall view of the likelihood of a major war, and its interest in improving its high-technology military capabilities has resulted in giving priority to the development of the Navy, Air Force and the Second Artillery.[4,5]

Beijing perceives the primary challenges to its territorial integrity and sovereignty coming from the East and South, in particular its conflicting claims with multiple countries over rocks and islands in the East and South China Seas, underscoring the continuing relevance of this priority listing of development. The government claimed in its 2013 White Paper on the Diversified Employment of China's Armed Forces that China is a 'major maritime country as well as a land country' and that it is essential to build China into a maritime power.[6] To deal with this threat perception the PLA has placed greater emphasis on the joint employment of sea, air and missile forces than 30 or more years ago when the primary threat was perceived to be from the North. Nonetheless, as a 'land country' bordering 14 countries, the development of the PLA Army is still important and the Army remains the largest of all the services.

In November 2013 the Third Plenary Session of the eighteenth CCP Central Committee decided to 'optimize the size and structure of the army, adjust and improve the proportion between various troops, and reduce non-combat institutions and personnel', adding 'that joint operation command authority under the [Central Military Commission], and theater joint operation command system, will be improved'.[7] At the time of writing, further details have not been announced officially.

At its simplest, China's military modernization programme aims to build a smaller, more technologically advanced military capable of fighting local wars mostly along its periphery while gradually expanding its operational areas into the western Pacific and beyond. It seeks to build capabilities 'commensurate with China's international standing' and to 'meet the needs of its security and development interests'.[8]

The PLA maintains a strategically defensive posture, but acknowledges that offensive operations are essential for victory. Its military strategy of 'active defence' is often described as 'We will not attack unless we are attacked; if we are attacked, we will certainly counter-attack'. Doctrinally, the PLA considers the deterrence of war as of equal importance to warfighting, stating, 'The more powerful the warfighting capability, the more effective the deterrence.' While emphasizing caution in initiating combat, PLA doctrine allows for pre-emptive attacks if the enemy is perceived to have already fired the 'first shot on the plane of politics and strategy'.[9]

Without specifically using the terms, Defence Minister Chang Wanquan emphasized the principles of active defence and deterrence in a press conference with the U.S. Secretary of Defense in April 2014. When asked if the PLA was training to conduct a short, sharp war with Japan in the East China Sea, General Chang responded:

> If you came to the conclusion that China is going to resort to force against Japan, that is wrong. I would like to reiterate to that, on our Chinese side, we will not take the initiative to stir up troubles. Second, we are not afraid of any provocation.[10]

The reference to China not taking the initiative to use force reflects its strategy of active defence while the statement that the PLA is 'not afraid of provocation' signals that China has the military capabilities to respond to what it perceives as aggression against its sovereignty or territorial claims. The Minister's intention was to display a level of confidence sufficient to deter future military escalation of the situation.

Political components of military modernization

The 'absolute leadership' of the CCP is the most fundamental management principle for the armed forces, making it a Party-Army. Party leadership is exercised through the CCP's Central Military Commission (CMC), led by Party General Secretary Xi Jinping, who is CMC Chairman as well as President of the PRC. This dictum goes back to Mao Zedong's 1938 essay, 'Problems of War and Strategy', which states, 'Every Communist must grasp the truth, "Political power grows out of the barrel of a gun." Our principle is that the Party commands the gun, and the gun must never be allowed to command the Party.'

The Party's senior-most decision-making body is the Central Committee's Politburo Standing Committee. Xi and six other senior civilians on the Standing Committee manage national-level issues through a system of 'leading' or 'small' groups that span the various functions including, but not limited to, reform, finance and the economy, state security, foreign affairs, relations with Taiwan, internet security, ideology and propaganda, and party building.[11] These informal groups act with other elements of the Party, government ministries and organizations, and the National People's Congress to formulate and implement policies based on the collective leadership decisions of the Standing Committee. The National People's Congress announces national budget decisions annually around its session in March. On the other hand, the CMC is a formal party and state organization, led by the senior party and government leader.

Since 1999 when the Party has groomed a successor to the general secretary/chairman of the CMC, it has made a second civilian (first Hu Jintao and then Xi Jinping in 2010) the senior vice-chairman of the CMC. The PLA's two most senior officers serve as vice-chairmen with the Minister of Defense and seven other senior generals/admirals as members. The CMC develops strategic policy and exercises operational command over the armed forces.

Significantly, since 1997 no uniformed military officer has served on the Politburo Standing Committee, and only two senior PLA officers serve on the full Politburo. Since 1987, PLA representation on the roughly 25-member Politburo has been limited to two or fewer members.[12] About 300 members of the PLA and People's Armed Police (PAP) are designated as delegates to the 2,270-member National Party Congress.[13] Likewise, around 300 members of all ranks from the PLA and PAP are delegates to the roughly 3,000-member National People's Congress (NPC), where they take part in NPC sessions, work groups and investigations as an additional duty and honour.

In November 2013, the CCP Central Committee formed a National Security Commission (NSC) led by the president, premier, and the chairman of the NPC Standing Committee. The NSC is subordinate to the Politburo Standing Committee and is responsible for 'decision-making, deliberation and coordination on national security work', including *both external and internal* security matters.[14] The exact structure and operating procedures for NSC have not been announced, but it is expected to improve coordinated security policy planning and implementation among China's bureaucracies, since lateral communication among government agencies has long been considered a shortcoming within the Chinese government. Consistent with the uniformed military's subordinate role to the Party leadership, no military officers have been identified in the NSC's senior leadership. The PLA and PAP will undoubtedly be represented within the NSC, but their duties have not yet been defined.

Party control is exercised at every level of command of the armed forces. In every military unit from company on up, responsibility for everything the unit does or does not do is shared by a commander and political commissar of equal grade. Nearly all (if not all) PLA officers and most non-commissioned officers (NCOs) are Party members and, as such, share the Party's values, world outlook and objectives. Within the PLA, the General Political Department controls a system of political officers who share responsibility with commanders in all units from company level up. Each unit has a Party committee led by a political commissar who ensures that Party policy is understood and consensus is achieved on matters both political and military. Every soldier takes an oath pledging first to obey Party orders, then to serve the people. The soldiers' oath does not mention loyalty to the Chinese government.

Political and ideological training within the military has the same level of priority as tactical training. Depending on the type of unit, units spend either 20 or 30 per cent of their overall training time in political education.[15] This training inculcates all personnel on the need for loyalty to the Party, reinforces CCP values and objectives, and warns of the dangers China faces from internal and external forces. For example, the theme of stopping any attempt to 'separate the military from the Party, depoliticize, or nationalize the military' has been a consistent element of political training for over a decade.[16] Perhaps the most dangerous scenario for the armed forces to confront would be a divided Party leadership.

The Chinese military does not have its own foreign policy and obeys the orders from its Party superiors in its dealings with foreign countries. An example of the military's role in foreign policy is the establishment of the East China Sea Air Defense Identification Zone (ADIZ) in November 2013, which overlaps with an existing Japanese ADIZ in the area. Although the Ministry of National Defense (MND) announced the creation of the ADIZ and is responsible for its implementation, the decision to do so was made by the PRC government, in which the military was but one agency involved in the process.[17] An important factor in this PRC government decision was probably the October announcement by the Japanese Defense Ministry that it was working on protocols for 'necessary measures' to include shooting down drones flying in the vicinity of the Senkaku islands.[18] Beijing's decision to create its own ADIZ was probably in part intended to establish a legal justification for its aircraft to fly in the international airspace covered by both the Japanese and Chinese ADIZs. The controversy over the ADIZ is one of many tit-for-tat political and military measures which both countries have taken over the Senkaku islands, an area of contention which heated up in 2012.

Likewise, as part of China's larger foreign policy efforts, the PLA works with civilian law enforcement agencies, such as the newly formed China

Coast Guard, to patrol the waters in the three near seas. The creation of the Coast Guard in early 2013 was intended to improve command and control over several disparate agencies (the former Maritime Surveillance Force under the State Oceanic Administration, the Ministry of Public Security Border Defense/PAP Maritime Police/Coast Guard, the Fisheries Bureau under the Ministry of Agriculture, and the Customs Anti-Smuggling Bureau) and to better coordinate their activities with other central and local government organizations. With regard to asserting China's territorial claims over the Senkaku islands, the PLA is clearly in a supporting role as Coast Guard vessels and aircraft perform the primary patrolling duties in disputed waters. The PLA Navy is more active in patrolling the South China Sea, though there too it also cooperates with other civilian agencies.

Economic components of military modernization

The subordination of military modernization to economic development is the fundamental economic principle guiding PLA development:

> In the past three decades of reform and opening up, China has insisted that defense development should be both subordinated to and in the service of the country's overall economic development, and that the former should be coordinated with the latter. As a result, defense expenditure has always been kept at a reasonable and appropriate level.[19]

For the first decade and a half of military modernization, this relationship resulted in minimal defence budgets as the civilian economy grew rapidly. Throughout the 1980s and early 1990s, with the force numbering between three and four million personnel, the military budget hovered in the single or low double-digit billions of US dollars. During this time frame, as the economy was blossoming, the PLA was encouraged to enter into commercial ventures to augment its budget. However, by the late 1990s this experiment with profit-making activities was determined to be detrimental to the overall health of the military by allowing too much temptation for graft and corruption while diverting emphasis away from training and readiness. As a result, in 1998 CMC Chairman Jiang Zemin directed the PLA to divest itself of its commercial enterprises.

Following divestiture, the central government sought to reimburse the PLA for some of its monetary losses by increasing the defence budget. At the same time, soldier pay and benefits began a series of enhancements, and more funds were allotted for new equipment and training and to offset the effects of inflation. Nonetheless, even as the defence budget began to grow at double-digit rates in the 1990s, the amount of resources dedicated to the military did not impede other elements of China's economic development programme. The defence budget for 2014 amounted to about US$132 billion[20] and, when compared to 15 years ago, that figure more accurately, but not completely, represents the total funding available to the PLA. Even as defence budgets have increased, PLA leaders have encouraged thrift to allow for the still limited funding for such a large force to be better spent.

Like the rest of Chinese society, corruption remains a major problem in the armed forces, though the degree to which it affects readiness is uncertain. As the PLA has downsized over the past decade, it has offered excess property for sale or rent. The sale or rental of land, along with the development that usually takes place after the PLA has vacated that land, provides opportunity for graft. The most scathing examples of corruption are the dismissal of Lieutenant General Gu Junshan and the Party's decision to expel former vice-chairman of the CMC General Xu Caihou for suspected bribery.[21] Xu, who also had served as director of the General Political Department, is the highest ranking military officer to be caught up in the anti-corruption campaign. Gu had been a deputy director of the General Logistics Department and had previously been in charge of managing 'capital construction and barracks' work for the entire PLA. Gu was charged with embezzlement, bribery, misuse of state funds and abuse of power in March 2013.[22] The PLA now audits senior officers before they are promoted or retire.[23] Other measures have been ordered to limit drinking, lavish banquets and staying in luxury hotels.[24]

Due to organizational changes implemented in the late 1990s, it is much more difficult now for military officers to benefit personally from arms sales than it may have been decades before. In April 1998, the Commission of Science Technology, and Industry for National Defense (COSTIND), the PLA headquarters organization that managed policy for the civilian defence industries, was dissolved and its functions turned over to a new civilian bureaucracy with the same name. In 2008, COSTIND was superseded by the State Administration of Science, Technology and Industry for National Defense (SASTIND). Thus, for approximately two decades the PLA has had no direct control over the actions of the defence industrial sector.

Some progress in the quality of weapons and equipment has been made since these reforms were implemented, particularly in some electronics and computer systems and some missiles, but many technological gaps remain. As a result, the PLA continues to purchase selected advanced weapons and technology from foreign sources, primarily Russia. As one recent study observed:

> Although the Chinese leadership's stated goal is for defense technological self-sufficiency, this is a long-term over-the-horizon political aspiration and the near- to medium-term (5–10 years) reality is of continuing heavy reliance on foreign sources for technology and knowledge, although combined with increasing levels of domestic input....

China's defense enterprises are likely to continue to provide the [PLA] with increasingly advanced technologies (continuing on a more linear, production model approach) but not with radically innovative, disruptive, technologically integrated or innovatively engineered systems.

The PLA and the ten large civilian defence industrial enterprise groups that manufacture weapons and equipment operate in a buyer and seller relationship. The PLA may provide some research and development money to the industries in support of specific projects, but military officers do not have direct input to the daily decisions made in the defence industrial sector. A limited number of PLA officers are assigned to factories throughout the defence industrial sector in Military Representative Offices (MRO) and are responsible for overseeing quality control and conducting liaison with the forces. As the PLA purchases new weapons and equipment from the industries, it seeks to preserve good relations with the manufacturers. In particular, PLA units frequently send personnel to factories to learn how to operate and maintain new equipment, and it is common for the industries to dispatch technical personnel to PLA units to assist in training and maintenance. This type of civil–military integration was described by the 2010 White Paper:

> The state takes economic development and national defense building into simultaneous consideration, adopts a mode of integrated civilian–military development. It endeavors to establish and improve systems of weaponry and equipment research and manufacturing, military personnel training, and logistical support, that integrate military with civilian purposes and combine military efforts with civilian support.[25]

This sort of interaction between the military and civilian enterprises is not confined to weapons manufacturers. Currently the military and civilian sectors aim to work in conjunction with each other in economic and security functions. This programme is run under the auspices of the National Defense Mobilization Committee (NDMC) system. The premier of the PRC is chairman of the State NDMC with vice-premiers of the State Council and vice-chairmen of the CMC as NDMC vice-chairmen. The NDMCs found at military regional, provincial and lower government levels are tasked with the preparation for and execution of the mobilization of the armed forces, national economy and transportation, civil air defence, and national defence education.[26] Local NDMCs and their corresponding PLA headquarters conduct surveys and inventories of civilian assets, such as food supplies, transportation assets and repair/maintenance facilities that may be utilized by the military during periods of emergency or while in training. By using civilian resources to support the military whenever possible, the government seeks to conserve funds spent on defence.

One of the most important logistical reforms of the past 15 years is the 'socialization' of support functions in which PLA units contract with civilian entities to provide services such as managing dining halls, sharing bulk storage facilities, providing fuel and vehicle maintenance, and augmenting PLA air, sea and land transportation capabilities. In theory, this practice will free up troops for operational assignments and should cost the Chinese government less than if the PLA performed all these functions. For example, these policies allow PLA vehicles to get fuel and receive repairs at civilian facilities when operating away from their home bases or local communities to provide food to troops passing through, all reimbursed through 'smart cards' connected to the banking system. To date, 'socialization' of support has been provided mostly within China itself. However, as the PLA finds itself conducting operations outside of China, it is also experimenting with obtaining civilian logistics support in other countries.

The entire scope of civilian–military integration is aimed primarily to better use the funds and resources allotted to the PLA and the civilian defence industries and is not an effort to 'privatize' or 'civilianize' the military. Its goal is to improve efficiency, to avoid duplication of effort, and to create a structure for cooperation between the military and civilian sector when needed. It seeks to create synergies between the military and civilian sectors but does not 'outsource' the PLA's basic mission of external defence and deterrence.

International components of military modernization

One of the biggest changes the PLA has undertaken over the past decade is the diversity of missions it now executes outside of China. Operating beyond China's borders is a relatively new experience for the modern PLA and managing these new missions has been a learning experience for all services.

Since 2002, the PLA has participated in well over 60 training exercises with more than 30 other countries and international organizations, such as the Shanghai Cooperation Organization and ASEAN.[27] Units from all services and military regions in the PLA (except for the Second Artillery) and the PAP have been involved in these exercises, some of which have been conducted in China and others beyond its borders. Most of the exercises focus on non-traditional security missions such as anti-terrorism, and often include special operations forces, and disaster relief operations. Exercises outside of China also give the PLA and PAP experience in moving units over long distances, mostly using rail and air transport, and require some degree of accompanying logistics support. Nonetheless, the number of personnel involved in these exercises is a small percentage of the total PLA and PAP force. Although some exercises include thousands of personnel, most are much smaller, with only dozens or hundreds of personnel involved. Perhaps the most important aspect of this training is that headquarters from all over the PLA and PAP, not just the most advanced units, are exposed to the complexities of operations beyond China's borders.[28] Since 2005, the PLA has also invited foreign observers to several of its brigade and division-size exercises inside China.

Likewise, increased unilateral naval and air training further from China's borders has been a point of emphasis for the past several years for the PLA Navy and Air Force. In 2009, the Commander of the Navy, Wu Shengli, told reporters: 'In the past decade the People's Navy has organized more than 30 combat operations group campaign exercises at sea.'[29] These numbers suggest that each of China's three fleets conducted one such exercise annually. Four years later the number of these exercises had more than doubled to seven, at least two per fleet annually.[30] As the PLA Navy extends its operations further at sea it is practising for *both* warfighting and non-traditional security missions, which involves cooperation with other countries. The PLA Air Force is also conducting larger, more complex exercises as it transitions from a mainly territorial defensive posture to a force capable of both offensive and defensive operations.[31] Some of these exercises are carried out over water, but most large PLA Navy and Air Force exercises are single-service events to improve their own functional capabilities as the overall force is still exploring how to command and control joint operations that employ all of its new capabilities.

The PLA has engaged in a number of overseas non-traditional security missions. The most prominent of these missions is China's participation in multiple UN peacekeeping operations and in anti-piracy escort missions in the Gulf of Aden. Currently, China has deployed over 2,100 military personnel and police officers to ten UN missions.[32] Responsibility for providing personnel and equipment for these missions is shared among multiple military regions. The vast majority of troops provided are engineers and logistics (medical and transport) personnel, with only two instances of the deployment of combat forces for local security. Starting in December 2008, China has dispatched naval forces to participate in escort operations off Somalia. The navy task forces usually consist of two combatants and a logistics support ship, a few helicopters, and some special operations forces from the PLA Navy. Responsibility for the provision of these forces has rotated among the three fleets, allowing each headquarter to be exposed to the complexities of long-distance deployments. The supply ships have established support relations with ports in Salah, Oman, Aden, Yemen and Djibouti.[33] Ships from these task forces have also used this opportunity to make port calls to many countries en route and on return to China, as well as to conduct exercises with other nations' forces on patrol or while in transit. In addition, PLA forces have contributed to a few foreign disaster relief efforts (often flying in supplies) and the evacuation of Chinese citizens from Libya in 2011 when four Il-76 aircraft and a frigate were sent to the area. The hunt for Malaysian Air Flight 370 in

March and April 2014 involved PLA Navy and Air Force assets working in conjunction with Chinese civilian government entities.

Social components of military modernization

Traditionally, the PLA has prided itself on being a people's army in addition to being a Party army. This relationship was sullied in June 1989 after the PLA obeyed Party orders to clear civilian demonstrators out of Beijing. Over the following two decades the military has tried hard to restore its relationship with the people and improve its domestic image. As it seeks to create a smaller, more technologically advanced military these efforts are even more important as the armed forces attempt to attract qualified personnel into its ranks. The military must compete with opportunities in the civilian world for young people graduating from high school. Based on the percentages of soldiers, NCOs and officers in the force, every year the PLA and PAP must conscript or encourage over half a million young adults to join the service out of about ten million males and nine million females reaching military age annually.[34] Attracting and keeping qualified personnel in the armed forces is easier to do when the military has a good image among the population of the country.

Recruitment has been complicated by the decades-old 'one-child family' policy. A frequent complaint heard from commanders is that many 'one-child family' youth are 'too soft', not disciplined or physically fit enough to enter the forces. Many better-educated military-age people living in cities prefer to find civilian employment than join the military. On the other hand, many rural youth in depressed parts of the country are looking for ways to improve their situation, with one option being to join the military, but they may not be as well-educated and technologically savvy as urban youth. Accordingly, depending on their circumstances, some parents and young people may be inclined to attempt to bribe officials either to get their children into or out of military service.

Because operating and maintaining the PLA advanced equipment requires greater technical skill than previous generations of equipment, the PLA is attempting to attract more college students (while they are still in school) and graduates to enlist. In order to encourage students to join the military, the central and local governments are offering monetary incentives, tax benefits, bonus points on school entrance exams, and preferential job placement after leaving the service.[35] These inducements have been successful to a degree, especially for graduates who have problems finding civilian work after leaving college, but they also result in large numbers of volunteers being rejected for failing to meet health and physical standards.[36] As a result, physical fitness standards for joining the military have been lowered twice. So far, the PLA's experiment with recruiting college students and graduates is not complete and more adjustments are likely.

The Chinese government has undertaken a national defence education programme to inform the civilian public, including high school and college students, about what the military is and does. Since 1985 China has implemented a system of military training in senior middle schools and institutions of higher education.[37] Since 2002, all students have been required to undergo military instruction sessions, often comprised mainly of basic drill and ceremonies, which provide them with only a small sample of what military life and discipline is like. As might be expected, reportedly a large percentage of students attempt to avoid this training.[38]

PLA and PAP units routinely participate in local projects, coordinated by local governments, to provide medical assistance and infrastructure construction support. This sort of activity is considered beneficial to the local economy. Local governments and business enterprises frequently reciprocate by providing gifts, such as fruit and other local products, to PLA and PAP units in their areas, though the practice of exchanging gifts may be limited in the future as part of the battle against corruption.

The purpose of all of this activity is to create good relations between the military and the population. The PLA leadership understands that it is dependent upon the public to provide it with qualified people and to support its peacetime and warfighting operations. The better its relationship with the Chinese population, the more likely the PLA will perform its missions successfully. Although domestic security is considered among the non-traditional security tasks to which the PLA may be assigned, as an institution it does not desire to perform the internal security mission except when absolutely necessary. Instead, internal security is the primary task of the civilian police force and PAP, while the PLA concentrates on its military missions, focused mainly on external threats or armed insurgency or terrorist groups, and other non-traditional security tasks. Managing the details of preparing for and executing those missions is left to the PLA chain of command.

Military components of military modernization

Within the parameters prescribed by the defence budget, missions assigned by the Party and government, and applicable national policies, PLA modernization is managed almost exclusively by its professional, uniformed military leadership. The Central Military Commission provides strategic leadership and policy direction. Although the civilian chairman is at the head of this organization, the CMC performs its duties through collective leadership and consensus decision-making. The CMC has a small staff to assist the principal officers.

PLA organization and command chain

The CMC commands the PLA through the four General Departments and service headquarters in Beijing down to units distributed throughout the country in seven military regions (MR). Each of the four General Departments, the General Staff Department (GSD), General Political Department (GPD), General Logistics Department (GLD) and General Armaments Department (GAD), oversees its own functional system that extends from Beijing to headquarters at regimental level throughout the PLA.

The four General Departments act as the national-level service headquarters for the Army, while also providing policy guidance for the other services. The PLA Navy, Air Force and Second Artillery (technically an 'independent branch' of the Army, but usually treated as a service on a par with the Navy and Air Force) all have national-level headquarters in the Beijing area. The Ministry of National Defense is not in the chain of command from the CMC to operational units and is primarily responsible for relations with other government agencies, foreign militaries and international organizations. The Minister of Defense's influence mainly comes through his seat on the CMC.

Below the four General Departments, the Academy of Military Science (AMS), the National Defense University (NDU) and the National University of Defense Technology (NUDT) are directly subordinate to the CMC. The AMS is primarily a research centre for the study of military strategy, operations and tactics; military systems; military history; and foreign militaries. The AMS does not focus on student education, but rather is the premier doctrinal research and development institute for the entire PLA. The NDU and NUDT form the pinnacle of the PLA professional military education system.

The PLA Navy, Air Force and Second Artillery are at the same organizational level as the seven military regions. The Navy exercises command of its operational forces through three fleet headquarters (North, East and South Sea Fleets), while the Air Force commands its units through seven regional air forces (one in each of the military regions). The Second Artillery commands six missile launch bases (numbered 51 through 56). In peacetime these forces report to their service headquarters and, during war or emergency, elements of the services may be assigned to war zone headquarters, which would be formed around the existing military region headquarters.

The Navy comprises five branches: submarine (nuclear and conventional), surface, naval aviation, coastal defence, and marine forces. Each of the three fleets has division-, brigade- and regimental-level units for each of the first four branches. The South Sea Fleet commands the PLA's only two Navy marine brigades. Within the marine brigades are small Special Operations Force (SOF) units. The Navy is currently deploying the second generation of nuclear-powered ballistic missile submarines (SSBNs), with a second generation of submarine-launched ballistic missiles (SLBMs) expected to be operational in the near future.

The Air Force has four branches: aviation (fighters, ground attack, fighter-bombers, bombers, reconnaissance aircraft, helicopters and transport aircraft), surface-to-air missile, anti-aircraft artillery and airborne forces, organized into division-, brigade- and regimental-level units. Specialized troops, consisting of radar, communications, electronic countermeasures and chemical defence troops, support the branches. The majority of the PLA's airborne capacity is found in the Fifteenth Airborne Corps, which comprises three airborne divisions, an SOF unit and a helicopter unit. Strategic airlift, however, is a long-standing shortfall and only a portion of the airborne force could be transported by all of the PLA's long-distance transport aircraft in a single lift. Air defence responsibilities are shared among all three services, the reserves and the militia.

The six Second Artillery launch bases are each structured differently to command nuclear and conventional ballistic and/or ground-launched, land-attack cruise missile brigades. According to the U.S. Department of Defense, the force is composed of roughly 1,100 short-range ballistic missiles (SRBM), 75 to 100 medium-range ballistic missiles (MRBM), including an anti-ship ballistic missile (ASBM) under development and being deployed, five to 20 intermediate-range ballistic missiles (IRBM), 50 to 75 intercontinental ballistic missiles (ICBM), and 200 to 500 land-attack cruise missiles.[39] While the trend over the past 15 years in the other PLA services has been to downsize their force structures, the number of launch brigades in the Second Artillery has roughly doubled to currently about 26 or 27 brigades. The force is being further upgraded with more survivable and more accurate mobile missiles of all types, though the total number of missiles has basically levelled off over the past five years. Second Artillery forces are commanded directly by their headquarters in Beijing, which control all nuclear operations, but may temporarily assign conventional units to support operations commanded by war zone or military region headquarters.

Managing the PLA's new joint doctrine

In 1997 CMC Chairman Jiang Zemin announced a 500,000-man reduction in the size of the PLA. Shortly thereafter new personnel policies were introduced, such as shortening the conscription period to two years and the creation of a professional non-commissioned officer corps, and a new fighting doctrine promulgated. As the personnel size of the PLA has been reduced, the force structure has also changed dramatically through the simultaneous processes of mechanization and intelligence. For instance, the Army transformed and created units that have the potential to be used beyond China's borders as well as in its territorial defence. Two infantry divisions were transformed into amphibious mechanized divisions. These two units, in addition to the previously established amphibious armoured brigade, give the Army more amphibious capability than the two marine brigades in the Navy (Naval amphibious lift remains a shortfall, however, and the PLA would have to use civilian forces to conduct amphibious operations larger than a division in size). The Army has expanded its Army

Aviation (helicopter) Corps so that one or two brigades or regiments are found in each MR, though the number of airframes, estimated to be around 760, is very small for such a large force.[40] The Army also created and deployed SOF units in every MR; however, these units are not supported by the same array of special mission support aircraft, ships and intelligence support found in other militaries, particularly the US military. As a result, based on reports and photos of their training, most PLA Army SOF units resemble foreign commando or Ranger units in capabilities and missions more than they do higher level (Tier One units in US parlance) special operations units.

In order to conduct both defensive and offensive cyber warfare operations, the PLA has established 12 operational bureaus subordinate to the GSD Third Department and 16 technical reconnaissance bureaux in the military regions and the services. These organizations conduct 'communications intelligence, direction finding, traffic analysis, translation, cryptology, computer network defence, and computer network exploitation' in support of military operations. In addition, electronic warfare units are found in all services and are part of the GSD Fourth Department's operational system.[41] A few PLA reserve units and militia units have also been assigned cyber and electronic warfare duties. Perhaps the most important doctrinal concept about cyber, electronic and other forms of information war (such as the 'Three Warfares' of media, psychological and legal war) is found in *The Science of Military Campaigns* (and other texts), which states, 'Information warfare itself is not a goal but a means'.[42] Or, as found in *The Science of Military Strategy*: 'information operation capability is an important constituent of battle effectiveness.'[43] Accordingly, information, electronic and cyber operations are to be employed throughout a conflict to multiply or enable the effects of other combat systems; they are not considered an end in themselves.

Through these processes the PLA has received significant amounts of new and upgraded equipment, though modern equipment probably amounts to less than half the inventory of most units, resulting in a mix of old and new equipment in much of the force. Concurrently, a new joint warfighting doctrine, based on the strategy of active defence and the principles of People's War updated for the 21st century, has been introduced. All units are engaged in rigorous training to learn how to integrate the new force structure, equipment, and doctrine into actual capabilities. Perhaps the single word to best describe the current PLA training regime is *experimental*.

The key to the PLA's joint doctrine is incorporating all capabilities from all services in the force, using both old and new weapons, equipment and technology, with paramilitary and civilian support in a coordinated effort. This is not as easy or as straightforward as it sounds, since units throughout the country have different mixes of weapons and equipment and are at different stages in their own modernization. Traditionally Army leaders have commanded PLA operations within or near China's borders, but in the future naval and air campaigns are more likely to be needed to defend China's interests. Consequently naval and air officers must now be prepared to command offensive and defensive joint operations beyond China's periphery. At the same time as it gets smaller but more technologically advanced, the Army is practising how to shift forces within China from region to region using both military and civilian land, air and sea transportation assets. Leaders are urged not to forget the PLA's heritage of use of speed, stratagem and deception as they plan for the employment of their modernized forces.

The PLA has not conducted a true joint campaign since the operation to seize Yijiangshan Island from Kuomintang troops, which culminated in a relatively small-scale amphibious assault in January 1955. With no modern joint combat experience, the PLA is literally learning by doing, albeit on the training ground.

Units in all services are learning how to command, control, execute and sustain joint and combined arms operations through a process of trial and error. Many training events are organized to identify problems to be overcome in subsequent training and annual evaluations are conducted to measure progress. An after-action review, which focuses more on shortcomings than on strong points, follows every major training event.

Most importantly, PLA leaders understand that new weapons and equipment (and money) by themselves will not result in advanced capabilities. Therefore, at all levels, units focus on training commanders, political officers and their staffs in leadership, innovation and initiative, as well as emphasizing the basic functional proficiencies of NCOs and conscripts. They realize that, as an organization, the PLA has no experience in modern combat and, though the PLA may have studied the wars of others, it must determine its own methods of operations suitable for the unique circumstances in China.

When speaking to foreigners, senior PLA leaders often say that their forces lack the capabilities of advanced militaries by 20 to 30 years. Among themselves in their military media, PLA authors often refer to this condition as the 'main contradiction in army building' or the 'two incompatibles' explained as follows: 'the level of our modernization is incompatible with the demands of winning a local war under informatized conditions, and our military capabilities are incompatible with the demands of carrying out the army's historic missions in the new century and new stage.'[44] This self-assessment was attributed to CMC Chairman Hu Jintao and has been prevalent in the Chinese internal Party and military media since January 2006 (but not in the English-language versions of their publications). In late 2013 a new slogan, 'the two inadequacies and two big gaps' ('there are big gaps between the level of our military modernization and the requirements for national security and the level of the world's advanced militaries; our military's ability to fight a modern war is inadequate, the ability of our cadres at all levels to command modern war is inadequate), began to be included in the literature.⁴⁵

These general assessments of PLA shortcomings are backed up by many specific functional evaluations of the inabilities of some commanders and staff to command and control joint and combined arms operations, a shortage of properly trained NCOs, problems in training (such as 'fakery' and formalism), organizational shortcomings (such as lack of staff personnel at battalion level to control combined arms operations), lack of interoperability of equipment, shortfalls in many elements of logistics support, and the general lack of combat experience in the force. This self-awareness suggests that many senior military officials may not be as 'hawkish' in their counsel to Party leaders as they are frequently portrayed in the foreign media.⁴⁶ The judgements of senior and operational PLA leaders often contrast with the more publicized, often aggressive words of numerous military pundits such as Luo Yuan, Yin Zhuo and Dai Xu, who speak for themselves and may reflect the nationalistic sentiments of much of the public and maybe even some other military personnel. Most active or retired PLA officers, like Luo and Yin, who are frequently quoted by the foreign media, have not been assigned to operational units recently, but have rather spent the majority of their careers in academic capacities.

Nonetheless, if ordered by the Party leadership, the PLA will attempt to use whatever capabilities are available, both military and civilian, to accomplish the missions assigned. The likelihood of success depends on where future battles are fought (the closer to China, the better for the PLA), who is the enemy and whether he is supported by allies (like the United States), the international security environment (what else is going on in the world that may distract national leaders and military forces), the perennial questions of the weather and the battlefield environment, and the reasons all parties are fighting.

Conclusion

Managing military change in China is subject to the country's unique circumstances, its national strategy of economic development, and threat perceptions. As a result, the organization of the force and the methods it employs will differ from the organizations and methods adopted by other militaries. PLA leaders consistently acknowledge the military's lack of experience in modern combat, but this does not mean that they are looking for a fight to prove themselves in battle. They and their CCP leaders prefer to achieve China's development objectives peacefully, without fighting, a tradition going back to Sun Tzu. A strong military, however, is essential to achieving that goal.

There is little doubt that, because of its economic growth, the Chinese government could choose to spend much more on military modernization than it has in the past. However, even with the increases to the defence budget over the past 15 years, the amount China spends on its military remains relatively small for such a large force. The size of the force, in fact, mandates that new equipment cannot be deployed to all units quickly. Unless the PLA undergoes further reductions in the number of its personnel and the size of its force structure (which appear to be forthcoming, but will likely take several years to implement), most of its units will be composed of a mix of advanced and legacy weapons for some years to come. Training and education of PLA leaders and troops will remain a priority even as new equipment enters the inventory.

Realistically speaking, the transformation of the PLA into a modern fighting force is a decades-long, generational mission. This timeline is often minimized as reports of new weapons under development capture the headlines of the foreign press. Nonetheless, the Chinese leadership appears to recognize the nature of the task ahead and is prepared to give the military the time and resources it needs to achieve its modernization objectives (assuming the Chinese economy continues its general expansion, albeit at a slower rate of growth than in past decades). If the Chinese economy were to suffer a severe crisis, it is likely that the military budget would also feel the impact of the situation.

In the meantime, between now and 2049, if an external threat to China's sovereignty or territorial integrity arises, the PLA along with other elements of the Chinese government will act, at the order of the Party and government, to defend China's core interests. This has been demonstrated in recent years in the military's actions in conjunction with other PRC government agencies in the East and South China Seas, actions often labelled 'assertive' or 'aggressive' by others. How effective the PLA will be on a modern kinetic battlefield, however, depends not only on its new equipment but more so on the capabilities of its personnel and on who, when, where and why the PLA fights.

Notes

1 Information Office of the State Council (IOSC) of the People's Republic of China, 'China's National Defense in 2006', 2006. Information Office of the State Council (IOSC) of the People's Republic of China, 'China's National Defense in 2008', 2009.
2 Information Office of the State Council (IOSC) of the People's Republic of China, 'The Diversified Employment of China's Armed Forces', 2013.
3 *PLA Daily*, 'Strengthen and Accelerate the Sense of Responsibility for National Defense and Military Modernization', 16 April 2013, http://chn.chinamil.com.cn/gd/2013–04/16/content_5301118.htm (accessed 16 September 2014).
4 Information Office of the State Council (IOSC) of the People's Republic of China, 'The Diversified Employment of China's Armed Forces', 2013. 'China's National Defense in 2004', 2004.
5 The Chinese armed forces consist of the active and reserve forces of the PLA, People's Armed Police (PAP) and militia. The active duty PLA officially numbers 2.3 million personnel, with about 1.6 million in the Army, 235,000 in the Navy, 398,000 in the Air Force and 100,000 in the Second Artillery, responsible for

most nuclear and conventional ballistic and cruise missiles (IOSC, 2013). This active duty total includes an unknown number of uniformed PLA civilian cadres, who mostly perform a variety of non-combat functions. The size of the PLA reserve force was given as 510,000 personnel in 2009, but is probably increasing as new reserve units are being formed to support the Navy, Air Force and Second Artillery (*Xinhua*, 'National Defense Reserve Strength', 22 September 2009, http://news.xinhuanet.com/mil/2009–09/22/content_12098695.htm). The size of the PAP was reported as 660,000 in 2006 (IOSC, 2006), considerably smaller than most previous foreign estimates.
6 IOSC, 2013.
7 'China to Optimize Army Size, Structure: CPC Decision', *PLA Daily*, 16 November 2013, http://eng.chinamil.com.cn/news-channels/china-military-news/2013–11/16/content_5651458.htm (accessed 16 September 2014).
8 IOSC, 2013.
9 Guangqiang Peng and Youzhi Yao (eds), *The Science of Military Strategy* (Beijing: Military Science Publishing House, 2005), p. 426.
10 U.S. Department of Defense, 'Joint Press Conference with Secretary Hagel and Minister Chang in Beijing, China', 8 April 2014, www.defense.gov/Transcripts/Transcript.aspx?TranscriptID=5411 (accessed 16 September 2014).
11 Alice Miller, 'More Already on the Central Committee's Leading Small Groups', *China Leadership Monitor*, no. 44, 28 July 2014, p. 6, www.hoover.org/research/more-already-central-committees-leading-small-groups (accessed 16 September 2014).
12 Alice Miller, 'The Politburo Standing Committee under Hu Jintao', *China Leadership Monitor*, summer 2011, no. 35, p. 6, www.hoover.org/research/politburo-standing-committee-under-hu-jintao (accessed 16 September 2014).
13 Minnie Chan, 'PLA Names 300 Delegates to Party Congress', *South China Morning Post*, 24 August 2012, www.scmp.com/article/1014453/pla-names-300-delegates-party-congress (accessed 16 September 2014).
14 *Xinhua*, 'Xi Jinping to Lead National Security Commission', 24 January 2014, http://news.xinhuanet.com/english/china/2014–01/24/c_133071876.htm (accessed 16 September 2014).
15 IOSC, 2009.
16 *People's Daily*, 'No Nationalization of Military in China: Senior PLA Officer', 21 June 2011, http://english.peopledaily.com.cn/90001/90776/90786/7415343.html (accessed 16 September 2014).
17 *Xinhua*, 'Statement by the Government of the People's Republic of China on Establishing the East China Sea Air Defense Identification Zone', 23 November 2013, http://news.xinhuanet.com/english/china/2013–11/23/c_132911635.htm (accessed 16 September 2014).
18 *Asahi Shimbun*, 'Defense Ministry Working on Protocol to Shoot Down Encroaching Drones', 2 October 2013, http://ajw.asahi.com/article/behind_news/politics/AJ201310020040 (ccessed 16 September 2014).
19 IOSC, 2009.
20 *Xinhua*, 'China Defense Budget to Increase 12.2 pct in 2014', 5 March 2014, http://eng.mod.gov.cn/TopNews/2014–03/05/content_4494719.htm (accessed 16 September 2014).
21 *Xinhua*, 'PLA Supports Graft Probe into Xu Caihou', 1 July 2014. http://news.xinhuanet.com/english/china/2014–07/01/c_133451742.htm (accessed 16 September 2014).
22 Kenji Minemura, 'Corruption Among Top Military Officials Kept under Wraps in China', *Asahi Shimbun*, 7 February 2012, http://ajw.asahi.com/article/asia/china/AJ201302070003 (accessed 16 September 2014). *Xinhua*, 'Former Senior Military Officer Faces Corruption Charge', 31 March 2014,

23 *Xinhua*, 'Chinese Military Officials to be Audited before Promotion', 24 September 2012, http://news.xinhuanet.com/english/china/2013-09/24/c_132746834.htm (accessed 16 September 2014).
24 *Xinhua*, 'Chinese Military Bans Luxury Banquets', 21 December 2012, http://news.xinhuanet.com/english/china/2012-12/21/c_124131621.htm (accessed 16 September 2014).
25 Information Office of the State Council (IOSC) of the People's Republic of China, 'China's National Defense in 2010', (2011).
26 Information Office of the State Council (IOSC) of the People's Republic of China, 'China's National Defense in 2002', (2002).
27 *Global Times*, 'China Stresses its Defensive Military Policy', 2 August 2013, www.globaltimes.cn/content/801085.shtml#.Un-0qqX7oYV (accessed 16 September 2014).
28 Dennis Blasko, 'People's Liberation Army and People's Armed Police Ground Exercises With Foreign Forces, 2002–2009', in Roy Kamphausen, David Lai and Andrew Scobell (eds), *The PLA at Home and Abroad* (Strategic Studies Institute, U.S. Army War College, 2010), pp. 377–428.
29 *Xinhua*, 'Navy Enhances Overall Operational Capabilities through Military Training', 15 April 2009, http://news.xinhuanet.com/mil/2009-04/15/content_11191605.htm (accessed 16 September 2014).
30 *PLA Daily*, 'Top Ten Pieces of PLAN Bews in 2013', 2 January 2014, http://eng.chinamil.com.cn/news-channels/china-military-news/2014-01/02/content_5717538.htm (accessed 16 September 2014).
31 IOSC, 2009.
32 United Nations (UN) Peacekeeping, 'UN Mission's Summary Detailed by Country', 31 January 2014, www.un.org/en/peacekeeping/contributors/2014/jan14_5.pdf (accessed 16 September 2014).
33 Daniel J. Kostecka, 'Places and Bases: The Chinese Navy's Emerging Support Network in the Indian Ocean', *Naval War College Review*, vol. 64, no. 1 (2011), pp. 59–78.
34 Central Intelligence Agency (CIA), China, The World Factbook, 2014, www.cia.gov/library/publications/the-world-factbook/geos/ch.html (accessed 16 September 2014).
35 *Xinhua*, 'Over 200,000 College Students Apply for Joining Military', 21 August 2013, http://china.org.cn/china/2013-08/21/content_29781313.htm (accessed 16 September 2014).
36 *China Daily*, 'Students Fail Army Fitness Standards', 13 August 2013, http://english.people.com.cn/90786/8361647.html (accessed 16 September 2014).
37 IOSC, 2002.
38 *Global Times*, 'Students Dodge Military Training', 27 September 2012, www.globaltimes.cn/content/735806.shtml (accessed 16 September 2014).
39 Office of the Secretary of Defense (OSD), 'Annual Report to Congress: Military and Security Developments Involving the People's Republic of China 2012', 2012, p. 29. Office of the Secretary of Defense (OSD), 'Annual Report to Congress: Military and Security Developments Involving the People's Republic of China 2013', 2013, p. 5.
40 James Hackett (ed.), *The Military Balance 2014* (London: The International Institute for Strategic Studies, 2014), p. 233.
41 Mark A. Stokes, Jenny Lin and L.C. Russell Hsiao, 'The Chinese People's Liberation Army Signals Intelligence and Cyber Reconnaissance Infrastructure', Project 2049 Institute, 2011, pp. 6–13.

42 Yulang Zhang (ed.), *The Science of Campaigns* (Beijing: National Defense University Press, 2006), p. 151.
43 Peng and Yao, 2005, p. 344.
44 *PLA Daily*, 'Strengthen and Accelerate the Sense of Responsibility for National Defense and Military Modernization', 16 April 2013, http://chn.chinamil.com.cn/gd/2013–04/16/content_5301118.htm (accessed 16 September 2014).
45 *China Daily*, 'Always Insist on Being Able to Fight and Win as the Fundamental Focus', 17 November 2013, http://chn.chinamil.com.cn/jwjj/2013–11/17/content_5651790.htm (accessed 16 September 2014). *China Daily*, 'Focus on Strengthening the Military Objective of Deepening National Defense and Military Reform', 18 November 2013, http://chn.chinamil.com.cn/jwjj/2013–11/18/content_5652134.htm (accessed 16 September 2014).
46 Dennis Blasko, 'The "Two Incompatibles" and PLA Assessments of Military Capability', Jamestown Foundation China Brief, 9 May 2013, www.jamestown.org/single/?tx_ttnews%5Btt_news%5D=40860&no_cache=1#.VBhZm0u0wTt (accessed 16 September 2014).

3 Military change in Japan

National Defense Program Guidelines as a main tool of management

Isao Miyaoka

Japan is now at a strategic crossroads. In late 2013, the Diet enacted a bill to establish the National Security Council (NSC) in the Cabinet Secretariat as well as a secret protection bill, both of which had been submitted by the second Shinzo Abe administration. The newly established NSC immediately approved Japan's first National Security Strategy (NSS), based on the following recognition:

> Maintaining the peace and security of Japan and ensuring its survival are the primary responsibilities of the Government of Japan. As Japan's security environment becomes ever more severe, Japan needs to identify its national interests from a long-term perspective, determine the course it should pursue in the international community, and adopt a whole-government approach for national security policies and measures in order to continue developing a prosperous and peaceful society.
>
> Japan has contributed to peace, stability and prosperity of the region and the world. In a world where globalization continues, Japan should play an even more proactive role as a major global player in the international community.[1]

A sense of international crisis brought about the formulation of the NSS, which may suggest a sign of a major military change at the grand strategic level. The document identifies a ' "Proactive Contributor to Peace" based on the principle of international cooperation' as Japan's fundamental principle of national security.

Surprisingly, the National Security Strategy finally replaced the Basic Policy for National Defense, which was approved by the Cabinet in May 1957. For more than half a century, the Basic Policy had been Japan's top-level government document for national security. Until very recently, Japan in the post-war era could be best characterized as having no major formal change at the grand strategic level.

Since the mid-1970s, the Government of Japan has managed military change mainly at the defence-strategic level by formulating National

Defense Program Guidelines (NDPG).[2] In the words of the NDPG approved in 2010, 'These Guidelines provide the vision for our defense forces for approximately the next decade, to promote innovation of the defense forces.'[3] All of the NDPG documents are similar in content and share common elements, such as the NDPG's objective, international situation (or the security environment surrounding Japan), the buildup concept and role of the defence forces, and the posture and structure (organization, equipment and disposition) of the Self Defense Forces (SDF).[4] The NDPG documents were approved by the Security Council (or the National Defense Council until June 1986) and then the Cabinet. In these documents, the government has announced nearly all important decisions on the buildup and maintenance of the defence forces. There are only a few noted exceptions; for example, the Security Council and Cabinet decision of 19 December 2003 'On Introduction of Ballistic Missile Defense System and Other Measures'.

The NDPG also focuses on the operation of the defence forces, although there are two caveats. First, until the 2004 NDPG, Japan maintained the Basic Defense Force Concept that placed emphasis on preventive measures and deterrence against invasion through the existence, rather than operation, of defence capabilities. Second, the Japanese government has managed military change at the operational or doctrinal level by formulating joint and service Defense and Security Plans, which are treated as top secret.[5]

What is noteworthy is that the intervals between NDPG documents have increasingly narrowed. The 1954 Defense Agency Establishment Act stipulated that a National Defense Council be established in the Cabinet to deliberate on important matters on national defence, such as the Basic Policy for National Defense and the NDPG. Although the National Defense Council was established in 1956, it took 20 years, until 1976, to adopt the first NDPG. Moreover, an updated version of the 1976 inaugural NDPG was never approved during the so-called new Cold War in the late 1970s and 1980s. In contrast, the Government of Japan has adopted four NDPGs since the end of the Cold War: in 1995, 2004, 2010 and 2013.

This chapter aims to examine how Japan has managed military change at the defence-strategic level in the post-war period and why Japan has formulated NDPGs increasingly more frequently. In the first section of the chapter, I describe the basis of Japanese defence policy as the context in which Japan has managed military change. In the next two sections I analyse Japan's management of military change at the doctrinal level through the lens of differing challenges, namely Cold War détente, post-Cold War uncertainty, the increasingly severe security environment, and the rise of China. For this purpose, I carefully examine the strategic logic of the five NDPG documents. Finally, I explain the increasing frequency of formulating NDPGs in recent years, before summarizing the main developments.

The basis of defence policy

In this section, I outline Article 9 of the Constitution of Japan and basic defence policies as the basis of Japanese defence policy.

The Peace Constitution

In Japan, the Constitution plays an important role in security policy-making. The Constitution of Japan enshrines pacifism as one of its three central pillars, together with sovereignty of the people and respect for fundamental human rights. In particular, Article 9 of the Constitution renounces war, the possession of war potential and the right of belligerency by the state as follows:

> Aspiring sincerely to an international peace based on justice and order, the Japanese people forever renounce war as a sovereign right of the nation and the threat or use of force as means of settling international disputes.
>
> In order to accomplish the aim of the preceding paragraph, land, sea, and air forces, as well as other war potential, will never be maintained. The right of belligerency of the state will not be recognized.[6]

Since the establishment of the Self-Defense Forces (SDF) in 1954, the Government of Japan has justified this armed organization, by pointing out that, as a sovereign state, Japan has an inherent right to self-defence.

The government has officially adopted the following official interpretation of the Constitution:

> [A]rmed force can be used to exercise the right of self-defense only when the following three conditions are met: 1) When there is an imminent and illegitimate act of aggression against Japan; 2) When there is no appropriate means to deal with such aggression other than by resorting to the right of self-defense; and 3) When the use of armed force is confined to the minimum necessary level.[7]

The government has also made an interpretation of the geographic boundaries of self-defence: it is beyond the minimum necessary level of self-defence and thus constitutionally impermissible to dispatch the SDF to the land, sea or airspace of other countries with the aim of using force. In a similar manner, the government has also held the view that, although Japan possesses the right to collective self-defence[8] as a sovereign state under international law, it is constitutionally impermissible for Japan to exercise this right because such exercises go beyond the limits of the minimum necessary level of self-defence.[9]

As for Paragraph 2 of Article 9, the government has also clarified its constitutional view on the non-possession of war potential and the non-recognition of the right of belligerency. First, Japan can possess the minimum necessary level of self-defence capability. The Ministry of Defense gives a supplementary explanation on this interpretation:

> But in any case in Japan, it is unconstitutional to possess what is referred to as offensive weapons that, from their performance, are to be used exclusively for total destruction of other countries, since it immediately exceeds the minimum level necessary for self-defense. For instance, the SDF is not allowed to possess ICBMs [intercontinental ballistic missiles], long-range strategic bombers or offensive aircraft carriers.[10]

Second, the constitutional exercise of the right of self-defence is conceptually distinguishable from the unconstitutional exercise of the right of belligerency.

Basic defence policies and the Yoshida Doctrine

The 1957 Basic Policy for National Defense is a short document that consists of fewer than 300 characters in Japanese. The following is an English translation of the original text:

> The objective of national defense is to prevent direct and indirect aggression, but once invaded, to repel such aggression, and thereby, to safeguard the independence and peace of Japan based on democracy.
> To achieve this objective, the following basic policies are defined:
>
> 1 Supporting the activities of the United Nations, promoting international collaboration, and thereby, making a commitment to the realization of world peace.
> 2 Stabilizing the livelihood of the people, fostering patriotism, and thereby, establishing the necessary basis for national security.
> 3 Building up rational defense capabilities by steps within the limit necessary for self-defense in accordance with national strength and situation.
> 4 Dealing with external aggression based on the security arrangements with the U.S. until the United Nations will be able to fulfill its function in stopping such aggression effectively in the future.[11]

The four basic policies may be categorized into diplomacy, domestic politics, military and the United States, respectively.[12]

The most important part of this document is the last two basic policies: 'building up rational defense capabilities by steps within the limit necessary for self-defense in accordance with *national strength and situation*" [italics added] and 'dealing with external aggression based on the security arrangements with the U.S.'. The original draft of the Defense Agency comprised these two basic policies only.[13] Moreover, the Defense White Papers of 1989 and 1990 call them the 'two pillars of defense policy'.[14] In this connection, it is noteworthy here that the term 'national strength' (*kokuryoku*) refers to national economic and other material power while 'national situation' (*kokujo*) is a set of contradictory ideational elements such as the Peace Constitution, anti-militarism and the guns-or-butter sentiments of ordinary people.[15] As for the national situation, Japan's management of military change has been strictly constrained by feelings of pacifism, anti-militarism and an emphasis on the economy.

The Basic Policy for National Defense may be considered to be the first official document that institutionalized the so-called Yoshida Doctrine.[16] Shigeru Yoshida, Prime Minister of Japan during the period from the mid-1940s to the mid-1950s, had the strategic vision that a weak Japan should focus on the reconstruction of its economy while constraining its rearmament in reliance upon the United States for its military security. During the Cold War period, the Yoshida Doctrine took hold as an unwritten grand strategy, which consists of three basic principles: Japan–US security arrangements, light armament and emphasis on the economy.

In addition, Japan has retained the four other basic policies that were formulated during the Cold War era, namely: (1) exclusively defence-oriented policy; (2) the pledge to not become a military power that could threaten other countries; (3) the Three Non-nuclear Principles; and (4) securing civilian control.[17] These four basic policies were reconfirmed by the 2013 NDPG, although the Basic Policy for National Defense was replaced by the NSS.

Cold War détente and post-Cold War uncertainty

In this section, I show how Japan managed military change by focusing on the international situation, the buildup concept and role of the defence forces, and the posture and structure of the SDF that are described in the 1976, 1995 and 2004 NDPG documents.

International situation

The 1976 NDPG views the international situation as stable because of the détente in the 1970s and the Japan–US security arrangements.

> Under present circumstances, though, *there seems little possibility of a full-scale military clash between East and West or of a major conflict* possibly leading to such a clash, due to the military balance – including mutual nuclear deterrence – and *the various efforts being made to stabilize international relations.*

Furthermore, while the possibility of limited military conflict breaking out in Japan's neighborhood cannot be dismissed, this equilibrium between the super-powers and *the existence of the Japan–U.S. security arrangement[s] seems to play a major role in maintaining international stability, and in preventing full-scale aggression against Japan* [italics added].

It is noteworthy that the then government considered full-scale aggression against Japan unlikely.

After the end of the Cold War and the collapse of the Soviet Union, the framers of the 1995 NDPG reviewed the international situation, but in the end fundamentally accepted the same perception as the italicized parts of the 1976 NDPG quoted above,[18] recognizing that 'the possibility of a global armed conflict has become remote in today's international community'. Moreover, the 2004 NDPG points out that 'a full-scale invasion against Japan is increasingly unlikely'.

On the other hand, the newly established NDPG following the end of the Cold War highlights new problems arising from the international security environment. The 1995 NDPG focuses on the increase of '*new kinds of dangers*, such as the proliferation of weapons of mass destruction including nuclear arms, and of missiles' and the occurrence of 'regional conflicts and other *various situations*'. Similarly, the 2004 NDPG emphasizes that 'the international community is facing urgent *new threats and diverse situations* to peace and security' [italics added], while it also makes a special mention of the 9/11 terrorist attacks on the United States and international terrorist activities in general.

Buildup concept and role of the defence forces

In this optimistic view of the international situation, the 1976 NDPG is based on the Basic Defense Force Concept (*kibanteki boeiryoku koso*), although the term itself is not mentioned in this document. This concept is expressed in the following part of the NDPG:

> the most appropriate defense goal would seem to be the maintenance of a full surveillance posture in peacetime and *the ability to cope effectively with situations up to the point of limited and small-scale aggression*. The emphasis is on the possession of the assorted functions required for national defense, while retaining *balanced [posture in] organization and deployment*, including logistical support [italics added].

It is also added that Japan's future defence capability 'will be standardized so that, when serious changes in situations so demand, the defense structure can be smoothly adapted to meet such changes'. The introduction of this concept was based on the assumption 'that the international political structure in this region – along with continuing efforts for global stabilization – will not undergo any major changes for some time to come, and that Japan's domestic conditions will also remain fundamentally stable'.

According to the 1977 Defense White Paper, this concept mainly considers the level of future defence force in peacetime.[19] The previous defence buildup plans were aimed to cope effectively with 'the entire generic category of "limited aggression" which includes conflict expanded beyond a "small scale".' The 1976 NDPG lowered the defence goal to only 'limited and small-scale aggression'. Such aggression, which attempts to create a fait accompli, is conducted as a surprise attack without wholesale advance preparations by the enemy.[20] Behind this change lay the improvement of the security environment in Asia, as exemplified by the Sino-American *rapprochement* and the end of the Vietnam War. The Fourth Defense Buildup Plan, approved in 1972, could not be completed in the final year, 1976, due to the oil crisis and rampant inflation.[21] Nevertheless, according to the less ambitious 1976 NDPG, 'the present scale of defense capability seems to closely approach the target goals of the above-mentioned concept'.

The 1995 NDPG considers it 'appropriate that Japan continue to adhere fundamentally to the Basic Defense Force Concept'.[22] The 1995 NDPG, however, does not have the following stipulation contained in the previous NDPG: 'Japan will repel limited and small-scale aggression, in principle, without external assistance.' According to the 2011 Defense White Paper, 'In consideration of the expanded role of the defense capabilities, this stipulation was considered inappropriate as it focuses solely on invasions of Japan.'[23]

The 1995 NDPG expands the role of the defence forces, which includes not only national defence but also two new pillars: 'Response to large-scale disasters and various other situations' and '[c]ontribution to creation of a more stable security environment'. The first pillar includes Japan–US defence cooperation in 'situations in areas surrounding Japan that will have an important influence on Japan's peace and security'. In the face of the nuclear crisis in the Korean Peninsula in 1994, it had become clear that the Japan–US alliance was not prepared for a Korean contingency.[24] As for the second pillar, Japan had already started to dispatch the SDF overseas for United Nations peacekeeping operations (PKO), international humanitarian relief operations, and international relief operations for large-scale disasters overseas.[25]

In conclusion, the 1995 NDPG considers it

> appropriate that Japan's defense capability be restricted, both in scale and functions, by streamlining, making it more efficient and compact, as well as enhancing necessary functions and making qualitative improvements to be able to effectively respond to a variety of situations and simultaneously ensure the appropriate flexibility to smoothly deal with the development of the changing situations.

The 2004 NDPG considers it necessary to retain 'those elements of the Basic Defense Force Concept that remain valid'. The following two elements were considered valid: '1) Japanese defense force should not directly counter military threats, and 2) in order to forestall and prevent invasions, Japan should maintain a defense force that takes into consideration the strategic environment and geographic characteristics.'[26] In conclusion, the main concept of future defence forces in this NDPG is 'multi-functional, flexible, and effective defense forces that are highly ready, mobile, adaptable and multi-purpose, and are equipped with state-of-the-art technologies and intelligence capabilities measuring up to the military-technological level of other major countries'. Emphasis is placed not only on deterrence but also on response and international cooperation.[27]

Posture and structure of the SDF

The Japanese government has managed military change by setting up the numerical targets of major organizations and equipment in the Annex Table of each NDPG. In a practical sense, the Annex Table is the most important part of the NDPG document because it offers the most concrete guidelines for defence buildup plans. In the post-Cold War period, the Annex Tables have played a vital role in rationalizing and streamlining Japan's defence forces, and especially Cold War-style equipment (see Table 3.1). The 1995 NDPG emphasizes 'streamlining, making it [Japan's] defence capability] more efficient and compact'. The 2004 NDPG states:

> we will modify our current defense force building concept that emphasized Cold War-type anti-tank warfare, anti-submarine warfare and anti-air warfare, and will significantly reduce the personnel and equipment earmarked for a full-scale invasion.

Table 3.1 NDPG comparison table

Category		1976 NDPG	1995 NDPG	2004 NDPG
GSDF	Authorized personnel	180,000	160,000	155,000
	Tanks	About 1,200*	About 900	About 600
MSDF	Destroyers	About 60	About 50	47
	Submarines	16	16	16
ASDF	Combat aircraft	About 430	About 400	About 350
	(Fighter aircraft)	(About 350*)	(About 300)	(About 260)

Source: Ministry of Defense (ed.), *Defense of Japan 2014* (Tokyo: Urban Connections, 2014),152.

Note
* Although not stated in the 1976 NDPG, it is listed here for comparison with the NDPG tables after 1995.

In the first 15 years after the end of the Cold War, there was a decreasing trend in the main equipment numbers of each service of the SDF: Ground Self-Defense Force (GSDF), Maritime Self-Defense Force (MSDF) and Air Self-Defense Forces (ASDF).

The increasingly severe security environment and the rise of China

In this section, I show how Japan managed military change by focusing on the international situation, the buildup concept of the defence forces, and the posture and structure of the SDF, which are all described in the 2010 and 2013 NDPG documents.

International situation

The 2010 NDPG recognizes the relative decline of the United States' power as follows: 'we are witnessing a global shift in the balance of power with the rise of powers such as China, India and Russia, along with the relative change of influence of the United States.' The 2013 NDPG takes it one step further by using the expression 'multi-polarization of the world'. In fact, the NDPG describes Japan's security environment as having 'become increasingly severe' since 2010. For example, it depicts North Korea's nuclear and missile development as 'a serious and imminent threat to Japan's security' for the first time.

In particular, the Government of Japan has become increasingly sceptical about China's military activities. The reference to China in the 2004 NDPG is limited to the following:

> China, which has a major impact on regional security, continues to modernize its nuclear forces and missile capabilities as well as its naval and air forces. China is also expanding its area of operation at sea. *We will have to remain attentive to its future actions* [italics added].

By contrast, the 2010 NDPG describes trends in China's military activities as 'of concern for the regional and global community'. Furthermore, the 2013 NDPG uses three times as many characters for China as the 2010 NDPG and demonstrates a stronger sense of caution against China:

> In particular, China has taken assertive actions with regard to issues of conflicts of interest in the maritime domain, as exemplified by its attempts to change the status quo by coercion. As for the seas and airspace around Japan, China has intruded into Japanese territorial waters frequently and violated Japan's airspace, and has engaged in dangerous activities that could cause unexpected situations, such as its announcement of establishing an 'Air Defense Identification Zone'

based on its own assertion thereby infringing the freedom of overflight above the high seas....

As Japan has great concern about these Chinese activities, it will need to pay utmost attention to them, as these activities also raise concerns over regional and global security.

Buildup concept of the defence forces

The 2010 NDPG attracted much attention by introducing the new concept of a Dynamic Defense Force (*doteki boeiryoku*). This concept is characterized by 'readiness, mobility, flexibility, sustainability, and versatility', and 'advanced technology based on the trends of levels of military technology and intelligence capabilities'. It was developed from the Multi-Functional, Flexible, and Effective Defense Force of the 2004 NDPG.[28] The 2010 NDPG also declares that 'Japan should no longer base its defense on the traditional defense concept, the "Basic Defense Force Concept", which places priority on ensuring deterrence through the existence of defence forces per se'.

The concept of the Dynamic Defense Force emphasizes dynamic deterrence through the operational use of the defence forces, as explained by the 2010 NDPG:

> In particular, comprehensive operational performance such as readiness for an immediate and seamless response to contingencies is increasingly important, considering shortening warning times of contingencies due to exponential advances in military technology. Clear demonstration of national will and strong defense capabilities through such timely and tailored military operations as regular intelligence, surveillance, and reconnaissance activities (ISR), not just maintaining a certain level of defense force, is a critical element for ensuring credible deterrence and will contribute to stability in the region surrounding Japan. To this end, Japan needs to achieve greater performance with its defense forces through raising levels of equipment use and increasing operations tempo, placing importance on dynamic deterrence, which takes into account such an operational use of the defense forces.

In short, the Dynamic Defense Force is about defence capabilities focused on high-tempo, responsive operations.

The 2013 NDPG introduces the concept of Dynamic Joint Defense Force (*togo kido boeiryoku*). This concept aims to build 'an effective one [defence force] which enables conducting a diverse range of activities to be dynamic and adapting to situations as they demand' with an emphasis on 'both soft and hard aspects of readiness, sustainability, resiliency and connectivity'. With an intention to formulate the 2013 NDPG, the government conducted capability assessments based on joint operations for the first time, with a focus on the functions and capabilities of the entire SDF in relation to various potential contingencies. In comparison with the Dynamic Defense Force Concept, according to the Ministry of Defense,[29] the concept of Dynamic Joint Defense Force is characterized by a greater emphasis on joint operations; the development of capacities to ensure maritime supremacy and air superiority, and the establishment of rapid deployment capabilities; the strengthening of command, control, communications, and intelligence (C3I) capabilities; and consideration of establishing a wide-ranging logistical support foundation.

Posture and structure of the SDF

With regard to the SDF structure, the 2010 NDPG pays attention to specific functions for the effective and efficient buildup of Japan's defence forces. Specifically, according to the document, 'the SDF will enhance its defense posture by placing priority on strengthening such functions as ISR, maritime patrol, air defense, response to ballistic missile attacks, transportation, and command communications, including in the southwestern region'. Nevertheless, the 2010 NDPG does not increase the level of main equipment except submarines and the Aegis-equipped destroyers that may also be used in ballistic missile defence (see Table 3.2).

It is noteworthy that Japan's number of submarines increased from 16 to 22 in 2010. The NDPG explains this increase as follows: 'The MSDF will maintain augmented submarine units so that it can effectively conduct regular underwater ISR on a broad scale in the seas surrounding Japan as well as patrolling activity in those seas.' Vice-Admiral (retired) Kobayashi Masao, former Commander of Fleet Submarine Force, estimates that eight submarines are necessary for continuous surveillance and patrol over Chinese submarines.[30] The number eight is the sum of six submarines for the chain of islands extending from southwestern Kyushu to northern Taiwan and two for the strait between Taiwan and the Philippines. The continuous deployment of eight submarines requires the possession of 22

Table 3.2 NDPG comparison table

Category		2004 NDPG	2010 NDPG	2013 NDPG
GSDF	Authorized personnel	155,000	154,000	159,000
	Tanks	About 600	About 400	About 300
MSDF	Destroyers	47	48	54
	(Aegis equipped)	(4)	(6)	(8)
	Submarines	16	22	22
ASDF	Combat aircraft	About 350	About 340	About 360
	(Fighter aircraft)	(About 260)	(About 260)	(About 280)

Source: Ministry of Defense (ed.), *Defense of Japan 2014* (Tokyo: Urban Connections, 2014), 151–152.

submarines in addition to two training platforms. Only one-third of all the submarines could be deployed for missions while the others were used for training and repair.[31] A sense of caution against China seems to have brought about the increase of submarines in the Annex Table.

The 2013 NDPG shows the first clear trend of increasing the SDF's main equipment during the post-Cold War period. For example, the MSDF's destroyers increased from 48 ships in 2010 to 54 ships in 2013 while the ASDF's combat aircraft increased from 340 in 2010 to 360 in 2013. The 2013 NDPG was formulated based on the recognition that the current quality and quantity of the defence force was not sufficient for the augmented activities required to deal with the increasingly severe security environment.[32] This perception led to the first clear trend of increasing the SDF's main equipment in the post-Cold War period. On the other hand, the GSDF's Cold War-style equipment, such as tanks and artillery, will be further reduced.

Basic sources of military change

It cannot be denied that the last two NDPG documents may have been politically driven. In 2009, the government, led by the Liberal Democratic Party (LDP), reviewed the 2004 NDPG. Due to a regime change as a result of the general election held in August 2009, however, the Democratic Party of Japan (DPJ)-led government performed the review process all over again and finally approved the new NDPG in December 2010. Then, another change of government from the DPJ back to the LDP took place in December 2012. The LDP government adopted the latest NDPG in December 2013. It had been only three years since the DPJ government had approved the second-to-last NDPG as a decade-long vision for Japan's defence forces.

The recent increasing frequency of revising NDPGs is, however, more attributable to Japanese officials' perception of significant changes in the country's strategic outlook. As the 2013 NDPG points out, 'In the times of an ever-changing security environment surrounding Japan, defense forces need to be constantly reviewed to adapt to the environment.' In this connection, Barry Posen offers a noteworthy hypothesis: 'Generally, anything that increases the perceived threat to state security is a cause of civilian intervention in military matters and hence a possible cause of integration and innovation.'[33] The approval of NDPGs by the Cabinet is one means of top-level civilian intervention in military matters.

Technology is another important factor for Japanese military change. For example, the 2004 NDPG states: 'In case there are significant changes in circumstances, Japan will review and, if necessary, revise the Guidelines in light of the security environment and technological trends at that time, among other things.' Technological trends influence the international situation. In the words of the 2013 NDPG, 'military strategies and military balance in the future are anticipated to be significantly affected by the progress and proliferation of technologies such as those related to precision guided munitions, unmanned vehicles, stealth capability and nanotechnology.' Moreover, the 2004, 2010 and 2013 NDPG documents emphasize that Japan will develop a defence force 'reinforced by advanced technology'. The 2010 NDPG introduced the Dynamic Defense Force Concept against the backdrop that 'comprehensive operational performance such as readiness for an immediate and seamless response to contingencies is increasingly important, considering shortening warning times of contingencies due to exponential advances in military technology'.

In the remainder of this section, I explain why Japan has formulated NDPGs increasingly more often by focusing on two factors: the erosion of cultural anti-militarism, and the increasing difficulty of buck-passing.

Erosion of cultural anti-militarism

Some constructivist scholars attribute the restrained nature of Japanese security policy to the domestic culture of anti-militarism both on the elite and popular levels. For example, Thomas Berger distinguishes anti-militarism from pacifism and translates the former into the expression 'aversion to military power'.[34] The culture of anti-militarism can explain many aspects of Japan's defence policies and activities, including its decision to not send the SDF to the Persian Gulf War in 1991. After the end of the war, Michael Armacost, then US ambassador to Japan, sent the Department of State a cable with the following comments:

> Pacifist sentiment in general and distrust of the Japanese military in particular remain very strong, even among many so-called 'conservatives' in the LDP [ruling Liberal Democratic Party].... While public opinion polls revealed growing support among the Japanese public for the multinational forces and for some Japanese physical presence, there was also strong reaffirmation of Japan's 'Peace Constitution' and of continued non-involvement of Japan in military activities abroad.[35]

Indeed, the SDF had never been sent abroad for operational missions.

As Berger himself puts it, however: 'Changes in the objective condition of any given country may trigger changes in its cultural system, leading to shifts in the normative and interpretative schemes of its political actors.'[36] In the Persian Gulf War in 1991, Japanese policy-makers learned the great lesson that non-military aid alone was not sufficient for international peace cooperation. Although Japan contributed US$13 billion to support the allied force, Japan was not recognized as a responsible member of the international community. It became clear to Japanese policy-makers that it was at least necessary for Japan to dispatch the JSDF overseas for international peace. As former Japanese diplomat Kazuhiko Togo puts it, 'the deep sense of crisis, which enveloped Japan after the "defeat in the Gulf", in turn, became the basis for future development'.[37]

After the end of the Persian Gulf War, Japan started to dispatch the SDF overseas for non-combat military missions. The following is a list of the first deployments in various categories: (1) UN peacekeeping operation (PKO) in Cambodia from September 1992 to September 1993; (2) international humanitarian assistance in the former Republic of Zaire from September to December 1994; (3) international disaster relief operation in Honduras from November to December 1998; (4) activities to respond to international terrorism in the Indian Ocean from November 2001 to February 2010; and (5) cooperation in efforts towards the reconstruction of Iraq from January 2004 to February 2009.[38] In January 2007, moreover, international peace cooperation activities and activities responding to situations in areas surrounding Japan were officially added to the primary missions of the SDF, which had consisted of only the defence of Japan and the maintenance of public order.[39]

The incremental implementation of these deployments seems to have eroded the culture of anti-militarism. Berger also recognizes the decline of anti-militarism.[40] According to the public opinion survey on the SDF and defence issues conducted by the Cabinet Office in January 2012, a record 91.7 per cent of respondents said that they had a 'good impression' of the SDF. This record seems to be due to the SDF's activities in the Great East Japan Earthquake.[41] Moreover, the survey also shows an increasing trend from the early 1970s in the favourable opinion of the SDF (see Figure 3.1). In the 1970s and 1980s, pacifism had already weakened to a great extent.[42]

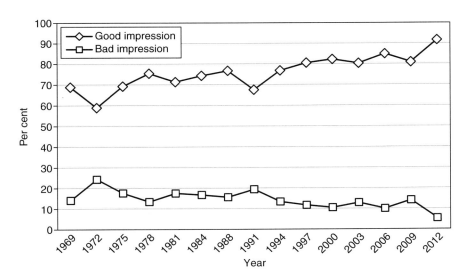

Figure 3.1 Impression of the SDF (source: '"Public Opinion Survey on the Self Defense Forces and Defense Issues" (excerpt) (Public Relations Office of Cabinet Office: as of January 2012)', cited in Ministry of Defense (ed.), *Defense of Japan 2014* (Tokyo: Urban Connections, 2014), 464–465).

Increasing difficulty of buck-passing

Realists challenge the constructivist argument for anti-militarism in Japan. For example, Eric Heginbotham and Richard Samuels consider Japanese foreign policy to be 'mercantile realism' with an emphasis on economic and technological security interests.[43] They pay attention to the strategic vision of Shigeru Yoshida. It is possible to interpret the Yoshida Doctrine as Japan's 'cheap ride on U.S. security guarantees'.[44] Similarly, Jennifer Lind argues that Japanese security policy may be attributed to a realist strategy of buck-passing to the United States.[45]

Lind's buck-passing theory is also useful in explaining Japan's recent sensitivity towards the international security environment. According to this theory, states will have to give up the preferable strategy of buck-passing when they face a growing external threat and a decline in the security protection provided by their powerful allies. Recent NDPG documents pay attention to the rise of China and the relative decline of the United States' power. In this situation, according to a balance-of-power theory, Japan has no choice but to strengthen its balancing behaviour by two means: 'internal effort (moves to increase economic capability, to increase military strength, to develop clever strategy) and external efforts (moves to strengthen and enlarge one's own alliance or to weaken and shrink an opposing one).'[46] Although the alliance remains the cornerstone of Japan's security, Japan can no longer afford to pass the buck to its ally.

As for internal efforts, the Government of Japan increased defence-related expenditures of 2013 in real terms for the first time in 11 years. Still, the growth rate of defence-related expenditures from previous years was just 0.8 per cent and the ratio of defence-related expenditures to GDP was still 0.96 per cent.[47] Although the defence budget increased again in 2014 for the second year in a row,[48] Japan is not likely to be able to afford a rapid increase, given internal budgetary pressures.

Budgetary constraints, however, may work in favour of military change. As the 2010 NDPG puts it:

> In order to deal with the increasingly difficult security environment, Japan needs to steadily build an appropriate-size defense force. In doing so, Japan will choose truly necessary functions on which to concentrate resources, and carry out structural reform of the defense forces, thereby producing more outcome with limited resources. To this end, *Japan will drastically rationalize and streamline the SDF overall through fundamentally reviewing, in light of its difficult fiscal condition, the equipment, personnel, organization and force disposition*, including the equipment and personnel that have been maintained as preparation to defend against a full-scale invasion [italics added].

In short, it appears that severe fiscal conditions may promote military change by creating the need to rationalize and streamline defence forces.

Conclusions

Since the mid-1970s, the Government of Japan has managed military change mainly at the defence-strategic level by formulating five NDPGs. On the basis of Japanese defence policy, each of these documents defines the buildup concept and role of the defence forces in response to the international situation, and then decides the posture and structure of the SDF. The 1976 NDPG document adopted the Basic Defense Force Concept and restrained defence force buildup in the context of Cold War détente. After the end of the Cold War, the next two NDPG documents of 1995 and 2004 retained the same defence force building concept and were instrumental in reducing the personnel and Cold War-type equipment earmarked for a full-scale invasion and, at the same time, preparing for a response to various situations and international cooperation. In contrast, the 2010 and 2013 NDPG documents presented the new concepts of Dynamic Defense Force and Dynamic Joint Defense Force given the changes in the security environment, and started to emphasize increasing levels of equipment use and operations tempo and, in the case of the 2013 NDPG, developing defence capabilities adequate not only in quality but also in quantity.

The reason why Japan has formulated NDPGs increasingly more often may be explained not only by strategic, political and technological factors but also by two other factors: the erosion of cultural anti-militarism and the increasing difficulty of buck-passing. In the past, Japan's management of military change was strictly constrained by the culture of anti-militarism, while Japan could afford to dismiss strategic thinking under the security protection provided by the United States. Therefore, the weakening of these constructivist and realist factors explains Japan's more prompt response to the changes in the security environment. In other words, a realist strategy of balancing is now a key to understanding Japan's military change in recent years. Tsuyoshi Kawasaki argues that Barry Posen's balance-of-power theory (1986) has a more explanatory power for the 1976 NDPG than domestic culturalist theory.[49] In my view, this argument is more applicable to the recent NDPG documents.

Notes

1 For the Japanese and English texts of the National Security Strategy, visit the Cabinet Secretariat website at www.cas.go.jp/jp/siryou/131217 anzenhoshou.html.
2 The Japanese title of the first and second NDPGs used to be translated by the words 'National Defense Program Outline'.
3 'National Defense Program Guidelines for FY 2011 and beyond', approved by the Security Council and the Cabinet on 17 December 2010 (hereafter cited as the 2010 NDPG). For the full texts of the NDPGs of 1976, 1995, 2004, 2010 and 2013 and the provisional English translation of the first two NDPGs, visit the website of 'The World and Japan' Database Project, Database of Japanese

Politics and International Relations, Institute of Oriental Culture, University of Tokyo at www.ioc.u-tokyo.ac.jp/~worldjpn/documents/indices/JPSC/index.html. For the provisional English translation of the NDPGs of 2004, 2010 and 2013, visit the website of the Ministry of Defense at www.mod.go.jp/e/d_act/d_policy/national.html.

4 The 2004 and 2010 NGDP documents also contain an embryonic form of national security strategy. They not only state the objectives of Japan's security policy but also mention a whole-government approach and non-military means such as Official Development Assistance (ODA). Since these elements developed into the NSS, the 2013 NDPG concentrates on the defence forces.
5 *Asahi Shimbun*, 26 September 2005, p. 1.
6 For the Constitution of Japan, see the website of Japanese law translation at www.japaneselawtranslation.go.jp/law/detail_main?id=174.
7 Ministry of Defense, ed., *Defense of Japan 2011* (Tokyo: Urban Connections, 2011), p. 137.
8 The Japanese government defines the right of collective self-defence narrowly as 'the right to use actual force to stop an armed attack on a foreign country with which it has close relations, even if the state itself is not under direct attack' (Ministry of Defense 2011, p. 138).
9 During a Cabinet meeting on 1 July 2014, the Abe administration approved the government's new view on security legislation, including a reinterpretation of Article 9 of the Constitution so that Japan may exercise the right of collective self-defence, 'when an armed attack against a foreign country that is in a close relationship with Japan occurs and as a result threatens Japan's survival and poses a clear danger to fundamentally overturn people's right to life, liberty and pursuit of happiness'. Cabinet Secretariat, 'Cabinet Decision on Development of Seamless Security Legislation to Ensure Japan's Survival and Protect its People', 2014, pp. 7–8, www.cas.go.jp/jp/gaiyou/jimu/pdf/anpohosei_eng.pdf (accessed 31 July 2014).
10 Ministry of Defense, 'Constitution of Japan and Right of Self-Defense', 2013, www.mod.go.jp/e/d_act/d_policy/dp01.html (accessed 20 November 2013).
11 Ministry of Defense, 'Basis of Defense Policy', 2013, www.mod.go.jp/e/d_act/d_policy/dp02.html (accessed 20 November 2013).
12 Kiyofuku Chuma, *Saigunbi no seijigaku* [Politics of remilitarization] (Tokyo: Chishikisha, 1985), p. 47.
13 Yomiuri Shimbun sengoshi han, ed., *'Saigunbi' no kiseki* [Locus of remilitarization] (Tokyo: Yomiuri Shimbunsha, 1981), p. 409.
14 *Boei hakusho* (Defense White Papers) are available at the following page of the Ministry of Defense website: www.clearing.mod.go.jp/hakusho_web/. Boeicho [Defense Agency]/Boeisho [Ministry of Defense], ed., *Boei hakusho* [Defense White Paper] (Tokyo: Okurasho Insatsukyoku/Zaimusho Insatsukyoku/Gyosei, annually).
15 Yomiuri Shimbun sengoshi han, 1981, p. 417.
16 Masaru Honda, *Nihon ni kokka senryaku wa arunoka* [Does Japan have a national strategy?] (Tokyo: Asahi Shimbunsha, 2007), p. 215.
17 Ministry of Defense, 2011, pp. 139–140.
18 Ministry of Defense, 2011, p. 151.
19 Defense Agency, ed., *Defense of Japan 1977* (Tokyo: Defense Agency, 1977), pp. 46–82.
20 Defense Agency, 1977, p. 54.
21 Akihiko Tanaka, *Anzenhosho – Sengo 50nen no mosaku* [Security – searching in the post-war 50 years] (Tokyo: Yomiuri Shimbunsha, 1997), p. 260.
22 In the 1995 NDPG, the Basic Defense Force Concept is:

defined as possessing the minimum necessary defense capability for an independent nation so that it would not become a source of instability in the surrounding regions by creating a vacuum of power rather than building a capability directly linked to a military threat to Japan.
23 Ministry of Defense, 2011, p. 143, n. 1.
24 Masahiro Akiyama, *Nichibei no senryaku taiwa ga hajimatta* [The Japan–US strategic dialogue began] (Tokyo: Aki Shobo, 2002), p. 50.
25 Isao Miyaoka, 'Japan's Dual Security Identity: A Non-combat Military Role as an Enabler of Coexistence', *International Studies*, vol. 48, nos 3 and 4, 2011, p. 245.
26 Ministry of Defense, 2011, p. 144, n. 1.
27 Ministry of Defense, 2011, p. 151.
28 Ministry of Defense, 2011, p. 160.
29 Boeisho (Ministry of Defense), *Aratana boei keikaku no taiko/chuki boeiryoku seibi keikaku* [The new NDPG/Mid-Term Defense Program], 2014, p. 21, www.kantei.go.jp/jp/singi/kaiyou/ritou_yuusiki/dai10/siryou.pdf (accessed 29 July 2014).
30 Masao Kobayashi, 'Shin boei taiko ni miru "sensuikan 22 seki taisei" no kaijo boei' [Maritime Defence by the 22 Submarine Structure in the New NDPG], *Gunji kenkyu* [Japan military review], vol. 46, no. 12, 2011, pp. 28–39.
31 According to Kobayashi, during the Cold War period, Japan developed an operational concept to block the Soviet naval forces from advancing into the Pacific Ocean at the three Straits of Soya, Tsugaru and Tsushima. It was necessary to maintain 16 submarines (and two training submarines) for the deployment of two submarines to each strait. Kobayashi, 2011.
32 *Boeisho*, 2014, pp. 20–21.
33 Barry R. Posen, *The Sources of Military Doctrine: France, Britain, and Germany between the World Wars* (Ithaca, NY: Cornell University Press, 1984), p. 79.
34 Thomas U. Berger, *Cultures of Antimilitarism: National Security in Germany and Japan* (Baltimore, MD: Johns Hopkins University Press, 1998), p. 1.
35 National Security Archive, 'The Gulf War: Impact on Japan and U.S.–Japan Relations', Secret, Cable, 14 March 1991, *Japan and the United States, 1977–1992*. JA01680.
36 Berger, 1998, p. 18.
37 Kazuhiko Togo, *Japan's Foreign Policy 1945–2003: The Quest for a Proactive Policy* (Leiden, the Netherlands: Brill, 2005), p. 77.
38 Ministry of Defense, 2011, pp. 518–519.
39 Ministry of Defense, 2011, pp. 347, 555.
40 Thomas U. Berger, *Redefining Japan and the U.S.–Japan Alliance* (New York: Japan Society, 2004), p. 24.
41 *Yomiuri Shimbun*, 11 March 2012, p. 2.
42 Peter J. Katzenstein, *Cultural Norms and National Security: Police and Military in Postwar Japan* (Ithaca, NY: Cornell University Press, 1996), p. 120.
43 Eric Heginbotham and Richard Samuels, 'Mercantile Realism and Japanese Foreign Policy', *International Security*, vol. 22, no. 4, 1998, pp. 171–203.
44 Richard J. Samuels, *Securing Japan: Tokyo's Grand Strategy and the Future of East Asia* (Ithaca, NY: Cornell University Press, 2007), p. 32.
45 Jennifer Lind, 'Pacifism or Passing the Buck? Testing Theories of Japanese Security Policy', *International Security*, vol. 29, no. 1, 2004, pp. 92–121.
46 Kenneth N. Waltz, *Theory of International Politics* (New York: McGraw-Hill, 1979), p. 118.
47 Ministry of Defense, ed., *Defense of Japan 2013* (Tokyo: Erklaren Inc, 2013), pp. 118, 327.
48 *Yomiuri Shimbun*, 25 December 2013, p. 11.

49 Tsuyoshi Kawasaki, 'Japan and Two Theories of Military Doctrine Formation: Civilian Policymakers, Policy Preference, and the 1976 National Defense Program Outline', *International Relations of the Asia Pacific*, vol. 1, no. 1, 2001, pp. 67–93.

4 Garuda rising?
Indonesia's arduous process of military change

Benjamin Schreer

Indonesia has long been the sleeping strategic giant in Southeast Asia. Yet for decades its armed forces (Tentara Nasional Indonesia, TNI) focused on its domestic role as the guardian of authoritarian governments. TNI also suffered from a lack of strategic direction, chronic underfunding and ageing defence equipment. However, Indonesia's positive political and economic development over more than a decade has led to a renewed ambition towards military change. Efforts are underway to modernize TNI's education, doctrine and equipment. Nevertheless, major impediments to military change persist. Indonesia's institutional capacity to formulate strategy and to manage military change remains limited. Furthermore, defence policy remains low on the political priority list. Finally, the defence acquisition process remains ad hoc and often not driven by strategic considerations but parochial interests and high levels of corruption. In sum, Indonesia's military change remains very much unfinished business and there exists a significant gap between strategic ambitions and reality.

The issue of Indonesia's military reform is important for regional stability in the Asia-Pacific region. Indonesia is located in the centre of key maritime chokepoints in Southeast Asia – such as the Malacca, Lombok and Sunda Straits – through which most maritime trade to and from the region has to transit. Moreover, its population and economy should make it 'the natural leader of Southeast Asia: its economy, in purchasing power parity terms, constitutes about one-third of the ASEAN (Association of Southeast Asian Nations) states' aggregate economies, and its population almost 40% of their aggregate populations'.[1] A population of over 250 million people already makes Indonesia second to none population-wise in Southeast Asia and the fifth largest country globally; and its population is expected to grow to over 300 million by 2050.

Indonesia has also experienced a remarkable political and economic transition in recent years. While there is still some level of uncertainty about whether the process of democratization upon which Indonesia embarked since the fall of the Suharto Regime in 1998 is irreversible, most analysts regard a return to authoritarianism as a very unlikely scenario.[2] As will be discussed below, one milestone in Indonesia's democratization has been the relative success of creating healthy civil–military relations (e.g. the subordination of the military to civilian rule). This has enabled political and military elites to pay more attention to other elements of TNI reform.

Moreover, for quite some time Indonesia has enjoyed solid economic growth rates. During the global financial crisis, Indonesia outperformed its neighbours and joined China and India as the only G20 members experiencing economic growth in 2009. In 2011, Moody's and Fitch upgraded the country's credit rating. Moody's further upgrade in 2012 was based on the assessment that Indonesia's 'resilience to large external shocks points to sustainably high trend growth over the medium term'.[3] The OECD 2012 Economic Outlook stated that Indonesia could surpass China by 2020 in terms of the highest growth rates among major countries.[4] A Pricewaterhouse-Coopers study predicted that it could even overtake G7 economies such as Germany, France and the United Kingdom by 2050.[5]

One should beware such predictions of long-term uninterrupted, linear economic growth. Nevertheless, the emergence of a much more powerful Indonesia over the coming decades is a distinct possibility. Military power will be one variable and Indonesia has formulated a very ambitious reform agenda for its armed forces. The 2010 'Strategic Defence Plan' announced the goal to develop a 'Minimum Essential Force' (MEF) by 2024. The wish list included new warships, submarines and combat aircraft to provide the backbone of a 'green-water navy'; a more capable air force; and a more mobile and agile land force. In addition, the capability development process is to be supported by a viable domestic industrial base.[6] The goal is to turn the TNI into a modern and deployable force, capable of addressing a range of external and internal security challenges.

Against this background, this chapter analyses the current process of military change in Indonesia. It starts by outlining the magnitude of the challenge to reforming Indonesia's military. This is followed by a discussion of the main domestic and external drivers behind Indonesia's current military modernization. The third part looks at the progress made, as well as the ongoing impediments to turning the TNI into a more modern defence organization. The fourth and final section provides an outlook on the future prospects for TNI reform.

The legacy of history and strategic culture

Before analysing current trends in Indonesia's military modernization, it is important to consider the impact of history and strategic culture upon the country's defence policy. Strategic culture comprises a nation's particular historical, geographical and cultural experiences and conditions that impact upon its strategic behaviour. It provides the political-societal context for strategic decision-making and is resistant to major changes.[7]

In Indonesia's case, the country's emergence from a guerrilla-style independence struggle against Dutch colonialism in the 1940s left a deep imprint on its strategic thinking. The major challenge for any hostile power to occupy and control Indonesia's vast territory and population is a significant advantage for Indonesian defence planning. In combination, strategic history, geography and culture made Indonesia a continental power – despite being an archipelago comprising about 13,000 islands. In defence policy terms, this manifested itself in an army-centric 'Total People's Defence' concept which is based on nonlinear guerrilla warfare against a potential invasion by external powers. The last Defence White Paper of 2008 still embraced this concept, despite the fact that Indonesia's external environment had changed significantly.

Indonesia's political history until the fall of President Suharto in 1998 also favoured an inward-looking defence policy as the armed forces' core role was to function as the guardian of the authoritarian state. Indeed, under the long, military-dominated authoritarian 'New Order' era of Suharto between 1967 and 1998, the TNI focused on maintaining its 'hegemonic grip on state institutions, the economy and society',[8] as well as addressing challenges of separatism. The result was a land-centric, manpower-intensive approach to 'protect the people', rather than protecting national sovereignty against external risks and threats. This further strengthened the dominant role played by the Indonesian land forces (Tentara Nasional Indonesia Angkatan Darat, TNI-AD) in the defence policy process, and a military doctrine that emphasized a defensive use of military power. It also prevented Indonesia from developing a strong maritime awareness and a coherent maritime strategy.

Finally, in the absence of a perceived external threat and faced with a small economic base, defence spending over decades was very low as well. Thus, when the regime finally collapsed in 1998, TNI was characterized by run-down military equipment (particularly in the air and maritime domains), low levels of military professionalism and readiness, high levels of corruption, and a major role in almost every level of politics and society. In addition, the new elites had very little institutional mechanism and intellectual support (such as independent defence policy think-tanks) to formulate, let alone execute, an effective strategy. In short, the task to reform TNI into a modern organization that served the young democracy was enormous. Consequently, efforts to reform TNI post-1998 started largely from scratch.

TNI reform in the early 2000s: the primacy of civil–military relations

Given this domestic context, Indonesia's young democracy understandably focused on 'putting the military back in the box' in the immediate aftermath of Suharto's demise. That is, the emphasis of military reform was on depoliticizing the military and establishing first-order principles of democratic civil–military relations.[9] Consequently, 'first-generation' military reforms included

> extensive changes to the country's institutional framework, judicial system, electoral mechanisms, composition of representative bodies, and the responsibilities of security agencies.[10]

This process of political marginalization of the TNI yielded remarkable results during the first decade of the 2000s. In line with Indonesia's successful democratic transition in other sectors, a high degree of civilian control over the military was established. As well, TNI lost its dominant position in internal security to the Indonesian National Police (INP). While it retained a strong role in national disaster relief missions and anti-separatist campaigns in Papua province, for instance, domestic counter-terrorism tasks were largely transferred to the INP.

However, the civil–military reforms were far from perfect. In particular the old, army-heavy territorial command structure (KODAMs) remained largely untouched – providing the TNI with an ongoing influence over local and provincial politics. As well, TNI continued its practice of raising off-budget funds to address the problem of low salaries of junior and mid-ranking officers.[11] As will be discussed below, these deficiencies continue to make current defence procurement projects very vulnerable to corruption.

The focus of the first-generation reform on restoring civil–military relations also meant that less effort was put into enhancing TNI's military capabilities. Simply put, this dimension of TNI modernization was a second-order priority for political elites. Moreover, significant investments in capability development also fell short due to insufficient funding. Ambitious force modernization plans formulated during the 1990s became a victim of the 1997 to 1998 Asian Financial Crisis which put additional stress on an already small defence budget. In 1997, the total defence budget amounted to approximately US$2.5 billion. In 2001, that figure dropped even lower to an estimated US$1.9 billion for an active force of 297,000 personnel.[12] To be sure, the TNI's 'unofficial budget' had always been higher due to corruption and the military's involvement in local economies. However, this 'self-financing' did not go into military modernization programmes (new platforms, etc.) but was mainly used to boost the low salaries of soldiers.

As a result, despite significant progress in civil–military relations, by the mid-2000s the overall status of the TNI as a modern force was generally poor. It still faced an urgent requirement to replace ageing military systems and continued to suffer from very low levels of readiness as well as professionalism. Critical 'enablers' such as logistics, command and control (C2), as well as command, control, communications, computers, intelligence, surveillance and reconnaissance (C4ISR) systems were basically non-existent. Finally, the military-industrial base to sustain and maintain the force was very weak.

It was only during the first decade of the 2000s when conditions for more concerted efforts to address some of these shortfalls improved. The ongoing process of democratic consolidation and economic growth provided new scope for TNI modernization beyond civil–military relations. The prospects of a military coup appeared very remote and political elites gained confidence in TNI's willingness to stay out of politics and focus to a greater degree on external security challenges. Ongoing economic growth meant that more money was available to tackle TNI's chronic underfunding. Table 4.1 demonstrates that between 2003 and 2013 Indonesia's defence spending indeed increased significantly, commensurate with a rising Gross Domestic Product (GDP).

Growing defence spending permitted an increase in TNI personnel pay and allowances. It also allowed greater investment in modern defence equipment, particularly with regard to air and naval systems. Moreover, largely stripped of its role in domestic politics and internal security, the military leadership had to focus more on keeping the TNI relevant as an instrument of a democratically elected government. The military's role in internal security was even further diminished after a wave of communal violence in Maluku, Kalimantan and Sulawesi subsided by around 2003, and after the resolution of the separatist conflict in Aceh in 2005. In the absence of a major outbreak of communal or separatist violence in the archipelago, the TNI was forced to direct more energy towards addressing external challenges to Indonesia's sovereignty.

Indonesia's external strategic environment

The TNI's most pressing task is to secure Indonesia's massive maritime Exclusive Economic Zone (EEZ). Its archipelago comprises 13,000 islands which stretch over nearly two million square kilometres. Indonesia's coastline is 54,716 kilometres long and it has land borders with Malaysia (1,782 kilometres), Papua New Guinea (820 kilometres) and Timor Leste (228 kilometres). As mentioned already, Indonesia also sits in the middle of Southeast Asia's most important maritime chokepoints. While its geography makes a hostile attempt of invasion extremely unlikely, it also greatly complicates the TNI's ability to exercise control over its territorial waters. Currently, the TNI struggles to protect even 12 of Indonesia's outermost islands.[13] Weak maritime capabilities also make the combat of non-traditional security challenges within the EEZ, such as illegal fishing, human and drug trafficking, and piracy, a very difficult undertaking. Furthermore, the TNI faces significant problems in effectively controlling the country's air space.

Indonesia's political elites also recognized a need to strengthen the military instrument to address a changing external security environment

Table 4.1 Indonesia defence spending, 2003 to 2013

Year	2003	2004	2005	2006	2007	2008	2009	2010	2011	2012	2013
Rupiah (trillion)	18.2	21.4	23.9	23.6	32.6	32.8	33.6	42.9	51.1	72.5	81.8
US$ (billion)	2.12	2.39	2.47	2.59	3.57	3.40	3.25	4.70	5.82	7.74	8.37
Real growth (%, US$)	33.0	13.0	3.0	5.0	38.0	−5.0	−4.0	45.0	24.0	33.0	8.13
Percentage of GDP (US$)	0.99	0.93	0.88	0.75	0.82	0.67	0.60	0.67	0.69	0.86	0.88
GDP (US$ billion)	214	258	281	346	433	511	544	704	846	894.9	946

Source: figures based on various issues of International Institute for Strategic Studies, *The Military Balance*, London, issues from 2004 to 2014.

in the broader Asia-Pacific region. In essence, regional power shifts pose new questions for Indonesia's defence policy. One is the need to respond to military modernization trends in neighbouring Southeast Asian countries which invest in sophisticated conventional submarines, air combat capability as well as intelligence, surveillance and reconnaissance (ISR) platforms.[14] From an Indonesian perspective, the military modernization trends in Malaysia and Singapore are specific points of reference. Malaysia is viewed by many Indonesian elites as a military 'peer competitor', a perception aided by the fact that both countries are still in a dispute over the jurisdiction of Ambalat in the Sulawesi Sea and Malaysian incursions into Indonesian territorial waters. Consequently, some TNI procurement plans have been explicitly justified with reference to a (perceived) need to keep up with Malaysia in this area. For instance, the TNI argued the case for new conventional submarines by pointing to Malaysia's commission of two French *Scorpène*-class submarines in 2009.

Singapore is upgrading its submarine and air combat capability, which is second to none in Southeast Asia. The Singaporean Air Force (SAF) already operates US F-15SG long-range strike aircraft and recently upgraded its fleet of F-16 C/D fighter aircraft. Moreover, it is an open secret that the SAF will acquire fifth-generation US Joint Strike Fighter (JSF) aircraft, the only Southeast Asian nation to do so. Indonesian elites envy this capability and argue that the TNI needs to minimize this air combat capability gap. In addition, despite military cooperation (for example, in the area of submarine rescue), the strategic relationship has not been without friction. For instance, high-ranking Indonesian defence officials 'skipped' the 2014 Singapore Air Show as a result of diplomatic tensions when the TNI named a new frigate after two Indonesian marines executed in 1968 for a 1965 bombing in Singapore, killing three and wounding 33 civilians. Indonesian sources also claimed that Singapore has helped Australia with spying activities.[15]

Therefore, political and military leaders in Indonesia increasingly emphasize the need for TNI to keep up with Southeast Asia force modernization. Yet, 'prestige' and 'status' thinking is arguably the key driver behind such arguments, as opposed to strategic necessity. Almost all armed forces in Southeast Asia (with the exception of Singapore) remain seriously imbalanced and underfunded, despite their modernization efforts. None is in a position to strategically pose a challenge to Indonesia's sovereignty and it is hard to see which of those nations would have any intention of doing so.

Indonesian strategic decision-makers are also concerned about the strategic implications of major power dynamics in Asia.[16] The consequence of China's rise for Indonesia's foreign and security posture is the key challenge. At this point the Indonesian government does not perceive China as a threat and prefers to maintain good trading relations. After all, China has become Indonesia's second-largest trading partner. Moreover, in recent years Jakarta has concluded a number of defence technology deals with Beijing. For example, under a technology transfer agreement, Indonesia now produces the Chinese C-802 anti-ship cruise missile and talks are underway for a similar arrangement regarding the C-705 anti-ship cruise missile.

That said, it grows harder for the Indonesian strategic establishment to ignore the challenge posed by China's rise. Beijing's expansive maritime claims in the South China Sea comprising the so-called 'nine-dashed line' demonstrate China's ambition to establish a hegemonic position in maritime Southeast Asia, thereby undermining Indonesia's desire to maintain the regional status quo. The line also runs close to Indonesia's Natuna Islands, and Chinese and Indonesian maritime patrol vessels have already encountered each other in several hostile incidents in the past. As a result, Indonesian officials have become increasingly concerned that China could contest Indonesia's territorial integrity by seeking to establish its influence around Natuna Island; yet they have stopped short of turning it into a public dispute with Beijing.[17] But, as will be discussed below, the TNI has sought to strengthen its capability for possible contingencies in these areas.

Moreover, the debate on whether to 'hedge' against China's rise, including through closer strategic relations with the United States, has gained traction among Indonesian elites. Whether this translates into actual defence policy changes will depend very much on China's strategic behaviour in the South China Sea and its willingness to openly contest parts of Indonesia's EEZ. In any event, China's trajectory has reinforced the need to increase the TNI's capabilities to respond to the emerging strategic environment in the South China Sea.

In sum, while Indonesia's threat perception still focuses predominantly on internal and non-traditional security challenges (such as terrorism, separatism, natural disasters, people smuggling and illegal fishing), external security challenges are becoming more prominent in Indonesia's strategic assessment. They include border disputes and the protection of outer islands, violations of sea lines of communications (SLOCs) and possible conflicts over energy resources in Indonesia's EEZ. Indonesia needs to pay greater attention to the possibility of hostile actions by foreign military powers, particularly around islands in its territorial waters (such as Riau and Riau Islands), and in maritime chokepoints such as the Malacca and the Lombok Straits. As well, Indonesian strategic planners are concerned about a possible foreign intervention in separatist conflicts such as Papua, and border conflicts with Malaysia.

TNI capability development: towards a 'Minimum Essential Force' (MEF)

The MEF has served as a conceptual framework for TNI military modernization based on what it called 'Flash-point Defence'. It is essentially a geographically based approach which requires the TNI to develop a capability to project and employ 'forces in the areas of potential conflict – most of which are part of the outer islands'.[18] In other words, the defence of Indonesia's vast archipelago and the adjunct waters became a central mission. One concrete result has been the upgrade of military facilities on Natuna and Riau Islands to accommodate combat aircraft.[19]

In addition, the TNI began to address other key MEF components.[20] First, the Indonesian Navy (Tentara Nasional Indonesia Angkatan Laut, TNI-AL) moved towards what it calls a 'green-water' navy. The ambition is not to develop an expeditionary regional navy or even a 'blue-water' navy. Instead, the role of a green-water navy is to focus on littoral warfare and the defence of territorial waters by investing in anti-access capabilities in order to fortify a nation's maritime approaches. This includes capabilities such as anti-ship missiles, fast-attack craft, submarines, land-based missiles, land-based tactical combat aircraft, sea mines and amphibious operations.[21] The TNI-AL's ambition is no small feat: by 2024 it aims to possess 110 modern surface combatants, 66 patrol vessels, 98 support ships, plus 12 new diesel-electric submarines. This is very ambitious indeed, especially when considering that fewer than half of its current fleet of 213 vessels are deemed seaworthy and those that are were mostly commissioned during the Cold War.

Nevertheless, the TNI-AL made some progress towards achieving this objective. It upgraded its existing six frigates with modern anti-ship missiles and radar technologies. It also acquired four new 1,700-tonne Dutch *Sigma*-class corvettes. As well, it plans to build two modular 2,300-tonne *Sigma*-class ships by 2017/2018. In addition, it has invested in a number of smaller ships optimized for coastal defence, such as locally produced small patrol boats and fast-attack guided missile boats. Moreover, the TNI-AL upgraded its amphibious capability by acquiring four 7,300-tonne landing platform docks (LPD) and more than 50 Russian amphibious fighting vehicles.

However, at this stage, other critical capability areas remain significantly underdeveloped. For example, Indonesia seeks to replace its two 1980s 1,200-tonne *Type 209* submarines with three modern 1,400-tonne *Type 209* boats from South Korea, with the third one to be assembled in Indonesia. In the future, the total number of submarines is set to grow to 12. However, given the limited financial reasons and enormous challenges involved in building and sustaining a complex military system such as a modern conventional submarine, huge doubts remain over Indonesia's capacity to operate this capability by 2024. Furthermore, TNI-AL still lacks significant investments in offshore patrol vessels, maritime patrol aircraft, mining, counter-mining or anti-submarine warfare. It is far from being a 'balanced fleet' which could effectively control most of Indonesia's territorial waters. This shortfall is compounded by the absence of a maritime strategic mind-set guiding Indonesian defence planning.

Second, Indonesia's Air Force (Tentara Nasional Indonesia Angkatan Udara, TNI-AU) has announced major investments in air combat capability, tactical airlift and unmanned aerial vehicles (UAVs). Modernization is urgently needed, since it is estimated that fewer than half of its fleet are currently operational and largely outdated. Yet, some of its recent procurement decisions when it comes to air combat lack focus. This is because the TNI-AU opted to acquire largely incompatible fighters from the United States (24 F-16 C/D) and Russia (16 Sukhois of different variants), adding up to about two fighter squadrons; it appears unrealistic that TNI-AU will operate 10 fighter squadrons by 2024.

In addition, it lacks airborne early warning and control (AEW&C) systems which are critical in modern air combat operations, as well as airborne refuelling to extend the range of its tactical fighter aircraft. Its approach to addressing the problem of an ageing, poorly maintained tactical airlift capability has also been somewhat piecemeal. Indonesia took up Australia's offer to provide up to nine refurbished C-130H transport aircraft. Yet, apart from that the TNI-AU currently only invests in locally produced, less capable CN-235 medium transport aircraft. As a result, it has modestly increased its capacity to patrol Indonesia's air space and to provide tactical airlift for operations within the archipelago. However, it continues to lack key assets for modern air operations and it is highly likely that it will remain a second-class air force for the foreseeable future.

Finally, the Indonesian Army (Tentara Nasional Indonesia Angkatan Darat, TNI-AD) has retained its prominent role in the defence policy process. It has announced the goal to transform into a more agile, rapidly deployable force to play a greater role in the defence of Indonesia's islands. However, it remains at heart focused on maintaining its place as the dominant service and preserves a continental strategic mind-set. This has been partly evident in its recent procurement of major weapon platforms, which appear to be driven more by status than by operational utility. For example, its decision in 2012 to buy 103 surplus Leopard 2A6 main battle tanks (MBT) and 50 infantry fighting vehicles from Germany has been described by defence officials as a 'cornerstone' of the MEF and vital to strengthening 'border security'. However, MBTs are ill-suited for operations within the archipelago, and the mountainous border with Malaysia also makes their operational utility highly questionable with regard to border protection.

Furthermore, the Army leadership has resisted several attempts since 1998 to abolish its 13 regional territorial commands (KODAMs) and to replace them with a command structure more suitable for expeditionary deployments. The KODAMs main function is to allow the TNI-AD ongoing involvement in local politics as well as off-budget financing activities. That said, the TNI-AD has taken some modest steps to improve army aviation. For example, it acquired 18 medium-lift transport helicopters from Russia to enhance its airlift capability for national contingencies such as disaster relief. It also bought six surplus Russian air-attack helicopters. In August 2013, Indonesia announced the purchase of eight US Apache attack helicopters from the United States, a world-class military system. Configured for littoral operations, it could indeed contribute to TNI's overall capability in this area. Yet, it remains to be seen if the TNI-AD leadership decides to choose this option; in a purely land-attack role the Apaches would amount to very expensive 'flying main battle tanks' with limited operational utility.

The land force will play a major part in Indonesia's intention to increase its contribution to international peacekeeping operations (PKO). In 2011, Indonesia established a new TNI Peacekeeping Centre in Sentul, Java. In March 2012, the government of President Susilo Bambang Yudhoyono (SBY) announced the goal of creating a standing PKO force of about 10,000 personnel to contribute to United Nations (UN) PKO; at present, Indonesia has a contingent of approximately 4,000 troops.[22] Most of those force elements will come from the TNI-AD.

Therefore, the TNI-AD has taken some measures towards a more agile and deployable force. But overall it remains largely non-deployable due to ineffective training schemes, as well as lack of financial resources and KODAM reform. Until the TNI-AD leadership addresses the problem of a land force that is too large and expensive, TNI modernization as a whole will remain challenging.

Obstacles to TNI reform

The previous section has shown that while TNI has made some progress towards a more agile and deployable military, reform efforts have been piecemeal and incoherent. Apart from the land-centric strategic culture, there are a number of additional factors that are likely to obstruct defence reform efforts over the coming years.

At the top of the list is the absence of clear political guidance by Indonesian governments about the TNI's future role in Indonesia's foreign and security policy. Arguably, they have struggled to provide a clear linkage between Indonesia's strategic interests and military objectives. It remains to be seen whether the shifts in Indonesia's external environment will lead to a readjustment of defence doctrine away from the army-centric 'Total People's Defence' concept to guide TNI reform. The next defence White Paper, which at this stage was due to be released in late 2014, will provide a good indication in this regard.

Another key challenge is to develop better institutional capacities to formulate and execute defence policy. Coordination between the major bureaucracies involved in defence affairs is still very limited. As well, government and parliamentary committees overseeing defence procurement decisions have only limited access to independent expert advice from civilian staff, think-tanks or journalists; this expertise and interaction is only slowly developing. This shortfall often disallows policy-makers from reaching an informed decision on procurement projects. In combination, these factors contribute to a defence procurement process which is incoherent, ad hoc and subject to significant levels of corruption.

Moreover, despite a growing awareness of a changing external threat and risk environment, defence policy ranks relatively low on the list of political priorities. Notwithstanding rhetoric about the need to develop a strong TNI, Indonesia's political elites continue to prioritize non-military services such as health, education and infrastructure. This is not surprising. After all, Indonesia remains a developing country and its long-term economic growth is far from guaranteed. Moreover, as highlighted above, neither force modernization trends in Southeast Asia nor China's strategic rise are (yet) seen as a major military challenge. Absent a dramatic shift in Indonesia's external threat environment, the focus in Indonesia's security policy on domestic factors is likely to persist into the foreseeable future.

One consequence is that financial resources to back TNI reform remain constrained. As mentioned before, Indonesian defence spending has been rising almost continuously for more than a decade. In addition, senior government officials repeatedly promised even further growth. Former President SBY, for example, promised in 2010 to increase defence spending to 1.5 per cent of GDP in 2014. Assuming that the Indonesian economy continues to grow by 6 to 7 per cent annually, the defence budget could potentially increase to about US$14 to 15 billion, potentially surpassing Singapore as the highest defence spender in Southeast Asia. However, defence spending as a proportion of GDP has remained persistently below 1 per cent, and SBY's pledge of 1.5 per cent in 2014 did not materialize. As a result, the TNI still had only relatively little to spend on a very large force of about 395,500 active personnel in 2014. It is difficult to envisage major increases in Indonesia's defence spending in the near to medium future that would bring it much closer to the investments of advanced regional militaries such as Australia, which currently spends over US$25 billion on its armed forces.

Moreover, budget increases do not automatically lead to increases in TNI capability, particularly with regard to modern weapons systems. Ample evidence exists that the TNI faces a problem similar to other emerging military powers: buying expensive military platforms without regard for the future costs of through-life support (e.g. maintaining and sustaining the capability). Simply put, the TNI is at risk of investing in new platforms without the ability to keep them in service.[23] In all likelihood the funding base to acquire many of the planned military platforms by 2024 will thus remain insufficient. This is also because investments in defence equipment compete with equally pressing efforts to increase military professionalism. In order to develop a more professional and deployable force, the TNI would need to increase the money spent per individual soldier on wages, housing, health and education. In an ideal world, shrinking overall TNI size – particularly the land force – in return for fewer but more deployable units would be one obvious measure. Yet, it remains uncertain whether the political and military leadership are ready for such a step; particularly the TNI-AD is likely to oppose it.

Finally, the SBY government pushed for the promotion of a capable defence industrial base. The Parliament approved a national defence industry bill in 2012 which sets out a framework for financial assistance to domestic defence companies, prioritizes the acquisition of indigenous military systems, and creates offset guidelines. The declared goal is to reduce Indonesia's reliance on foreign arms suppliers and to establish a 'strategic' defence industry. However, these plans could actually impede TNI modernization. Local defence industries have become more capable of building smaller, less sophisticated military systems such as patrol boats or the CN-235 maritime patrol aircraft. But indigenously producing complex systems such as modern submarines or frigates will pose major challenges given the poor state of the nation's defence industrial capacity. Besides, it would be very hard to achieve economies of scale. The effect on TNI modernization would be largely negative.[24]

Conclusion and outlook

Indonesia's military change remains a work in progress. It is important to consider the reform of the TNI in the context of the political transformation taking place after the fall of the Suharto regime in 1998. The primary driver for military reform during the first decade of the young Indonesian democracy was the establishment of civil–military relations. While the geopolitical changes in the Asia-Pacific region did factor in Indonesian strategic thinking at that time, they were clearly subordinate to efforts to create a firm civilian control over the military. Only once relative success was achieved in putting the military 'back in the barracks' did the Indonesian political elites turn their focus more towards the country's evolving external environment. Not surprisingly, particularly the Navy and the Air Force were supportive of strengthening the military's capabilities to deal with external challenges in Indonesia's maritime approaches.

Indonesia's political and military elites recognized the need to respond to the geopolitical changes in the Asia-Pacific region. Challenges to stability in maritime Southeast Asia became the key concern in this regard. Unresolved territorial disputes in the South China Sea, China's claims in the 'nine-dash line' as well as force modernization trends in neighbouring countries made Indonesia aware of the potential vulnerability of its vast EEZ against foreign interference and the TNI's limited capacity to effectively safeguard the maritime approaches.

Reflecting Indonesia's continental, strategic mind-set, TNI reform aims to improve its littoral warfare capacity to defend the archipelago and the EEZ. Strategically, this makes sense given Indonesia's strategic location at the epicentre of maritime chokepoints in Southeast Asia, its limited resources and its defensive strategic culture. The operational doctrine of (mostly) maritime 'flash-point defence', the navy's 'green water' ambitions, and efforts to upgrade facilities on some outer islands most clearly demonstrate this shift in strategic thinking. Moreover, some procurement programmes – particularly in the maritime and air domain – will enable the TNI to complicate a potential external aggressor's operations in Indonesia's littoral.

Nevertheless, military reform remains very much unfinished business. Overall, the TNI is still very much an imbalanced force, lacking critical military enablers to implement its operational objectives within the archipelago. In the absence of a clear and present external military threat, political and financial support for a more efficient and effective military modernization process has been lacklustre. Corruption takes its toll, and limited institutional capabilities prevent the making of defence strategy. As a result, the ambition to establish the MEF in a decade from now will most likely be only partially implemented.

A critical question for the coming years is how much effort the new President Joko Widodo (Jokowi) will invest in modernizing the TNI, also given that he is the first incumbent with a non-military background. A priority will certainly be to uphold democratic civil–military relations and to focus on economic and social reform. These are the fundamental pillars to prevent the return of military authoritarianism. However, and somewhat paradoxically, precisely because the new President has a civilian background, he could be pushing for a more coherent defence strategy and policy in order to position Indonesia in a changing Asia-Pacific strategic landscape. Otherwise, and absent a significant deterioration in Indonesia's external environment, the TNI's process of modernization is likely to remain arduous. Still, elements of the TNI will become more capable of conducting modern military operations. As a result, Indonesia is bound to become a more capable military power in Southeast Asia.

Notes

1 Damien Kingsbury, 'Two Steps Forward, One Step back: Indonesia's Arduous Path of Reform', *ASPI Strategy*, Australian Strategic Policy Institute, Canberra, 2012.
2 Edward Aspinall, 'From Authoritarian to Democratic Models of Post-conflict Development: The Indonesian Experience', *The Korean Journal of Defense Analysis*, vol. 29, no. 4, 2012, pp. 449–463.
3 Quoted in Novrida Manurung and Berni Moestafa, 'Indonesia Regains Investment Grade at Moody's After 14 Years', *Bloomberg*, 19 January 2012.
4 OECD 2012, *Economic Outlook*, p. 214.
5 PricewaterhouseCoopers, *The World in 2050. Beyond the BRIC's: A Broader Look at Emerging Market Growth Prospects*, 2008. www.pwc.com/gx/en/world-2050/pdf/world_2050_brics.pdf (accessed 2 October 2013).
6 Dzirhan Mahadzir, 'Indonesia's Military Modernization', *Asian Military Review*, 1

November 2012, www.asianmilitaryreview.com/indonesias-military-modernization/ (accessed 25 September 2013).
7 Colin S. Gray, 'Strategic Culture as Context: The First Generation of Theory Strikes Back', *Review of International Studies*, vol. 25, no. 1, 1999, pp. 49–69.
8 Marcus Mietzner, 'Overcoming Path Dependence: The Quality of Civilian Control of the Military in Post-authoritarian Indonesia', *Asian Journal of Political Science*, vol. 19, no. 3, 2011, p. 271.
9 See Rizal Sukma, 'Civil–Military Relations in Post-authoritarian Indonesia', in Paul Chambers and Ariel Croissant (eds), *Democracy Under Stress: Civil–Military Relations in South and Southeast Asia* (Bangkok: Institute of Security and International Studies, 2010), pp. 149–169.
10 Marcus Mietzner, 'The Politics of Military Reform in Post-Suharto Indonesia: Elite Conflict, Nationalism, and Institutional Resistance', *Policy Studies*, no. 23, East-West Center, Washington, DC, 2006, pp. vii–viii.
11 Mietzner, 2011.
12 The estimated figures for defence expenditure are based on the Stockholm International Peace Research Institute, *Military Expenditure Database*, 2014, http://milexdata.sipri.org/files/?file=SIPRI+milex+data+1988–2012+v2.xlsx (accessed 5 April 2014). Figures are in constant (2011) US$. The figure for TNI's active personnel in 2001 is from the International Institute for Strategic Studies, *The Military Balance* (London, 2002), p. 149.
13 Rizal Sukma, 'Indonesia's Security Outlook and Defence Policy, 2012', *NIDS Joint Research Series*, no. 7, 2012, p. 8.
14 International Institute for Strategic Studies, *Regional Security Assessment 2014: Key Developments and Trends in Asia-Pacific Security* (London: IISS, 2014), ch. 13.
15 Luke Hunt, 'Indonesia boycotts Singapore Airshow', *The Diplomat*, 12 February 2014, http://thediplomat.com/2014/02/indonesia-boycotts-singapore-airshow/ (accessed 25 March 2014).
16 Kanupriya Kapoor and Jonathan Thatcher, 'Indonesia Military Worries over Asia Arms Race, Territorial Tensions', *Reuters*, 3 April 2014, www.reuters.com/article/2014/04/03/us-indonesia-military-idUSBREA320GD20140403 (accessed 13 April 2014).
17 Daniel Novotny, *Torn between America and China: Elite Perceptions and Indonesian Foreign Policy* (Singapore: Institute of Southeast Asian Studies, 2010).
18 Evan A. Laksama, 'The Enduring Strategic Trinity: Explaining Indonesia's Geopolitical Architecture', *Journal of the Indian Ocean Region*, vol. 7, no. 1, 2011, p. 103.
19 Ridzwan Rahmat, 'Indonesia to Station Su-27, Su-30s on South China Sea Islands', *Jane's Defence Weekly*, 1 April 2014.
20 Unless otherwise noted, the following sections on Indonesia's capability development are largely based on Benjamin Schreer, 'Moving Beyond Ambitions? Indonesia's Military Modernisation', Australian Strategic Policy Institute, Canberra, 2013, ch. 3.
21 Robert C. Rubel, 'Talking About Sea Control', *Naval War College Review*, vol. 64, no. 3, 2010, pp. 45–46.
22 ABC Radio Australia, 'Indonesia to Significantly Boost UN Peacekeepers', 21 March 2012, www.radioaustralia.net.au/international/2012–03–21/indonesia-to-significantly-boost-un-peacekeepers/473422 (accessed 2 July 2013).
23 Tim Huxley, 'Australian Defence Engagement with Southeast Asia', *Centre of Gravity Paper*, no. 2, 2012.
24 S Rajaratnam School of International Studies, *Transforming the Indonesian Armed Forces: Prospects and Challenges* (Singapore, 2011).

5 The management of military change
The case of the Singapore Armed Forces

Bernard Fook Weng Loo

The Singapore Armed Forces (hereafter SAF) has undergone at least three processes of evolutionary change in its brief history. Singapore's defence strategy has been characterized by a number of analogies: from a 'poisonous shrimp', to a 'porcupine', and more recently to a 'dolphin'.[1] Coincidentally, with the emergence of the idea of the third-generation SAF (hereafter 3G SAF), it is also possible to associate the three analogies with the three generations of evolution of the SAF. The 'poisonous shrimp' analogy applied to the first-generation SAF and sought to impose upon a potential aggressor a 'Stalingrad style of close combat' in urban areas. By raising the human and material costs of aggression against Singapore to unacceptable levels, this was hopefully sufficient deterrence against this putative aggressor.[2] The second-generation SAF of the 1980s was likened to a 'porcupine', developing an increasing capacity to project at least limited military power (the porcupine's quills) at some distance from its shores. By the late 1990s, however, as the SAF began to leverage increasingly on information technologies and experiment with such 'transformational' concepts as network-centric operations and the Revolution in Military Affairs (hereafter RMA), a 3G SAF was evidently emerging, analogous to a 'dolphin', in that intelligence, speed and manoeuvrability would become the principal hallmarks of its style of military operations. As this chapter will subsequently demonstrate, this is a SAF increasingly willing to project power – albeit in support of UN-sanctioned and coalition operations as well as disaster relief operations – further afield from the immediate environment of Singapore.

A rational military organization is, presumably, one that manages to remain relevant to the principal security challenges that the state expects to face. Of course, it is impossible to predict the security challenges that the state will face in the future. Nevertheless, this is a challenge that all policy-makers and strategic planners have to address. For the SAF to remain strategically relevant, therefore, it has had to grapple with the fundamental questions of what its mission is, the security challenge it is likely to face (at any given moment), and the context (both external security and domestic political) in which the SAF will find itself. How the

SAF has answered these questions has been driven by a number of considerations: the nature of the security environment, and how it is likely to change in the medium- to long-term future; the sorts of capabilities the SAF will likely be required to deal with the projected conditions of the security environment; the relative and projected health of the economy, and therefore, how much the military organization can afford to buy. Finally, it is also necessary to ask exactly what operational missions the SAF has actually had to undertake. It is incorrect to assume that the answers to this third question derive exclusively from the nature of the security environment. How this process of evolutionary change was managed, balancing considerations of the series of questions posed above, is the focus of this chapter.

The strategic environment and the strategic culture discourse of vulnerability

The Melian Dialogue, coming from Thucydides' *The Peloponnesian War*, is the classic statement of the inherent vulnerabilities that plague small and weak countries. Its perspective of how weak states must suffer the vicissitudes of their position reflects the dominant concerns of Singapore's defence policy-makers. There is no specific country that threatens Singapore; rather its strategic problems stem from the simple fact of its small size and the absence of a self-defence capability. In a speech to Parliament on 23 December 1965, Goh Keng Swee, Singapore's first Minister for Defence, noted, 'It is no use pretending that ... the island cannot be easily overrun by any neighbouring country within a radius of a thousand miles.' Interestingly, there was no mention of who this aggressor might be or why this aggressor would want to resort to military force against Singapore. It is as if such considerations are irrelevant, that Singapore's very nature as a very small state, in an international politics that privileges the power principle, means that the island is structurally and naturally very vulnerable.

From Singapore's independence and through the 1970s, the Cold War was an ever-present consideration, manifesting itself in Southeast Asia, first in the form of the Vietnam War and subsequently in the Cambodian conflict that ended in 1989. From the immediate neighbourhood, there was the history of *Konfrontasi* with Indonesia[3] and occasional threats from Malaysia to cut off the island's water supply. It is difficult not to conclude that the dominant image was one of potential trouble that could directly or indirectly affect the security of Singapore.[4] Up until 1970, British troops were stationed in Singapore to safeguard the island's external defence, but the SAF had only two infantry battalions 'for [her] protection in normal times'.[5] The SAF, clearly, was not in a position by itself to do very much in terms of defending the island against external aggression. However, when the British government announced on 15 January 1968 its plan to withdraw all British forces east of Suez by December 1971, the Singapore government had to formulate plans to build up the SAF, since the island would no longer come under a British defence umbrella. As a consequence, a central element in the Singaporean strategic culture was a pervasive discourse of vulnerability:[6]

> If you are in a completely vulnerable position anyone disposed to do so can hold you to ransom, and life for you will become very tiresome.
> (Goh Keng Swee, speech to Parliament, 13 March 1967)

This realist view of the region has persisted. Subsequent prime ministers have enunciated similar world views:

> To have permanent peace, all Singaporeans must be ready, operationally ready, to keep out threats from any direction. The sharper our defensive skills, the higher the chances of our being left alone to progress and prosper in peace, to work and play.
> (Goh Chok Tong, speech to Parliament, 1 November 1983)

The current Prime Minister, Lee Hsien Loong, in a speech in 1984, noted:

> So we need a policy which says: 'If you come I'll whack you, and I'll survive.' This is a workable strategy. I may not completely destroy you, but you will have to pay a high price for trying to subdue me, and you may still not succeed.[7]

Geography underpins this strategic culture of vulnerability. In contrast to its immediate neighbours, Singapore lacks geo-strategic depth, stemming from its small size. Huxley also makes the point that Singapore's territorial waters are completely surrounded by the territorial waters of Indonesia and Malaysia.[8] Furthermore, Singapore's position lies astride the trade routes connecting the Pacific and Indian Oceans.[9] This position places Singapore in a position to control maritime trade in Asia, and this makes the island a potentially attractive target. If nothing else, the location of Singapore meant that any instability arising out of its weaknesses would strengthen the temptation of external actors to intervene:

> Small states are likely to be a great source of trouble in the world if they cannot look after themselves. If the management of their domestic affairs is so bad as to invite civil war and disorder, there is always the risk that larger states may be tempted to intervene.
> (Goh Keng Swee, speech to Parliament, 13 March 1967)

There is no reason to conclude that his perceptions have changed since then. That being said, it seems clear that Singapore policy-makers have viewed the various actors in Southeast Asia – both regional and extra-regional – differently. The United States has been regarded as among the more trustworthy of the great powers and is the principal source of modern military technologies for the SAF, in particular in the area of air combat technologies. With the exception of the Hawker Hunters, which the SAF inherited from Singapore's British colonial masters, the other combat aircraft – A-4 Skyhawk, F-5E, F-16 and F-15 aircraft – have all come from the United States, whether directly or indirectly. There is increasing speculation that Singapore will eventually acquire the F-35B.[10] The US–Singapore bilateral relationship was not always this positive. Initially, the United States evinced a certain strategic distrust of Singapore; Singapore's first acquisition of F-16A/B aircraft in the 1990s faced some resistance from the United States Congress, on the grounds of allegations of illegal reverse engineering that the Singapore defence industries had previously engaged in. Furthermore, notwithstanding this recognition of the strategic importance of the United States, especially in the realm of defence modernization, there have been occasional hiccups in the broader political relationship: an example was the furore generated in the United States when the Singapore legal authorities caned an American citizen, Michael Fay, for having been found guilty of engaging in acts of public vandalism. Nevertheless, the fact that Singapore participates in the F-35 Joint Strike Fighter programme is testimony to the strategic relationship between the United States and Singapore.

With China and India, the attitudes of Singapore policy-makers have been somewhat more problematic. China has had a problematic status in Singapore policy-makers' sense of history, given that China previously supported the indigenous communist movements in Southeast Asia, something that China officially disavowed only in the 1990s. Furthermore, Singapore's relationship with Taiwan, although not official, remains a bugbear in Singapore's relations with Beijing. In 2004, before assuming the position of Prime Minister, Lee Hsien Loong paid a private visit to Taiwan, an act that generated strong protests from Beijing.[11] Currently, while defence relations with China are cordial, there is no substantive cooperation between the two countries and certainly nothing comparable to defence cooperation with the United States. Singapore's defence relationship with India has been similarly problematic. In attempting to establish the SAF, India was regarded as a potential source of advice until Singapore was rebuffed. Current defence relations are cordial, with the signing of a defence cooperation agreement in 2003. But, as Kuik noted, 'Singapore's defence cooperation with [China and India] paled in comparison to its high-level and broad-ranging military ties with [the United States].'[12]

Indeed, if the SAF is now the most modern and well-equipped armed force in Southeast Asia, this realist sense of vulnerability has not dissipated. If any, the growing potential power of the SAF has merely facilitated a changing strategy, one that has increasingly emphasized the need to achieve a 'swift and decisive victory over aggressors'.[13] Then Minister of

State for Defence Mathias Yao even spoke of the need to give the aggressor a 'knock-out punch in round one'.[14] The pre-emptive posture envisaged by the 'porcupine' meant that in the event of hostilities, the SAF would strike first, establish the front line in the likely enemy's territory and prevent the enemy from being able to bring fire power to bear on the population and economic centres of Singapore.

Managing the pace of technological change: challenges for the SAF

Given the geo-political and geo-strategic conditions in which Singapore exists, and layering on top of these conditions the small population base, the SAF has always seen technology as a key force multiplier and solution to many, if not all, of its extant strategic problems. That being said, technology is a double-edged sword that needs to be managed carefully by balancing the requirements of a technologically advanced force structure with the competing elements of increasing technological costs and a mature economy that generates lower rates of growth.

Technology and the universality of the RMA

Technology – and specifically, technological change – is part of the broader phenomenon of what Barry Buzan and Eric Herring refer to as the arms dynamic, understood as the 'entire set of pressures that make actors (usually states) both acquire armed forces and change the quantity and quality of the armed forces they possess'.[15] Its pervasive influence over strategy and war[16] means that as military technologies change, the machinery of war must also undergo periodic change. Military capabilities change over time and, on occasions, it may be necessary to review the modus operandi of the military organization, especially if new types of capabilities avail themselves and these new capabilities require radically different modus operandi for their effects to be maximized. Given the 'technophiliac' nature of the SAF, technological change has implications for force structure decision-making, as will be discussed below.

The argument that the RMA was born during the height of the Cold War to answer the Cold War problem of the Warsaw Pact's superiority in heavy armoured forces means it is perfectly legitimate to question the applicability of the RMA to other states. In other words, the fact that the RMA technologies are relatively diffuse in nature does not rule out the question of whether the military organizations of Southeast Asia can and should engage with this RMA. The specific geo-political origin of this RMA means that it is perfectly reasonable to question its universality today. The questionable universality of the RMA is reinforced by the culturally specific nature of the RMA. Nevertheless, it is obvious that this RMA has had quite broad-based appeal – witness the RMA-inspired modernization programmes of various military organizations in Europe as well as the Asia-Pacific region. In the case of Singapore, the SAF has adopted an RMA-inspired operating doctrine: the Integrated Knowledge-based Command and Control (hereafter IKC2).[17]

However, is the RMA a universally applicable concept? Should the SAF have embraced this strategic concept in the first place? This is an important issue, since the degree to which the United States dominates the arms market, and the extent to which it drives this RMA, will shape the acquisitions policy of the SAF. As it stands, the United States and its European partners have dominated the acquisitions policy of the SAF.[18] This is especially the case in the area of high-value, big-ticket items, such as advanced air combat or naval platforms. In the realm of air combat systems, with the exception of the first-generation combat aircraft that the SAF acquired from the United Kingdom, the United States has been the principal supplier: A-4 that were subsequently upgraded to A-4SU with avionics from the United States, F-5Es, F-16A/B and subsequently the C/D variants, and F-15SGs. Much of its advanced air combat training is done in collaboration with the United States Air Force (USAF). When the old Paya Lebar International Airport was abandoned in favour of the current international airport and Changi, the Paya Lebar Airport was converted into an air base for the RSAF, precisely because of the perceived benefits from allowing regular stopovers by USAF aircraft.[19]

The potential problem for Singapore, however, is that the United States' military-industrial complex has been moving almost inexorably towards increasingly exotic technologies. What this also hints at, which is another difficulty that the United States military-industrial complex poses for Singapore's military modernization ambitions, is the issue of costs (an issue that will be examined below), and whether or not the Singaporean military will be able to remain within the US military-industrial technological universe.

The accelerating pace of military change: organizational and doctrinal implications for the SAF

The problem with having a technology-driven force structure decision-making policy is that technological change is accelerating. Indeed, as a global phenomenon, it is fair to argue that ever since the 20th century, an increasingly important driver of military change has been technological in nature. Certainly, extrapolating from Moore's Law, it is reasonable to expect that technological change will, if anything, accelerate even faster in the future. With the SAF then fully embracing the RMA – the dominant school of thought in the RMA emphasizes the central role of advanced technologies[20] – it will have to face up to the accelerating pace of technological change.[21] One challenge then is to manage change and transformation within the SAF as the technological base changes – an issue that will be examined in greater detail below.

A second challenge will be to manage the impact of this RMA-inspired transformation on strategy and operational planning.[22] In particular, the RMA facilitates the implementation of concepts of joint warfare by networking the entire military organization into a holistic fighting entity that generates a *potential* net strategic effect far greater than the sum of its parts. In so doing, the traditional distinctions between the services and the areas where the services' strategic effects can be felt will disappear.[23] The net effect of the RMA is an increasingly lethal (but also complex) military organization, capable of unprecedented levels of precise destruction.[24] A necessary precondition for this transition to joint warfare is a mature, conventional warfighting capability. At face value, the SAF qualifies to undertake this RMA-inspired transformation, having adopted conventional warfighting doctrines for a very large part of its history. That being said, the fact that the SAF has never had to undertake combat operations means that its conventional warfighting capability is hypothetical at best, never having been actually tested in war.

There is also a doctrinal issue to transformation, and navigating this doctrinal path will not be easy. RMA advocates have insisted that wholesale changes – flattened organizational structures, network-centric operations and effects-based operations doctrines in the tactical realm – in the military organization are absolutely necessary for the promise of Singapore's RMA to be fully realized.[25] Such radical departures are, for a variety of reasons, difficult to accept, not least because radical departures require great leaps of faith into the unknown.[26] This 'vision' of the transformed military organization then comes into potential conflict with the roles and functions that the SAF performs. At one level the SAF may more or less resemble its European counterparts, albeit superficially, in terms of its internal organization and nominal responsibilities.[27] At another level however, the roles that the SAF assumes exceed those of conventional Western military organizations.[28]

As a start, the flattened and open structures that the RMA apparently requires challenge the hierarchical structures that the SAF continues to manifest. For one, the doctrinal concept that underpins the SAF's transformation programme is called Integrated Knowledge-based Command and Control.[29] What is interesting about the SAF documents is the importance placed on the training and education of its personnel.[30] This suggests that the SAF is fully aware of the doctrinal complexities associated with its RMA-inspired third-generation SAF transformation agenda. Automatically, however, this transformation agenda potentially runs into trouble with the conscription policy that the SAF maintains. Conscription policy in Singapore mandates that able-bodied male Singapore citizens and permanent residents are required to serve up to two years in the armed forces, police or civil defence force; of the three the vast majority of conscripts are deployed in the SAF. Upon completion of the two years of full-time training, these conscripts transition into Operationally Ready National Servicemen, which form the bulk of the combat forces.

Conscription is not likely to be dismantled, at least not for the immediate to medium-term future. In March 2013, the Singapore Minister for Defence, Dr Ng Eng Hen, announced the establishment of a Committee to Strengthen National Service, the aim of which is to study and recommend measures to strengthen public commitment to conscription policy.[31] An independent survey was commissioned, undertaken by the Institute of Policy Studies, and the survey results were subsequently published in October 2013.[32] As long as conscription remains the bedrock of manpower policies for the SAF, these challenges will need to be addressed.

Can the SAF afford the RMA?

The technological developments discussed above are already realities, and their impact will grow. These technologies are very often the product of, or have applications in, the civilian world. Moreover, in many cases – whether it is optics, robotics, nanotechnology or supercomputing – the basic research, product development and availability are of commercial interest. This means that these technologies are increasingly available 'commercial off the shelf' (COTS), which allows military organizations to avoid the large and often long investment in research and development. The widened interest and applications in these advancements mean wider availability of both the knowledge and the capabilities themselves. This, then, ought to make the RMA more affordable.

However, this is only one perspective. Another perspective suggests otherwise, namely that this RMA promises to be expensive and, for smaller military organizations with fewer resources, potentially prohibitively expensive.[33] For the SAF, this increasing cost of new technologies may result in a form of structural disarmament: simply put, this means that when the SAF replaces an existing platform with new technologies, the replacement platform will almost certainly be lesser in number than its predecessor. This is especially the case with air power.[34] A case in point is the Singapore Air Force's decision to replace its existing fleet of over 40 F-5Es with 12 F-15SGs. One way by which the related issues of rising costs and structural disarmament might be managed is through off-the-shelf purchase (often accompanied by licensed production) and collaborative procurement or weapons research development, such as Singapore's participation in the F-35 Joint Strike Fighter programme. Such international collaborations appear to be an attractive option, but can be problematic.[35]

As Singapore committed herself to a policy that sustained high yearly military spending to make the SAF a potent military force in Southeast Asia, this placed great demands on Singapore's financial resources. At the same time, however, policy-makers have maintained a careful balance between necessary defence spending and economic imperatives; simply put, defence spending remained economically supportable, and building up the SAF should not bankrupt the economy.

We are a small nation and our financial resources are limited. We therefore cannot afford to raise and maintain large defence forces on a permanent basis.

(Goh Keng Swee, speech to Parliament, 30 December 1965)

At the end of 1968, Singapore's defence budget was announced to target 10 per cent of her gross national product (GNP),[36] or 6 per cent of gross domestic product (GDP) annually.[37] By 1972, Singapore's military expenditure was nearly compatible with that of Indonesia and Malaysia in terms of per capita figures.[38] Singapore's defence spending through the 1970s was maintained at 4.2 to 6.8 per cent of her GDP, and with the concurrent rapid growth of the Singapore economy, Singapore increased her military expenditure 114 per cent over the 1969 to 1978 period.[39] Without any parliamentary opposition until 1981, the PAP was free to sustain a high level of military expenditure year after year.[40] It is worth noting that even the advent of a parliamentary opposition did not result in any significant reduction in defence spending. As the World Bank's figures (see Table 5.1) show, certainly since the 1990s, the defence budget has never exceeded 5.3 per cent of Singapore's GDP. In 2013, the Singapore government allocated about S$12 billion of its budget to military spending; in contrast, its immediate neighbours, Malaysia and Indonesia, dedicated S$6 billion and S$10 billion to military spending respectively.[41]

This financial prudence was not grounded solely in concerns about economically unsustainable levels of defence spending. Singapore policymakers remain conscious of how the buildup of the SAF could have potentially negative ramifications for the wider strategic environment. Care was taken to match what challenges the strategic environment would pose on the one hand with the strategic requirements of the SAF on the other:

> [O]ur defence forces are adequate for our present needs. Those who study the allocation of funds in the annual estimates of expenditure

Table 5.1 Singapore defence spending, 1991 to 2010

Year	1991	1992	1993	1994	1995	1996	1997	1998	1999	2000
D/B	4.5%	4.5%	4.1%	3.8%	4.2%	4.3%	4.5%	5.3%	5.3%	4.6%
GDP	43b	49b	59b	69b	80b	94b	104b	95b	85b	95b

Year	2001	2002	2003	2004	2005	2006	2007	2008	2009	2010
D/B	4.9%	5.0%	4.9%	4.5%	4.4%	4.0%	3.7%	3.9%	4.2%	3.8%
GDP	91b	90b	93b	109b	123b	139b	168b	178b	194b	217b

Source: The World Bank: http://data.worldbank.org/indicator/MS.MIL.XPND.GD.ZS, http://data.worldbank.org/indicator/NY.GDP.MKTP.CD.

passed by Parliament will have noticed a levelling off in recent years of our defence expenditure.... This is as it should be. It does not make sense when everything in Southeast Asia is relatively calm and stable, to keep on spending more and more money on defence.
(Goh Keng Swee, speech to Officer Commissioning Parade, 12 July 1974)

As a consequence, the buildup of the SAF was undertaken with a constant eye towards how the strategic environment was evolving, the emergence of new strategic challenges, and the new capabilities that these new challenges would demand. As a consequence, the manner in which the SAF was built up could be described as systematic.[42] In other words, every new acquisition makes 'strategic sense'; it is meant to plug an existing strategic capability gap.

The emergence of a Singaporean defence industry

Finally – and this is something that is more often than not ignored – the 1990s also saw the emergence and consolidation of a credible defence-industrial base. The viability of such a strategy is contested. Bilveer Singh argues that 'defense industries are ... the political industries of a country', aiming to reduce dependency on foreign suppliers and to provide self-sufficiency.[43] Furthermore, the 'political' and the 'strategic' assumed primacy over the 'economic' in the decision to proceed with defence industrialization, that 'it was strategically and militarily a necessity'. Richard Bitzinger argues, however, that indigenous defence industries do not have a large enough market share and are not economically viable, and that such countries may be better served buying COTS instead.[44]

As a result, throughout the 1990s, spending on defence research and development increased substantially, from approximately 1 per cent of defence spending in 1990 to 4 per cent in 2000.[45] In real terms, this amounted to an increase of from US$20 million to US$160 million. A key strategy continues to be technology transfers through collaborative efforts with other local and international partners. Singapore's participation in the F-35 Joint Strike Fighter project is one example of this strategy. Starting out in 1999 as an observer in the demonstration phase and moving on in 2003 to become a security cooperation participant in the system design and development phase of the multilateral programme, the SAF seems certain to acquire the F-35 eventually.[46] Other collaborations have taken place with Sweden, in particular the latter's Defense Research Establishment, and with France, most recently in the acquisition of the *Formidable*-class modified *Lafayette* frigates.[47]

That being said, however, Singapore's indigenous defence industries will almost certainly never replace the current reliance on overseas suppliers for high-end military technologies across the land, sea and air dimensions. As the *Formidable*-class modified *Lafayette* frigates case indicates, when it comes to major combat platforms, at best the Singapore defence industrial base can only undertake licensed production. Furthermore, there has been some recognition of the need for Singapore's defence industries to be as economically viable as possible, and hence there is an increasing desire to seek international markets. Singapore's defence industries secured a £150 million deal to produce Bronco All-Terrain Tracked Carriers for the United Kingdom's Ministry of Defence. The Ultimax-100 light machine-gun that Singapore designed and built has been sold to 24 different armed forces.[48] Nevertheless, the point still remains that Singapore's defence industries exist to serve the SAF's operational requirements first; sales to other armed forces should therefore be regarded as a bonus rather than as a necessity.

The SAF remaining relevant?

Finally, there is the issue of whether or not the SAF remains relevant to the security challenges facing Singapore in the 21st century. This may very well be the most contentious issue in how military change is managed in the SAF. The SAF has not been immune to the challenges posed by the emergence of new security concepts, which are being manifested in new military operations undertaken by the SAF. There has been an ongoing debate regarding the security agenda in the 21st century for states and military organizations.[49] One side argues for such ideas as 'new' wars, namely fourth-generation (possibly even fifth) warfare and cyber warfare. The other side argues that the grammar of war (its specific manifestations) may change but the logic of war (its essence) will remain unchanged. However, on the assumption that the new security environment thesis is true, this is a development that military organizations cannot ignore. It means that the military organization, in the words of former United States Secretary of Defence Donald Rumsfeld, has to 'defend our nation against the unknown, the uncertain, the unseen and the unexpected'.[50] The challenge is to manage a security environment that contains known threats from traditional putative adversaries, as well as to be prepared for unforeseeable circumstances, and to do both at the same time.

Some even raise the question of the continuing relevance of military organizations *du jour*.[51] Rupert Smith succinctly summarizes the situation in which military organizations are now likely to find themselves: 'war as cognitively known to most non-combatants, war as battle in a field between men and machinery, war as a massive deciding event in a dispute in international affairs: such war no longer exists.'[52] Rather, if military organizations are called to action, this action will fall under the rubric of operations other than war (OOTW). Military organizations have been configured to deal with the traditional security challenges – what Stephen Biddle[53] calls 'major wars' – centred on the threat of invasion by other states, concerns of territorial integrity and sovereignty. Deploying the military organization for these new security concerns may be necessary, but this prospect may also be inherently problematic.[54] Furthermore, these new security concerns require different skill sets from those that conventional military operations demand, which typically do not occupy very much attention in the training regimes of modern militaries.[55] Of course, there is every possibility that the 'old' and the 'new' security agendas may clash. Charles Krulak paints a scenario where a soldier needs to decide on a course of action that may have implications at the strategic level.[56] In the new security landscape, military organizations need to be equipped with the skill sets that will allow them to move seamlessly from one challenge to the next.

The irony is that the SAF's RMA-inspired agenda of military transformation may therefore result in an organization increasingly at odds with the security missions it is likely to face. In the quest to maintain its technological edge, the SAF may eventually become irrelevant to the dominant security concerns of the time.

The challenge of OOTW

What is of particular interest is the maintenance of what is a standard defensive/deterrent mission together with the addition of non-traditional military operations in recent years. The global trend since 1945, enhanced further in the 21st century, is that conventional wars are being increasingly supplanted by insurgencies, civil wars and terrorist threats.[57] Clearly, military organizations around the world are facing these types of so-called 'low-intensity' conflicts more frequently than traditional conventional wars. Singapore, in this regard, may be no different.[58] In addition, an increasingly salient challenge facing many military organizations is that of OOTW,[59] which the SAF has gone into in a big way.[60] Given Singapore policy-makers' positive responses to the SAF's long participation in operations in Afghanistan,[61] it is not difficult to surmise that the SAF will respond positively to future opportunities to participate in other OOTW.

This study rejects the notion that conventionally structured military organizations lack the skill sets to be able to engage in OOTW. In counter-terror operations, for instance, inasmuch as terrorist facilities can be located, these facilities can be attacked by increasingly precise instruments of military power. Even the more passive counter-terror measures – such as the guarding of critical infrastructure – resonate with Singapore's military organization.[62] Tapping this reservoir of manpower resources for these operations therefore appears to make sense. Furthermore, even as a hypothetical proposition, the RMA can apply in the case of counter-terrorism operations, since RMA technologies can provide commanders with a clearer situational picture of the area of operation, facilitating a faster operational cycle.[63] Austin Long (Chapter 13, this volume) certainly demonstrates that this in the case of the United States' operations in Afghanistan and Iraq. When the alleged Jemaah Islamiya terrorist Mas Selamat Kastari escaped from a Singapore prison in 2008, the SAF was deployed to assist law enforcement and internal security agencies in Singapore in the latter agencies' attempt to locate and re-apprehend him.[64] That being said, the SAF adopts a supporting role here, given that counter-terrorism falls within the command and control of the National Security Coordinating Committee, which is chaired by the Deputy Prime Minister and Coordinating Minister for National Security, and involves the ministers and senior officials from Home Affairs, Defence, Foreign Affairs, Transport, and other related government agencies.[65]

Nevertheless, a caveat needs to be introduced. Great care must be taken to tailor these traditional skill sets to the unique conditions of the particular counter-terrorism operation. The application of military force against terrorist bases has to be very carefully calibrated, so as to not incur unnecessary levels of destruction, especially collateral damage. This, of course, is what the RMA excels in. However, terrorist bases are not so easy to locate and destroy; otherwise the problem would not be as intractable as it seems. Rather, counter-terrorism involves passive security measures which more resemble law enforcement and policing. In both law enforcement and counter-terrorism, the measure of success is reflected in the absence of incidents. Restraint in the use of force is desirable,[66] but this may run against the grain of the military mind-set. The principles of counter-terrorism are not entirely consonant with the principles of conventional warfare. The conventional military mind-set is reflected in two axioms: 'Never send in a man when a bullet will do'; and 'Firepower is cheaper than manpower'. However, fire power is notably more expensive in the highly politicized milieu of counter-terrorism, where the critical effort resides in so-called 'hearts and minds' measures.

Furthermore, today's terrorists are as computer-savvy as their military counterparts. Information technology is a double-edged sword. This is an era in which the bulk of military capabilities are increasingly civilian developed, relatively cheap, commercially available and globally distributed. Thomas Hammes argues that, as such, terrorists are also able to access technology that is commercially available to carry out operations against states that have strong military forces.[67] An increasing number of individuals and non-state actors, as well as states, can gain access to overhead imagery, night vision devices, biological weaponry, thermal image defeating materials, robotic vehicles (land, sea and air), systems integration software, micro-satellites, sophisticated communications and conventional weaponry of all kinds. The means to wage war are no longer the exclusive domain of states.

As for OOTW, military organizations have some skill sets and capabilities to react effectively in OOTW – organizational and logistics management skills, the heavy lift capabilities that allow men and material to be moved over long distances, for instance. But these are manpower-intensive operations.[68] Where does the RMA fit within the spectrum of OOTW? It is possible that OOTW present challenges that run against the grain of the skill sets that military organizations have traditionally focused on – namely in conventional military operations in defence of the sovereignty and territorial integrity of the state, or in conventional campaigns projected into distant theatres of operations in pursuit of the interests of the state concerned. Time is however the ultimate scarce resource, and military organizations may not have the capacity to train their personnel to perform both traditional and non-traditional missions. The point to note is that soldiers not trained for such missions will likely find 'learning on the job' at best a difficult experience, and certainly a less-than-ideal situation.

The increasing salience of OOTW has had an impact on force structure planning. The SAF acquired a fleet of locally designed and produced *Endurance*-class landing ship tanks that have been deployed in anti-piracy operations in the Gulf of Aden as well as in humanitarian relief operations in Indonesia and elsewhere. There is more recently a declared interest in acquiring a helicopter carrier-type vessel precisely for disaster relief operations. Such an option may represent a potentially successful attempt at achieving a balance between these conflicting interests and tendencies, but obviously this is a judgement that thus far cannot be made.

Conclusion

For the SAF, therefore, managing this process of military change has to strike a careful balance between almost contradictory considerations. Singapore's defence and security policy rests ultimately upon a number of mantras. The first mantra is a discourse of geo-political and geo-strategic vulnerability, which has never been seriously challenged. It is likely that there is widespread acceptance, at least within policy-making circles, of this perception of Singapore's geographical conditions. The second mantra, also widely accepted, is of technology as a strategic panacea. The SAF's desire to remain technologically ahead of its immediate neighbourhood counterparts – something that is culturally and geo-strategically understandable, even necessary – may be increasingly economically and politically insupportable. An issue that has not been examined in this chapter is the domestic political context in Singapore, which may increasingly undermine the old social contract where, in return for high economic growth rates and rapid improvements in standards of living, the Singapore electorate opted to not question defence and security policy. Recent unhappiness over the institution of military conscription and over the size of the defence budget relative to other government expenditures may suggest that this social contract is breaking down. There is also the question of what kind of military organization the SAF will become at the end of this current process of military change, and whether or not it can remain relevant to the main security challenges that Singapore faces.

There is no evidence to suggest that the SAF is undertaking a radical departure from its existing practices. While the SAF appears to acknowledge that its operational missions stray quite some distance from its stated mission – the defence of Singapore against some external threat – the fact that it has consciously remained focused on the mission of defending Singapore against another external, state-based actor suggests a mind-set which concludes that 'traditional' military skill sets are the most necessary, and that SAF personnel being deployed to support OOTW missions can acquire OOTW-specific skill sets 'on the job'. There is mixed evidence as to whether or not this is as yet the correct strategy.

Notes

1 See Bernard Fook Weng Loo, 'From Poisoned Shrimp to Porcupine to Dolphin: Cultural and Geographic Perspectives of the Evolution of Singapore's Strategic Posture', in Amitav Acharya and Lee Lai To (eds), *Asia in the New Millennium: APISA First Congress Proceedings, 27–30 November 2003* (Singapore: Marshall Cavendish Academic, 2004), pp. 352–375.
2 Richard A. Deck, 'Singapore: Comprehensive Security – Total Defence', in Ken Booth and Russell Trood (eds), *Strategic Cultures in the Asia-Pacific Region* (Basingstoke: Macmillan, 1999), p. 249.
3 Felix Chang, 'In Defense of Singapore',' *Orbis*, winter 2003, p. 108; Teik Soon Lau, 'National Threat Perceptions of Singapore', in Charles Morrison (ed.), *Threats to Security in East Asia-Pacific: National and Regional Perspectives* (Lexington, MA: Lexington Books, 1983), p. 122; Kuan Yew Lee, *From Third World to First, The Singapore Story: 1965–2000* (New York: HarperCollins, 2000), p. 20; Michael Leifer, *Singapore's Foreign Policy: Coping with Vulnerability* (New York: Routledge, 2000), p. 2.
4 See Lee, *From Third World to First, The Singapore Story*, p. 24; Leifer, *Singapore's Foreign Policy*, p. 53. See also Kin Wah Chin, 'Reflections on the Shaping of Strategic Cultures in Southeast Asia', in Derek da Cunha (ed.), *Southeast Asian Perspectives on Security* (Singapore: Institute of Southeast Asian Studies, 2000), p. 7; Alom Peled, *Soldiers Apart: A Study of Ethnic Military Manpower Policies in Singapore, Israel and South Africa* (Ann Arbor, MI: University Microfilms International, 1994), pp. 60–65; Bilveer Singh, 'Singapore's Management of its Security Problems', *Asia-Pacific Community*, no. 29, summer 1985, p. 85.
5 'Military Spending Modest and Purely for Defence', *The Straits Times*, 6 December 1967, p. 8.
6 See, for instance, Derek da Cunha, 'Defence and Security: Evolving Threat Perceptions', in Derek da Cunha (ed.), *Singapore in the New Millennium: Challenges Facing the City-State* (Singapore: Institute of Southeast Asian Studies, 2002), pp. 133–153; Chong Guan Kwa, 'Relating to the World: Images, Metaphors, and Analogies', in Derek da Cunha (ed.), *Singapore in the New Millennium: Challenges Facing the City-State* (Singapore: Institute of Southeast Asian Studies, 2002), pp. 108–132; Leifer, *Singapore's Foreign Policy*; Bilveer Singh, *The Vulnerability of Small States Revisited: A Study of Singapore's Post-Cold War Foreign Policy* (Yogyakarta: Gajah Mada University Press, 1999).
7 'A Conversation with BG (Reservist) Lee Hsien Loong', *ASEAN Forecast*, vol. 4, no. 10, 1984, p. 164.

8 Tim Huxley, *Defending the Lion City: The Armed Forces of Singapore* (Crows Nest, NSW: Allen and Unwin, 2000), p. 31.
9 See Deck, 'Singapore: Comprehensive Security – Total Defence', pp. 248–249; Tim Huxley, *Defending the Lion City: The Armed Forces of Singapore* (St Leonards, NSW: Allen & Unwin, 2000), pp. 31–33.
10 See: Carl Thayer. 'Why Singapore Wants the F-35', *The Diplomat*, 10 March 2014, http://thediplomat.com/2014/03/why-singapore-wants-the-f-35/ (accessed 12 May 2014). Kelvin Wong, 'F-35 Singapore's Next Generation Fighter', *Today*, 2 April 2013, www.todayonline.com/commentary/f-35-singapores-next-generation-fighter (accessed 13 May 2014). Leithen Francis, 'Singapore Leaning Toward F-35B Variant', *Aviation Week*, 3 April 2013, http://aviationweek.com/awin/reports-singapore-leaning-toward-f-35b-variant (accessed 12 May 2014).
11 'China Protests Against Lee Hsien Loong's Visit to Taiwan', *People's Daily Online*, 23 July 2004, http://english.peopledaily.com.cn/200407/12/eng20040712_149 197.html (accessed 13 May 2014).
12 Cheng-Chwee Kuik, 'Shooting Rapids in a Canoe: Singapore and Great Powers', in Bridget Welsh, James Chin, Arun Mahizhnan and Tan Tarn How (eds), *Impressions of the Goh Chok Tong Years in Singapore* (Singapore: NUS Press, 2009), p. 158.
13 Bilveer Singh, 'A Small State's Quest for Security: Operationalizing Deterrence in Singapore's Strategic Thinking', in Kah Choon Ban, Anne Pakir and Chee Kiong Tong (eds), *Imagining Singapore* (Singapore: Times Academic Press, 1992), p. 57.
14 Huxley, *Defending the Lion City*, p. 56.
15 Barry Buzan and Eric Herring, *The Arms Dynamic in World Politics* (Boulder, CO, and London: Lynne Rienner, 1998), p. 5.
16 See, for instance, Martin Van Creveld, *Technology and War*, 2nd edn (New York: The Free Press, Maxwell Macmillan Canada, Inc, 1991), p. 1.
17 See *Realising Integrated Knowledge-based Command and Control: Transforming the SAF*, Pointer Monograph no. 2, 2003.
18 Data accessed online: 'SIPRI Arms Transfers Database', Stockholm International Peace Research Institute, www.sipri.org/databases/armstransfers (accessed 16 September 2013).
19 In a land-scarce island, converting the old airport into a military air base meant sacrificing valuable land that could have been redeployed for commercial and residential purposes. As it stands, buildings in the vicinity of the Paya Lebar Air Base have very strict height restrictions. Interview with former Chief Defence Scientist Professor Lui Pao Chuen on 13 August 2010.
20 Frank Kendall, 'Exploiting the Military Technical Revolution: A Concept for Joint Warfare', *Strategic Review*, spring 1992, pp. 23–30; Michael J. Mazaar *et al.*, *Military Technical Revolution: A Structural Framework*, Final Report of the Study Group on the Military Technical Revolution, Center for Strategic and International Studies (CSIS), Washington, DC, March 1993, pp. 17–39.
21 See *Realising Integrated Knowledge-based Command and Control*; and *Integrated Knowledge-based Command and Control for the ONE SAF: Building the 3rd Spiral, 3rd Generation SAF*, Pointer Monograph no. 5, 2008. Interestingly, the United States Marine Corps has repudiated the effects-based operations concept. See General J.N. Mattis, 'USJFCOM Commander's Guidance for Effects-based Operations', 14 August 2008.
22 See Bernard Fook Weng Loo (ed.), *Military Transformation and Strategy: Revolutions in Military Affairs and Small States* (London and New York: Routledge, 2009).
23 Ryan Henry and C. Edward Peartree (eds), *The Information Revolution and International Security* (Washington, DC: Center for Strategic and International Studies, 1998).

24 Robert W. Chandler, *The New Face of War: Weapons of Mass Destruction and the Revitalization of America's Transoceanic Military Strategy* (McLean, VA: AMCODA Press, 1998).
25 Interestingly, the United States Marine Corps has repudiated the effects-based operations concept. See General J.N. Mattis, 'USJFCOM Commander's Guidance for Effects-based Operations', 14 August 2008.
26 See, for instance, Thomas G. Manhken and James R. FitzSimonds, 'Tread-heads or Technophiles? Army Officer Attitudes towards Transformation', *Parameters*, summer 2004, vol. 34, no. 2, pp. 52–72. While the Mahnken and FitzSimonds study addressed United States Army officer attitudes, it is reasonable to assume that such concerns may be replicated in other national armed forces.
27 Mark Beeson and Alex J. Bellamy, *Securing Southeast Asia: The Politics of Security Sector Reform* (London and New York: Routledge, 2008), p. 59.
28 Tim Huxley, 'The ASEAN States' Defence Policies 1975–81: Military Responses to IndoChina', *Strategic and Defence Studies Centre Working Paper* No. 88 (Canberra: Australia National University, 1984), p. 11.
29 See *Realising Integrated Knowledge-based Command and Control*; and *Integrated Knowledge-based Command and Control for the ONE SAF*.
30 Indeed, in the case of *Integrated Knowledge-based Command and Control for the ONE SAF*, this document dedicates almost one-third of its space to the issue of training and education.
31 Ministry of Defence, Singapore, 'Committee to Strengthen National Service Holds Fourth Meeting', 2013, www.mindef.gov.sg/imindef/mindef_websites/topics/strengthenNS/news/nr/08oct13_nr.html#.U1Eu-PmSyMI (accessed 18 April 2014).
32 IPS Report on Singaporeans' Attitudes to National Service, 2013, http://lkyspp.nus.edu.sg/ips/news/ips-report-on-singaporeans-attitudes-to-national-service-2013 (accessed 18 April 2014).
33 See, for instance, Ron Matthews, 'Managing the Revolution in Military Affairs', paper presented at the IDSS Conference on Revolutions in Military Affairs: Processes, Problems and Prospects, 22–23 February 2005.
34 David Kirkpatrick and P. Pugh, 'Towards the Starship Enterprise – are the Current Trends in Defence Unit Costs Inexorable?', *Aerospace*, May 1983. Kirkpatrick argues that this acquisition cost is increasing by approximately 10 per cent each year. See David Kirkpatrick, 'Starship Enterprise Revisited – Prospects for the 21st Century', *The Hawk Journal*, RAF Staff College (1995).
35 Ron Matthews, *European Arms Collaboration* (New York: Harwood Academic Press, 1992), ch. 3; Ron Matthews, 'International Arms Collaboration: The Case of Eurofighter', *International Journal of Aerospace Management*, vol. 1, no. 1 (February 2001), pp. 73–79.
36 Huxley, *Defending the Lion City*, p. 183.
37 Huxley, *Defending the Lion City*, p. 27.
38 Morrison and Suhrke, *Strategies of Survival*, pp. 183–184.
39 Paribatra and Samudavanija, 'Internal Dimensions of Regional Security in Southeast Asia', p. 79.
40 Kin Wah Chin, 'Singapore: Threat Perception and Defence Spending in a City-state', in Kin Wah Chin (ed.), *Defence Spending in Southeast Asia* (Singapore Institute of Southeast Asian Studies, 1987), p. 203.
41 Data acquired from 'The SIPRI Military Expenditure Database', Stockholm International Peace Research Institute, www.sipri.org/research/armaments/milex/milex_database (accessed 6 April 2014).
42 Bernard Fook Weng Loo, 'Maturing the Singapore Armed Forces: From Poisonous Shrimp to Dolphin', in Welsh *et al.* (eds), *Impressions: The Goh Years in Singapore*, pp. 188–197.

43 Bilveer Singh, 'Singapore's Defence Industries', *Canberra Papers on Security and Defence*, No. 70 (Canberra: Strategic and Defence Studies Centre, Australia National University, 1990), p. 38.
44 Richard A Bitzinger, *The Modern Defense Industry: Political, Economic and Technological Issues* (New York: Praeger, 2009).
45 Keynote address by Dr Tony Tan at the launch of Temasek Laboratories, MINDEF Internet Webservice, 6 September 2000.
46 Mike Yeo, 'Singapore and the F35B Joint Strike Fighter', *The Diplomat*, 18 October 2013, http://thediplomat.com/2013/10/18/singapore-and-the-f-35b-joint-strike-fighter/ (accessed 12 May 2013).
47 Joo Lin Tay, 'Formidable Voyage to Deliver First Stealth Frigate', *The Straits Times*, 26 May 2005.
48 'Ultimax 100 Light Machine Gun', Singapore Technologies Engineering, www.stengg.com/products-solutions/products/ultimax-100-light-machine-gun (accessed 12 May 2013).
49 For those who insist that a new security agenda has emerged, that 'traditional' wars have become less relevant, see, among others: Mary Kaldor, *New and Old Wars: Organized Violence in a Global Era* (Cambridge: Polity Press, 1999); Thomas X. Hammes, *The Sling and The Stone: On War in the 21st Century* (St Paul, MN: Zenith Press, 2004); and Bruce Berkowitz, *The New Face of War: How War Will Be Fought in the 21st Century* (New York and London: The Free Press, 2003). A counter-argument to these scholars is Colin S. Gray, *Another Bloody Century: Future Warfare* (London: Weidenfeld & Nicolson, 2005).
50 The Honorable Donald H. Rumsfeld, 'Transforming the Military', *Foreign Affairs*, vol. 81, no. 3 (May/June 2002), p. 23.
51 See Rupert Smith, *The Utility of Force: The Art of War in the Modern World* (New York: Alfred A. Knopf, 2007).
52 Rupert Smith, *The Utility of Force*, p. 3.
53 Biddle, *Military Power*.
54 Michael W. Doyle, 'Discovering the Limits and Potential of Peacekeeping', in Olara A. Otunnu and Michael W. Doyle (eds), *Peacemaking and Peacekeeping for the New Century* (Oxford: Rowman & Littlefield, 1998), pp. 8–12.
55 Charles Moskos *et al.* (eds), *The Postmodern Military: Armed Forces after the Cold War* (Oxford: Oxford University Press, 2000).
56 Charles C. Krulak, *The Strategic Corporal: Leadership in the Three Block War* (Marine Corps Association, 1999), pp. 14–17.
57 *SIPRI Yearbook 2002* lists 57 major armed conflicts in the period between 1990 and 2001, of which only three may be considered as inter-state conflicts. See also: Brian Reid, 'Enduring Patterns on Modern Warfare', *The Nature of Future Conflicts: Implications for Force Development* (Camberley, UK: Strategic and Combat Studies Institute, 1998); Steven Metz and Raymond Millen, *Future War/Future Battlespace: The Strategic Role of American Landpower* (Carlise, PA: Strategic Studies Institute, US Army War College, 2003); and Kalevi J. Holsti, *The Decline of Interstate War, or The Waning of Major War* (London: Routledge 2006). For a counter-view, see Meredith Reid Sarkees, 'The Correlates of War Data on War: An Update to 1997', *Conflict Management and Peace Science*, vol. 18, no. 1 (2000), pp. 123–144.
58 See, for instance: Arabinda Acharya, *Ten Years After 9/11: Rethinking the Jihadist Threat* (London: Routledge, 2013); Kumar Ramakrishna, *Radical Pathways: Understanding Muslim Radicalization in Indonesia* (New York: Praeger, 2009); Kumar Ramakrishna and See Seng Tan (eds), *After Bali: The Threat of Terrorism in Southeast Asia* (Singapore: World Scientific, 2003); Andrew T.H. Tan and Kumar Ramakrishna (eds), *The New Terrorism: Anatomy, Trends and Counter-strategies* (Singapore: Marshall Cavendish, 2002).

59 The SAF defines OOTW as including transnational terrorism, epidemic outbreaks and disaster relief, distinct from conventional war. See Lim Guan He, 'A Ready SAF: A Strategy for Tomorrow', *The Pointer*, vol. 38, no. 1 (2012), p. 41.
60 The SAF has almost from the time of its inception been sending personnel on overseas OOTW; see 'Overseas Operations', MINDEF, www.mindef.gov.sg/imindef/key_topics/overseas_operations.html (accessed 13 May 2013).
61 See 'Singapore Armed Forces Concludes Deployment in Afghanistan', MINDEF, 25 June 2013, www.mindef.gov.sg/imindef/press_room/official_releases/nr/2013/jun/25jun13_nr.html#.UmDMVSjHafQ (accessed 13 May 2013).
62 This is the mission of the second People's Defence Force, a component element of the SAF. See 'The 2 People's Defence Force', MINDEF, www.mindef.gov.sg/imindef/mindef_websites/atozlistings/army/ourforces/2PDF.html (accessed 13 May 2013).
63 Bruce Berkowitz, *The New Face of War: How War will be fought in the 21st Century* (New York: The Free Press 2003).
64 'Update on the Security Operations on the Search for Mas Selamat Kastari: Comments by Deputy Prime Minister and Home Affairs Minister Wang Kan Seng', Ministry of Home Affairs, 26 March 2008, www.mha.gov.sg/news_details.aspx?nid=MTE4OA%3D%3D-xCtuBeinhu8%3D (accessed 12 May 2013).
65 See 'National Security Coordinating Structure', NCSC, http://app.nscs.gov.sg/public/content.aspx?sid=23 (accessed 15 May 2014).
66 See Kumar Ramakrishna and Bernard Loo, 'The U.S. Military and Nonconventional Warfare: Is Firepower Cheaper than Manpower?', *IDSS Commentary*, no. 34, 21 July 2005; and Bernard Loo, 'The Military and Counter-Terrorism', *IDSS Commentary*, no. 89, 8 December 2005.
67 Hammes, *The Sling and the Stone*, p. 260.
68 When the Singapore Armed Forces responded to the humanitarian crisis in Banda Aceh arising out of the December 2005 tsunami, it used airlift assets like C-130s and Fokker-50 and CH-47 helicopters to move 1,200 military personnel with over one million tons of cargo over some 250 sorties, while heavy sealift assets moved engineering equipment such as bulldozers, excavators and cranes to establish beach landing points and clear supply routes from the coast to the devastated areas to respond rapidly to the growing crisis in Banda Aceh. See Gail Wan, 'Fast Aid', *Pioneer*, no. 328 (February 2005).

6 The sources of military change in India

An analysis of evolving strategies and doctrines towards Pakistan

S. Kalyanaraman

The security challenges posed by Pakistan and China have dominated Indian strategic thinking since Independence and the late 1950s, respectively. Both challenges have, however, acquired significant new features during the past 25 years. Pakistan developed nuclear weapons with Chinese assistance and used its nuclear shield to expand support for and sponsorship of terrorist groups against India.[1] China dramatically rose to become Asia's most powerful resident economic and military power, which inevitably translated into worrying levels of Chinese presence and influence around India's continental and maritime periphery.[2]

The nature of the security challenges posed by China and Pakistan to India is not reducible to a mere contestation over disputed border alignments or conflicting claims over adjoining territories. While it is true that the principal dispute between India and Pakistan is over the disposition of Jammu and Kashmir, the very fact that Pakistan was created as a separate 'nation' for the Muslims of British India conflicts with the idea of India as a secular polity that is home to the world's third-largest Muslim population.[3] This imparts an ideological character to India–Pakistan rivalry. As for China, although an ideological content has hitherto been absent and the situation across the disputed border flares up only occasionally, India is acutely concerned about China's unstinted military and strategic support for Pakistan and the prospect of a China-dominated Asia. Together, these aspects entail serious consequences for India's diplomatic and military strategies.[4] In sum, these characteristics of India's relationships with China and Pakistan indicate the persistent nature of rivalry.[5]

Military change in India has been a function of two factors: the evolving capabilities and postures of its long-term adversaries, Pakistan and China, and the limited resources available for modernization. In this chapter I will show how India's grand strategy has not fundamentally changed, with only long-term incremental adjustments being made to cope with the new reality of China's rise. I will then argue that with regard to Pakistan, strategy, doctrine and organization have changed in both an incremental and major fashion. The pursuit of victory and operational lessons learned drove India's military change on strategic, doctrinal and organizational levels. This was especially so after Pakistan's development of nuclear weapons, which led to the replacement of the pursuit of victory in full-scale war with gaining strategic advantage in limited war. Doctrines and structures have changed to reflect this reality.

Economic constraints and military change

India has had to either divert resources from developmental to defence purposes or defer military modernization during periods of economic difficulties. Thus, although the Indian political leadership perceived economic development as the foundation of defence capability and therefore favoured development over defence in the 1950s and 1960s, it did divert resources for military purposes in the wake of Pakistan's acquisition of modern platforms from the United States and the emergence of China as a threat.[6] During the acute economic constraints of the 1990s, the advice of retired military leaders that the three Services abandon 'pet theories, prejudices and grandiose notions such as a blue-water navy, total mastery of the skies and two-front war-waging capabilities' and instead aim for smaller force levels that are 'lean but mean' went unheeded.[7] Instead, the political and military leaderships resorted to stop-gap measures such as technological upgrades of military aircraft and the acquisition of small numbers of modern platforms for maintaining a reasonable degree of military preparedness while waiting for better times to continue with robust modernization.[8] Deferring military modernization during difficult economic circumstances continues to persist as a preferred option, as is evident from the recent decision to postpone by at least one year the purchase of an already identified frontline fighter aircraft for the Indian Air Force because 'there is no money left' for new acquisitions.[9]

Subtle shifts in grand strategy and organizational responses

The persistence of the twin security challenges from China and Pakistan as well as the intermittent process of military modernization, both due to the 'lack of fiscal space ... for capital expenditure'[10] and the more recent roadblock imposed by corruption scandals linked to defence imports,[11] have ensured that major military change in India has largely been incremental and adaptive rather than sudden and disruptive. For instance, the objectives of India's defence policy and by extension the goals of the Indian military have largely continued to be those articulated in May 1995:

- to defend national territory and maintain the inviolability of 'land borders, island territories, offshore assets and ... maritime trade routes';
- to defend against internal security challenges;
- to 'exercise a degree of influence' over countries in the 'immediate neighbourhood' of South Asia and to prevent their destabilization

through the possession of 'an effective out-of-the-country contingency capability';
- to 'effectively contribute towards regional and international stability' – an unstated but evident reference to India's long record of participation in UN-mandated peacekeeping missions.[12]

During the past two decades, these objectives have been modified to some extent at least in terms of expressed aspiration, even if not yet realized in actuality. Thus, during the past decade or so, political and military leaders have repeatedly referred to India's interests in the 'extended neighbourhood', a region stretching from the Suez Canal to the South China Sea,[13] as well as to the country's role as a 'net security provider' in the Indian Ocean Region.[14] The consequent expansion in the area of possible future operations has found a reflection in the doctrines published by the three services during the 2000s. But the fact remains that the Indian military is unlikely to acquire the necessary capabilities for undertaking purposeful missions in this wider geographical area for another decade at least.[15]

Another recent modification to military goals occurred in December 2009 when the political leadership directed the armed forces to prepare for a two-front war with China and Pakistan.[16] Accordingly, the Indian Army is in the process of raising the new XVII Corps to beef up its capabilities vis-à-vis China and the Indian Air Force has been repositioning assets and establishing additional facilities along the China front.[17] These measures are, however, unlikely to significantly alter the existing static defensive posture towards China for a decade at least, for two main reasons. First, the raising of XVII Corps is earmarked for completion only by 2018/2019.[18] More importantly, current financial constraints and the unclear prospects of sustained high economic growth rates are likely to limit appropriations for both equipping these new land formations with heavy-lift helicopters, gunships, mountain howitzers and modern communication systems and maintaining, let alone expanding, the steadily declining operational strength of the air force.[19]

Other incremental changes have occurred in organizational structures. As part of the measures undertaken to reform the national security system in the wake of the Kargil conflict, the Headquarters Integrated Defence Staff was established in 2001 to serve as the Secretariat of the Chiefs of Staff Committee and, in that capacity, to coordinate the framing of the annual and five-year defence budgets and to formulate joint doctrines and manage training. Significantly, this was an interim arrangement and it was resorted to because of resistance from both the air force and the principal opposition party to the appointment of a Chief of Defence Staff.[20]

A second incremental organizational change that occurred in 2001 was the creation of a tri-service command in the Andaman and Nicobar Islands. Such a command was deemed necessary because of China's growing presence in the Bay of Bengal and the rise of piracy, as well as the

Aceh separatist movement in the vicinity of the Malacca Straits.[21] Two years later, the Strategic Forces Command was formed to operationalize the nuclear deterrent and exercise control over warheads and delivery systems.[22] Plans are now being drawn up to establish three other tri-service commands: one to effectively employ the special forces capabilities possessed by the three services, and the other two to cater for challenges in the cyber and space domains.[23]

Individual services have also periodically effected changes in their organizational structure. The Indian Air Force established the Eastern Air Command in 1959 and the Central Air Command in 1962 in response to the Chinese threat, and later the Southern Air Command in 1984, in the wake of the growing presence and activities of superpower navies in the Indian Ocean and the onset of the ethnic conflict in Sri Lanka.[24] More recently, in 2005, the Indian Army created a new command – the South Western – to better enable the prosecution of a short, sharp war against Pakistan.[25]

China: constrained change and engagement

In the case of China, India has persisted with the view that a war with the country is 'unthinkable' and has therefore sought to deal with it 'through a mixture of engagement, diplomacy, trade, and positioning adequate forces along the borders'.[26] This is not, however, to mean that 'creating conditions advantageous for the *possible* [emphasis in original] application of military power' has been unimportant.[27] On the contrary, 'typically, this has been done through non-military means: building roads, establishing legal regimes permitting or denying certain activities, and creating political alignments that make up the political context in which military force is used or not used'.[28]

Thus, since the early 1990s, India has forged a series of confidence-building measures with China – the latest in 2013 – that lay down ground rules to avoid, or non-militarily deal with, border incidents.[29] At the same time, in the light of China's dramatic improvement of its communication infrastructure in Tibet,[30] its successive articulations about waging 'local wars under high-tech conditions' and 'local wars under informationalized conditions',[31] and the expansive nature of its recent territorial claims to Indian territory,[32] India has initiated measures to beef up its defence capabilities. These include: a strategic road-building programme;[33] the ongoing raising of the XVII Corps by the Indian Army; the repositioning of assets and establishment of new facilities along the China front by the Indian Air Force;[34] and stepping up diplomatic and military engagement with key countries in Asia including Japan, Australia and the United States, although none of these relationships has as yet acquired an explicit balance of power characteristic as was the case with the former Soviet Union between 1971 and 1991.[35] As for the actual use of force, India has, since the 1960s, adopted a static defence posture[36] involving the forward deployment of troops to deny China its territorial aims.[37] As noted earlier, this posture is likely to change only in the 2020s when the new infantry forces being raised are endowed with greater mobility and fire power and the now declining operational strength of the air force is not only arrested but reversed.

In contrast to this stands India's record of dealing with the military challenge posed by Pakistan, and the strategies and doctrines it has successively adopted in this regard.

Pakistan: India's changing strategies and doctrines

The overall tenor of India's strategy, whether in dealing with China or Pakistan, has been marked by a defensive posture.[38] But its strategy, especially towards Pakistan, has invariably contained an integral offensive element – what Clausewitz described as 'the flashing sword of vengeance' that delivers 'well-directed blows'.[39] Defence, in Clausewitz's conception, consists only in waiting for the adversary to initiate an attack and thereafter transitioning to the offensive at the suitable moment. Without such a transition to the offensive, defence would be tantamount to 'passive endurance'.[40]

Within this overarching conception of defence, two distinct phases may be discerned in the evolution of India's strategy towards Pakistan. In the first phase, between 1947 and 1997, India adopted the strategy of annihilation, whereby it sought to defeat Pakistan and compel the country to seek peace. In the second, ongoing phase, which began with Pakistan's acquisition of nuclear weapon capability in the 1990s, India has been pursuing the strategy of exhaustion.[41] Now, the objective is no longer the military defeat of Pakistan but the occupation of a portion of its territory for bargaining purposes.[42]

Within the context of these two strategies, the operational doctrines of the three armed services have also exhibited major changes. In the process, they have left distinct impressions particularly on the strategy of annihilation. Indeed, the change in doctrines, which first occurred during the 1971 war and became consolidated thereafter, neatly divides into two the phase during which India had adopted the strategy of annihilation. Before that war, the strategy of annihilation was pursued through the land warfare doctrine of attrition. Attrition involved a broad-front advance by units that engaged in a series of set-piece tactical battles in order to cumulatively wear down the Pakistan Army and literally push it back.[43] In the air, the corresponding concept to attrition is aerial combat – the 'direct contest for dominance' to attain a favourable air situation – which is what the Indian Air Force sought during this period.[44] And on the seas, the equivalent of attrition is the destruction of the adversary's fleet to secure 'a working control of the sea'.[45] But given the very limited strength possessed both by itself and by its Pakistani counterpart, the Indian Navy limited itself to coastal defence and the custody of maritime assets during the 1950s and 1960s.[46]

All this changed during the course of the 1971 war, where lessons learned combined with shifts in local conditions and political realities drove the army to adopt the doctrine of manoeuvre for attaining a quick victory.[47] In contrast to attrition, which is a test of opposing strengths, manoeuvre involved incapacitating the Pakistani military and causing its 'systemic disruption' by targeting weak points, severing lines of communications and destroying key nodes deep behind the front line.[48] For its part, the air force obtained command of the air not only through aerial combat but also through ground interdiction in East Pakistan. The navy exercised sea control in the Bay of Bengal through a tactical blockade,[49] while also participating in aerial missions over East Pakistan with its carrier-based aircraft and projecting power on to land through an amphibious operation. India's victory in the 1971 war led to the further consolidation of these doctrines through operational and procurement changes during the course of the next two decades. The army evolved capabilities to mount an armour-led blitzkrieg across Pakistan. The air force sought command of the air through both aerial and ground interdiction. And the navy aimed at exercising control over designated sea areas through a blockade to both prevent the Pakistan Navy from venturing out to undertake attacks on India's coasts or commerce and the plying of Pakistan's sea-borne trade.[50]

But with Pakistan's acquisition of nuclear weapons capability and the consequent shift in India's strategy to one of exhaustion, the armed services have been forced to suitably adapt their doctrines. In this regard, the army has found itself torn between pursuing attrition, manoeuvre and surprise as its organizing principle.[51] In contrast, the air force and navy have found it easier to tune their extant doctrines for the new strategy. While the navy continues to pursue the exercise of sea control over designated areas through the imposition of a blockade, the air force is reverting to its doctrine of attaining in-theatre air superiority.

Annihilation through attrition

The formative experience that determined India's defence policy and strategy was the conflict with Pakistan over Jammu and Kashmir (hereafter Kashmir). In late October 1947, Pakistan launched an irregular force to annex Kashmir[52] whose international legal status at that time was indeterminate.[53] In response to the plea for help from Kashmir's ruler and after obtaining his formal accession to the Indian Union, the Indian leadership despatched an armed rescue mission.[54] India's initial successes in not only thwarting the capture of Srinagar, the state's capital, but also in gradually pushing the irregulars back led Pakistan to induct its own regular forces into Kashmir. In the meantime, India had lodged a formal complaint with the United Nations Security Council on Pakistan's support for the irregulars, a measure which not only prevented the declaration of open war but also afforded an opportunity for the geopolitically minded officials of the United Kingdom to transform the Kashmir file into a formal dispute between India and Pakistan.[55] By the time the UN-brokered ceasefire came into effect on 1 January 1949, Indian forces had managed to regain control over approximately two-thirds of the state.

India could not evolve a clear strategy for the conflict due to three main reasons. First, it was confronted by three equally important imperatives: pushing the irregulars back, relieving the towns and garrisons they had besieged, and preventing them from occupying new areas. Second, even after Pakistan introduced its regular troops into the conflict zone, India could not declare a war because the United Nations Security Council had become aware of the issue. This constraint was further reinforced by the efforts made by British officers then commanding India's armed forces as well as by India's first Governor-General, Lord Mountbatten, to prevent the escalation of the conflict into an open war between two British Dominions.[56]

Unable to devise a clear strategy, the Indian Army, ably supported by the air force, undertook a series of offensives aimed at thrusting the irregulars as far back as possible and relieving besieged towns and garrisons.[57] The army's doctrinal concept in these operations was attrition – 'the set-piece battle, phased programmes to capture strong points – and venturing on further advance only after due regrouping and re-enforcement' – which it had imbibed by 'training and tradition' and consequently preferred.[58] For its part, the air force provided critical support for the army through interdiction of the irregulars, armed reconnaissance, close air support and transportation.[59]

Why the strategy of annihilation?

In the wake of this experience and based on a military assessment that the only probable future military contingency likely to confront India was a more elaborate invasion of Kashmir by a Pakistani irregular force built around a sizeable core of the Pakistan Army, India resolved upon a strategy of annihilation. The reasoning behind this decision was as follows. Because of the existence of good communication facilities between Pakistani territory and Kashmir, Pakistan could keep the irregulars adequately supplied and backstop them with its own army. This would not only tie down large numbers of Indian Army personnel in Kashmir's mountainous terrain, but also neutralize India's superior numbers in other sectors. Moreover, India would not be able to attain a decisive victory because of the sanctuary that the irregulars and regulars would enjoy in Pakistani territory. Under these circumstances, the only practical alternative was not to distinguish between irregulars and Pakistani troops or even 'to concede the possibility of another limited war in Kashmir'.[60] In other words, India should wage a general war in response to another irregular invasion of Kashmir with or without the participation of the Pakistan Army.

When this assessment was accepted by the Defence Committee of the Cabinet, contingency plans were devised for 'possible operations in Kashmir, Punjab, and Rajasthan' as well as 'precautionary measures' along the border with East Pakistan.[61] The concentration of military operations on Pakistan's western wing was inevitable. From a military point of view, Pakistan had to be forced into taking its eyes off Kashmir. This could be achieved only by posing a threat to its political and military centres of gravity, which lay in the western wing. The thrust of the plan evolved was to adopt the defensive in Kashmir and undertake the main offensive across the Punjab border in the form of an advance towards Lahore and Sialkot. In order to prevent the concentration of Pakistani forces in the main operational theatre, diversionary thrusts were planned towards Rawalpindi or Karachi. While the army's role would be decisive, the air force in particular but also the navy were to support the army's efforts.[62] The aim of the plan was to 'inflict a decisive defeat' on the Pakistan Army and occupy important centres such as Lahore.[63] India's end-game would not be the acquisition of Pakistani territory but merely the use of captured territory as a bargaining chip in the peace negotiations.

Indian strategy in the 1965 war

An adapted version of this contingency plan was implemented during the 1965 India–Pakistan War. When popular protests erupted in Kashmir in May 1965 over the arrest of the Kashmiri leader Sheikh Abdullah, Pakistan began to plan the organization of a full-fledged uprising. Codenamed Operation Gibraltar, the plan was implemented in early August 1965 with the dispatch of some 7,000 trained and armed infiltrators into Kashmir. The objectives of the infiltrators were the establishment of bases, widespread sabotage and the creation of conditions for Pakistan's open military intervention. However, neither did the expected rebellion break out nor did Kashmiris (except for a small faction) provide support for the infiltrators.[64] But what the planned uprising did result in was the initiation by both countries of a series of small-scale military actions along the Kashmir border including the occupation of advantageous tactical positions. This jockeying for advantage reached a critical point on 1 September when the Pakistan Army executed *Operation Grand Slam* – an armoured thrust against the town of Akhnur, whose capture would have enabled Pakistan to sever all land communications between Kashmir and the rest of India.[65]

In the face of this threat, the Indian Army was ordered to take the offensive across the Punjab border. The aim of the plan developed by the army for this purpose was two-fold. First, to pose a threat to Lahore and 'drive a wedge' between Pakistani forces deployed around Lahore and those based in Sialkot; and second, to destroy Pakistan's military potential.

Accordingly, the army carried out two offensive thrusts – one towards Lahore and the other towards Sialkot. In addition, limited offensives were carried out in Kashmir and across the Rajasthan border to, respectively, capture suitable ground and tie down Pakistani forces.[66] Although the army's two offensive thrusts scored some initial success and relieved Pakistan's military pressure on Kashmir, they soon lost momentum in the face of counter-attacks. Neither were the bridges on the Ichhogil Canal captured nor was Sialkot encircled.[67] Similarly, the objective of destroying Pakistan's war potential was also only partially successful, with a noticeable degree of attrition inflicted upon its armour formations.[68]

As envisaged, the air force played a supporting role to the army during the war. Nearly 79 per cent of its missions were related to the provision of fighter escorts and sweeps, close air support, and combat air patrol for air defence purposes; the remainder were bombing and counter-air strike (against air bases and other infrastructure) missions.[69] The service's doctrinal orientation was towards attaining a favourable air situation. This was a function of several factors. First, it was due to the force structure that had evolved with an emphasis on interceptors, driven by the political imperative of not targeting population centres. This has resulted in a fighter-pilot ethos pervading the service's operational culture, best reflected in the motto 'Touching the Sky with Glory'.[70] When the war broke out, the only bomber in the inventory was the Canberra, which moreover proved highly vulnerable to anti-aircraft fire.[71] Second, since the air force's inception in 1933, every new squadron has been formally assigned the 'army cooperation role'.[72] Yet, there was a complete absence of joint planning during the war. Each service pursued its respective objectives and cooperation was 'incidental' rather than 'well planned'.[73]

As for the Indian Navy, with its sole carrier undergoing repair and refit, the political leadership advised it to 'lie low' and play a defensive role.[74] And it did exactly that.

Limitations of attrition in a short war

The official history of the 1965 war details several reasons for India's limited success and its unwillingness to pursue the war to its logical conclusion. These include: deficient command and control, lack of strategic planning, inadequate understanding among military commanders about the type of war being waged (all-out or limited) and the objectives being pursued (capture of territory or attrition of Pakistani forces), the army chief's fearful cautiousness about Pakistan's military capability and his professional failure in not being fully aware of the large-scale availability of war reserves, and the faulty 'strategic concept of attacking ... at many places' instead of 'a few selected powerful thrusts'.[75]

But this list ignores the more fundamental reason: the mismatch between the doctrine of attrition and the time available to make it effective. Success in a war of attrition depends on the ability to absorb greater losses than the adversary and outlast the other in what is essentially a slugfest. A fundamental prerequisite for this is 'overall superiority in net attritive capacity'; that is, the possession and generation of greater resources than the adversary.[76] At the same time, this superiority in material resources needs time to make itself felt, exactly what is not available in a short war.[77] This limitation was clearly demonstrated during the 1965 war in which India enjoyed overall advantages in manpower and industrial capabilities as well as a bigger war machine and greater military resources. When the war ended, the army had used up only 14 per cent of its frontline ammunition and possessed twice the number of tanks than Pakistan did, while the air force 'started prevailing over' its 'smaller but more modern' counterpart.[78] In contrast, by the closing stages of the war, the Pakistan Army 'was short of supplies' and 'running out of ammunition'.[79]

Yet, India did not pursue the war beyond three weeks, even though 'the planning parameters for war were based on the expectation of a long war' lasting several months, as occurred in 1947/1948.[80] It agreed to an early ceasefire because of two factors – the prospect of China initiating hostilities and India's isolation in the international arena – which were expressly highlighted in a military assessment prepared by the then army chief.[81] If China were to initiate even only limited and localized action, India would get caught in a two-front situation for which it was ill prepared. Not agreeing to a ceasefire would only heighten such a prospect. Further, India stood isolated in the international diplomatic arena: the Soviet Union had adopted a position of studied neutrality, the United States and Britain had imposed an economic and military embargo, and the United Nations Security Council had passed a series of unanimous resolutions calling for a ceasefire.[82] Together, these two factors shrank the time available and made the pursuit of victory through attrition impossible to attain even if the war had been prolonged 'for some days', which was all the time the political leadership saw itself as being able to offer the Indian military for achieving 'a spectacular victory'.[83]

Impact of the 1965 war on service doctrines and plans

That they may have only a window of three weeks in future wars as well was an important lesson that civilian and military decision-makers learned during the 1965 war. Yet, this lesson did not fundamentally alter the plans or doctrines of the army and air force in the immediate aftermath of that war. The only change effected by the army was the development of a contingency plan for the East Pakistan front, which had been deliberately ignored up until then. Given, however, the possibility of China entering a future India–Pakistan war and in such an eventuality Pakistani forces attempting a link-up with Chinese forces coming down from the north, the army evolved a plan to undertake two limited thrusts to keep Pakistani forces bottled up within East Pakistan.[84] Similarly, the air force continued with its doctrinal orientation of attaining a favourable air situation, although it rearranged its priorities in the following order: air defence first, followed by close air support and ground interdiction.[85] Only the navy began to rethink its approach by envisaging a greater role for itself in the Indian Ocean.[86]

Annihilation through manoeuvre

India's embroilment in the civil war between Pakistan's western and eastern wings, which resulted in the flow of ten million refugees into Indian territory, necessitated a fundamental shift in service doctrines and plans.[87] The Indian government was determined not to absorb the refugees, as had been its practice since Partition.[88] Given Pakistan's determination to find a military solution at the expense of a political accommodation to resolve the crisis, which was only stiffened by the unstinted support extended by the United States, an Indian intervention became inevitable.[89] Despite insistent calls within the country for an immediate military intervention, neither the prime minister nor the army chief was enthusiastic about such a course of action.[90] Citing three reasons against an immediate intervention – China initiating a military diversion, the Indian Army's limited strength on the East Pakistan front, and the constraints that the soon-to-arrive monsoon would pose – the army chief advocated December 1971 or January 1972 as the ideal time to intervene. By that time, snow would block the Himalayan passes and significantly constrain Chinese military action, India would be able to build up its forces on the East Pakistan front, and the ground in post-monsoon East Pakistan would have dried up.[91]

If an intervention were indeed to be undertaken, what doctrines should govern the employment of the armed forces? Pakistan made it amply clear that even a limited Indian military offensive aimed at capturing a portion of territory and establishing a Bangladesh government there would mean all-out war including on the western front. Further, to ensure that such an Indian attempt did not succeed, it established the bulk of its forces in strong defensive positions centred on key towns nearer to the border.[92] This move nullified the contingency plan that the Indian Army had evolved for the East Pakistan front in the late 1960s, although this forward defence posture[93] adopted by Pakistan opened up the possibility of developing a new plan based on the doctrine of manoeuvre. At the same time, the Indian Army decided to adopt the defensive, albeit composed of limited offensives, on the western front.[94]

Why manoeuvre?

The army chose manoeuvre for the East Pakistan front because of several interrelated reasons. First, the employment of attrition would make the war a prolonged affair because not only had the Pakistan Army built up entrenched defensive positions but its forces also had the option of falling back on an inner defence line made up of 'great river obstacles'.[95] Overwhelming these positions would consume considerable time. In contrast, a campaign based on manoeuvre would not only bypass the entrenched Pakistani forces but also prevent them from falling back on the river obstacles, thus enabling India to gain a swift victory. A swift conclusion to the war was imperative because the additional forces required for the intervention in East Pakistan were drawn from the China front, which could not be kept denuded for too long.[96] Further, a prolonged war of attrition would also afford the international community an opportunity to intervene diplomatically and impose a ceasefire, which would transform Pakistan's civil war into an India–Pakistan dispute, perpetuate the status quo by freezing the birth of an independent Bangladesh, and leave India with the enormous challenge of dealing with millions of refugees. The United States, in particular, was expected to take the lead in this effort given its unstinting support for the Pakistani military regime even in the face of the brutal repression in East Pakistan.[97] In such an event, India expected Soviet support for its cause to falter, especially given the latter's reluctance to contemplate a breakup of Pakistan, scepticism about India's military capability, and advice against military action.[98]

The army's role

Eastern Command, which spearheaded the offensive into East Pakistan, drew up a plan composed of two elements: containing entrenched Pakistani forces located near the border, and undertaking three powerful, mobile and flexible thrusts to bypass fortified positions, cut lines of communication and rush along the path of least resistance towards the 'Dacca bowl' – East Pakistan's triangular heartland bordered by the rivers Jamuna and Meghna. Accordingly, whole brigade groups avoided roads, moved over paddy fields or were transported by helicopter across rivers, even as their logistics travelled by bicycles and rickshaws.[99] These thrusts isolated Dhaka from Pakistani forces deployed closer to the border and led to Pakistan's theatre commander losing 'contact with his divisional commands ... [and] the divisional commands with the brigade headquarters', while his troops became 'an aimless crowd'.[100] At the end of the war, a substantial part of Pakistani forces remained in being and surrendered en masse.

Roles of the air force and navy

In tune with its determination to support the ground offensive to its fullest extent, the Indian Air Force began by transporting a significant portion of the army's buildup and undertaking aerial reconnaissance. The specific tasks assigned to it during the course of the war were: to eliminate the

Pakistan Air Force at the earliest opportunity, to provide close air support to the army, to help the navy isolate East Pakistan, and to ensure air defence.[101] Accordingly, within the first two days of the war, the small Pakistan Air Force contingent in East Pakistan was neutralized both through aerial and ground interdiction. The resulting command of the air allowed greater focus on the provision of support for the ground offensive, well reflected by the fact that 68.1 per cent of the air force's sorties on the East Pakistan front were devoted to close air support and interdiction of Pakistani forces.[102]

The Indian Navy also had its share of air operations for which it employed its carrier-based aircraft.[103] Both naval aircraft and ships specifically targeted Chittagong, Cox Bazar and other ports. In addition, the navy imposed a blockade and seized a number of Pakistani merchantmen carrying contraband.[104] One other task it carried out, albeit marked by a mishap, was an amphibious landing to cut off the escape routes of Pakistani personnel to Burma.[105]

Post-1971 evolution of doctrines

Inspired by this spectacular victory through a war of manoeuvre, military thinking in India began to immediately shift towards offensive armour operations that would reach deep into (the now reduced) Pakistan.[106] In pursuit of this line of thinking, the political leadership appointed a three-member expert committee in 1975 to develop a 20-year perspective plan for the Indian Army. The committee, comprising three senior generals, recommended 'a force advantage of two corps' over Pakistan, increasing the number of tank regiments from 27 to 58, and creating two mechanized infantry divisions.[107] Accordingly, the army's armoured force was steadily built up during the course of the next several years.[108] This resulted in the reformulation of the army doctrine during the 1980s. An effort spearheaded by General Sundarji, hence the appellation Sundarji Doctrine, the new doctrine called for deep armoured thrusts into Pakistan with the aim of bringing about its systemic collapse.[109] Such a war would be waged in response to a Pakistani intervention not only in Kashmir but also in Punjab to aid Sikh militants – a scenario explicitly visualized during an 11-month-long military exercise named Brasstacks.[110] Sundarji also developed a perspective plan 'Army 2000', which envisaged the service attaining a force level of 45 divisions, including four armoured divisions, eight mechanized infantry divisions, seven Reorganized Army Plains Divisions and two Air Assault Divisions.[111] This latter aspiration could not, however, be translated into reality because of economic constraints during the 1990s.

The Indian Air Force's experience in the 1971 war made it realize the imperative of attaining command of the air in the early stages of a conflict. Consequently, it began to emphasize command of the air as its primary mission, with close air support for ground forces coming immediately thereafter in its list of priorities. During the 1970s and 1980s, the service acquired a range of Western and Soviet aircraft including those that provided strategic airlift capability. A similar spree of acquisitions for the navy during these years enabled it to firm up the doctrine of exercising control over designated sea areas both by blockading Pakistan to prevent its fleet from going out to sea and its commerce from plying and by projecting power on to Pakistani territory.[112]

Nuclear weapons and the strategy of exhaustion

With Pakistan's acquisition of nuclear weapons capability sometime between 1987 and 1990, India had to abandon the idea of attaining victory in a war.[113] In fact, the Indian political leadership went one step further in February 1999 by asserting the imperative of not only avoiding but also preventing a conflict with Pakistan through confidence-building measures in both nuclear and conventional 'concepts and doctrines'.[114] Immediately thereafter, the leadership was, however, compelled to craft a strategy to wage a limited border conflict with Pakistan in Kashmir. This reality subsequently led to the public proclamation by the then Defence Minister that the political leadership 'understood the dynamics of limited war' and its feasibility under the nuclear overhang, that it waged the Kargil conflict (see below) 'within this perspective', and that the capabilities, doctrines and force structures of the armed forces need to be re-tuned to wage, and succeed in, a war that must be 'kept below the nuclear threshold'.[115]

The Kargil conflict

During the winter of 1998/1999, the Pakistan Army occupied 'an area of about 130 square kilometres over a front of over 100 kilometres, and a depth ranging between seven to fifteen kilometres' in the Kargil district of Kashmir.[116] Its initial, minimal, aim of providing a fillip to the waning insurgency in Kashmir subsequently expanded to one of posing a combined Pakistani military–Kashmiri insurgency threat at the front and rear of Indian forces. Pakistani military planners calculated that India's likely decision to augment force levels in Kashmir would erode its overall force advantage for carrying out an all-out offensive, that even such an augmentation would not be able to deal with the combined challenge in Kashmir, and that such a circumstance would force India to the negotiating table where its weak hand would be exploited to seek 'a just and permanent solution to the Kashmir issue in accordance with the wishes of the people of Kashmir'.[117]

Contrary to this calculation, India was indeed able to orchestrate a strategy to deal with the Pakistani challenge. The broad contours of this strategy were the exercise of restraint in terms of neither declaring a war nor escalating the conflict by permitting the armed forces to cross the Line of Control, and the adoption of an assertive posture by all three armed services to prevent Pakistan from focusing only on the Kargil sector.[118] While the exercise of restraint earned India the support of the international community and the United States in particular, the adoption of an assertive posture enabled it to deter Pakistan.[119] In tune with the strategy approved by the political leadership, the army not only reassigned and built up its force levels, including artillery in Kashmir, but it also forward deployed in other theatres some elements of the strike corps along with their logistics.[120] In addition, formations along the International Border and Line of Control were instructed 'to exert pressure on the Pakistani military through forward deployment, active patrolling and surveillance'.[121] The air force activated its defences in the area covered by its Western Command (Kashmir to parts of Rajasthan), with the proviso that there should be 'no sudden or mass movement of aircraft'.[122] And to strengthen the air effort in the Kargil sector, it built up its aircraft capacity in Srinagar air base as well as at Awantipur.[123] For its part, the navy augmented its Western Fleet with some ships from the Eastern Fleet, transferred its amphibious brigade to the west coast, and, in a further effort to signal its 'readiness and resolve', shifted the venue of its 'Summerex' exercise from the Bay of Bengal to the Arabian Sea.[124]

Within this overall strategic posture aimed at deterring Pakistan from escalation, the doctrinal orientations of the army and the air force was the attrition and pushing back of the intruding forces through a combination of ground battles and the application of fire power from both land and aerial platforms. Thus, before the army commenced each of its tactical battles to physically capture occupied hill features such as Tiger Hill and Tololing, artillery fire power was applied to soften these targets and psychologically exhaust the intruders positioned there.[125] The air force too played a critical role in delivering fire power upon the camps, material dumps and supply routes of the intruders, with its most signal successes being the destruction of the supply depot at Muntho Dalo, and of the intruders' command-and-control bunkers on Tiger Hill. Overall, during the course of the war, the air force undertook a total of 7,631 sorties.[126]

The idea of a shallow armoured thrust

While the Kargil conflict highlighted the salience of attrition in mountain warfare, the imagination of the Indian Army has continued to be gripped by the idea of an armoured offensive not only for compelling Pakistan to vacate captured territory but also for punishing it for its support of anti-India terrorist groups operating from its territory. At the same time, however, the shadow that the nuclear bomb has cast on India–Pakistan relations since the late 1980s has indeed induced a degree of caution. Consequently, influential retired army officers began to advocate a shallower offensive that did not cross Pakistan's 'chemical/nuclear threshold' but still forced the Pakistan Army's reserves into battle, thus providing an opportunity for destroying them.[127]

That this mode of thought continued to prevail within the army became evident during the India–Pakistan crisis of 2001/2002, which was triggered by an attack carried out on the Indian Parliament by Pakistani terrorist groups. Its threshold of tolerance breached, the Indian government ordered general mobilization in an attempt to diplomatically coerce Pakistan into ceasing support for the India-targeting terrorist groups based in its territory.[128] The army evolved a plan for an immediate offensive across the Kashmir border with the aim of occupying those patches of territory that would help stem the infiltration of terrorists. Such a campaign in mountainous terrain would have necessarily involved it in a war of attrition. But in the absence of a prompt political go-ahead for this plan and given Pakistan's own military mobilization in the meantime, the Indian Army subsequently devised another plan to conduct an armoured offensive across the Rajasthan border. The army's calculation was that Pakistan would be forced to move both of its strike corps to meet the Indian offensive, thus presenting an opportunity for destroying them. Further, such an offensive would not only cause the destruction of Pakistan's centre of military gravity – its two strike corps – but also result in the capture of territory for use as a bargaining chip during peace negotiations.[129] But the political leadership was more focused on leveraging the military mobilization for diplomatic coercion rather than for initiating a conflict.

The Cold Start Doctrine

When the army conducted a post-mortem of Operation Parakram, the principal failure it identified was its own inability to undertake a quick offensive, which, in turn, afforded an opportunity for the political leadership to lose resolve and the international community to intervene. The army further diagnosed that its inability to undertake a quick offensive was a function of three interrelated factors. First, the defensive corps deployed close to the border did not have the capacity to adopt an offensive. Second, offensive power was concentrated in the strike corps but moving them forward to their concentration areas took time because they were bulky and located in the interior. And third, because of their size and location, the strike corps could not cause surprise and unsettle Pakistan, which could always find time to counter them through suitable redeployment.[130] In order to rectify these shortcomings, the army proposed a new doctrine called 'Cold Start', which seeks to creatively employ all three doctrinal features of attrition, manoeuvre and surprise.

In order to implement a war plan based on this doctrine, the army has begun to reorganize the structure and capabilities of the defensive corps. Re-designated as pivot corps, the defensive corps are now being assigned offensive capabilities in the form of division-sized integrated battle groups possessing integral armour, artillery and aviation assets to enable them to launch a quick offensive along multiple axes for exploiting Pakistan's unpreparedness.[131] In other words, the idea is to take Pakistan by surprise. This early offensive would subsequently be used as a pivot by the strike corps to effect a deeper armoured penetration, but without crossing Pakistan's nuclear red lines. During the course of these successive offensives, the army and air force would apply massive fire power to attrite Pakistani forces, and the portion of territory captured during such a limited war would be used as a bargaining chip for compelling Pakistan to cease support for anti-India terrorist groups on its soil.[132] In sum, the Cold Start seeks to catch Pakistan by surprise through an early offensive, capture territory by effecting a shallow armoured penetration, and attrite Pakistani forces in the process.

Since its promulgation in April 2004, various elements of this doctrine have been tested in frequent exercises, in which the air force has also participated.[133] But because of international diplomatic pressure, particularly from the United States, the Indian political leadership has not formally endorsed the doctrine during the past ten years.[134] This, in turn, forced an army chief to declare that the Cold Start is 'neither a doctrine nor a military term in our glossary' but only one among many 'contingencies and options'.[135] Nevertheless, the political leadership seems to have tacitly endorsed the doctrine, as is evident from the regular exercises that the army, with the participation of the air force, has been carrying out to validate the concepts involved.[136] In effect, the Cold Start is not a 'doctrine stillborn' but a work in progress.[137] Consequently, in the event of a future limited war, the army is likely to seek to implement the strategy of exhaustion through a creative doctrinal combination of surprise, manoeuvre and attrition. The air force would play a critical role in such a war through its doctrine of attaining in-theatre command of the air and, within that context, carry out aerial and ground interdiction. For its part, the navy is likely to aim at exercising control of designated sea areas to prevent its Pakistani counterpart from undertaking a surprise attack on the Indian coast, to raise the costs of its sea-borne commerce, and to distract the focus of its decision-makers from events on the land.

Conclusion

Military change in India has been a function of two factors: the evolving capabilities of long-time adversaries China and Pakistan; and the limited resources available for modernization. Consequently, most changes in military goals and organizational structures have been incremental and adaptive in nature rather than sudden and disruptive. Major change has occurred only in India's politico-military strategy towards Pakistan as well as in the doctrines adopted by the armed services for prosecuting wars against that country. These, in turn, have been determined by the imperative of prosecuting a quick war before military or diplomatic interventions of third parties deny India its defined objectives. Both minor and major changes have been formulated and implemented by the professional military in concert with the political leadership. Thus, politico-military synergy has been a hallmark of India's evolving military goals, the successive military strategies it has adopted to deal with adversaries, and the operational doctrines employed by its armed services in the various wars that have been prosecuted hitherto and are being contemplated now.

Notes

1 C. Christine Fair, 'The Militant Challenge in Pakistan', *Asia Policy*, no. 11 (National Bureau of Asian Research, January 2011), 105–137; Jessica Stern, 'Pakistan's Jihad Culture', *Foreign Affairs*, vol. 79, no. 6 (November/December 2000), 115–126.
2 For a concise description of the China challenge as perceived in India, see S. Kalyanaraman, 'Fear, Interest and Honour: The Thucydidean Trinity and India's Asia Policy', *Strategic Analysis*, vol. 37, no. 4 (July 2013), 381–382.
3 India's Muslim population is only a few hundred thousand less than Pakistan's total population; cited in Stephen Philip Cohen, *The Idea of Pakistan* (New Delhi: Oxford University Press, 2006), 65. For a treatment of how independent India conceived of itself, see Sunil Khilnani, *The Idea of India* (London: Hamish Hamilton, 1997).
4 Sujit Dutta, 'Managing and Engaging Rising China: India's Evolving Posture', *The Washington Quarterly*, vol. 34, no. 2 (spring 2011), 130.
5 Jonathan Holslag, 'The Persistent Military Security Dilemma Between China and India', *The Journal of Strategic Studies*, vol. 32, no. 6 (December 2009), 811–840; Stephen Philip Cohen, *Shooting for a Century: Finding Answers to the India–Pakistan Conundrum* (Noida: HarperCollins, 2013).
6 Raju G.C. Thomas, *The Defence of India: A Budgetary Perspective of Strategy and Politics* (Delhi: Macmillan, 1978), 125–135.
7 C.V. Gole, 'The IAF in 2001 AD', *Vayu Aerospace Review*, no. I (January 1994), 43.
8 On how the Indian Air Force dealt with the resource crunch during the early 1990s, see George K. Tanham and Marcy Agmon, *The Indian Air Force: Trends and Prospects* (Santa Monica, CA: RAND, 1995), 77–90.
9 'Rafale Fighter Deal Deferred for 2014–15, No Money Left: Antony', *India Today* (New Delhi), 6 February 2014, http://indiatoday.intoday.in/story/rafale-fighter-deal-deferred-for-2014–15-no-money-left-antony/1/342146.html (accessed 12 April 2014).
10 P. Chidambaram, 'India's National Security – Challenges and Priorities', K. Subrahmanyam Memorial Lecture, 6 February 2013, http://idsa.in/keyspeeches/IndiasNationalSecurityChallengesandPriorities (accessed 10 February 2013).
11 Deba R. Mohanty, 'India's Defense Sector Still Plagued by Corruption', *The International Relations and Security Network*, 13 February 2014, www.isn.ethz.ch/Digital-Library/Articles/Detail/?lng=en&id=176507 (accessed 13 April 2014); Jay Menon, 'India Slows Action On Procurement Deals', *Aviation Week and Space Technology*, 3 February 2014, www.aviationweek.com/Article.aspx?id=/article-xml/AW_02_03_2014_p40–658973.xml (accessed 13 April 2014).
12 'Towards a Clear Defence Policy', in *P.V. Narasimha Rao: Selected Speeches,*

Volume IV: July 1994–June 1995 (New Delhi: Government of India, 1995), 125–126.
13 David Scott, 'India's 'Extended Neighbourhood Concept', *India Review*, vol. 8, no. 2 (April–June 2009), 107–143.
14 Institute for Defence Studies and Analyses, *Net Security Provider: India's Out-of-Area Contingency Operations* (New Delhi: Magnum Books, October 2012), 12–13; Vinay Kumar, 'India Well Positioned to Become a Net Provider of Security: Manmohan Singh', *The Hindu*, 23 May 2013, www.thehindu.com/news/national/india-well-positioned-to-become-a-net-provider-of-security-manmohan-singh/article4742337.ece (accessed 5 May 2014).
15 Walter C. Ladwig III, 'India and Military Power Projection: Will the Land of Gandhi Become a Conventional Great Power?', *Asian Survey*, vol. 50, no. 6 (November/December 2010), 1162–1183.
16 Rajat Pandit, 'Army Reworks War Doctrine for Pakistan, China', *Times of India* (New Delhi), 30 December 2009, http://articles.timesofindia.indiatimes.com/2009-12-30/india/28104699_1_war-doctrine-new-doctrine-entire-western-front (accessed 30 October 2013); Rajat Pandit, 'Two-front War Remote, but Threat from China Real', *Times of India*, 12 October 2012, http://timesofindia.indiatimes.com/india/Two-front-war-remote-but-threat-from-China-real/articleshow/16775896.cms (accessed 19 April 2014).
17 Rajeev Sharma, 'India: China not Eyeing Attack', *The Diplomat*, 14 December 2011, http://thediplomat.com/2011/12/india-china-not-eyeing-attack/ (accessed 14 May 2014).
18 Rajat Pandit, 'Army Chief Reviews Mountain Strike Corps', *Times of India*, 8 May 2014, http://timesofindia.indiatimes.com/india/Army-chief-reviews-mountain-strike-corps/articleshow/34795843.cms (accessed 8 May 2014).
19 Sandeep Unnithan, 'Indian Army not Ready for War with China', *India Today*, 29 October 2011, http://indiatoday.intoday.in/story/indian-army-war-readiness-against-china/1/157763.html (accessed 30 October 2013); Pravin Sawhney and Ghazala Wahab, 'Cusp of Change: The IAF is the Only Force with the Capability to Provide Deterrence', *Force*, October 2011, www.forceindia.net/CuspofChangeOctober2011.aspx (accessed 19 April 2014); Manu Pubby, 'Will be Tough to Tackle Collusive Threat from Pak, China: IAF', *Indian Express* (New Delhi), 19 February 2014, http://indianexpress.com/article/india/india-others/will-be-tough-to-tackle-collusive-threat-from-pak-china-iaf/ (accessed 19 February 2014).
20 The tasks of the Chief of Defence Staff were identified as follows: providing 'single point military advice to the government', administering the strategic forces and the soon-to-be-established tri-service commands, enhancing 'the efficiency and effectiveness of the planning process through intra- and inter-service prioritisation', and forging links among the three services. See Satish Chandra, 'The National Security Set Up', *Agni: Studies in International Strategic Issues*, vol. 10, no. 4 (October–December 2007), 11.
21 Patrick Bratton, 'The Creation of Indian Integrated Commands: Organisational Learning and the Andaman and Nicobar Command', *Strategic Analysis*, vol. 36, no. 3 (May/June 2012), 446; Carolin Liss, 'The Roots of Piracy in Southeast Asia', *APSNet Policy Forum*, 22 October 2007, http://nautilus.org/apsnet/the-roots-of-piracy-in-southeast-asia/ (accessed 29 April 2014).
22 Harsh Pant, 'India's Nuclear Doctrine and Command Structure: Implications for India and the World', *Comparative Strategy*, vol. 24, no. 3 (2005), 280.
23 Rajat Pandit, 'Tri-service Commands for Space, Cyber Warfare', *Times News Network* (New Delhi), 18 May 2013, http://timesofindia.indiatimes.com/india/Tri-service-commands-for-space-cyber-warfare/articleshow/20115462.cms (accessed 20 May 2013).

24 Indian Air Force, 'IAF Commands', http://indianairforce.nic.in/ (accessed 29 April 2014).
25 Saikat Datta, 'The Army Streamlines. A New Command – South West – is Born', *Outlook* (New Delhi), 14 February 2005, www.outlookindia.com/article.aspx?226464 (accessed 19 April 2014).
26 Chidambaram, 'India's National Security'.
27 John W. Garver, *Protracted Contest: Sino-Indian Rivalry in the Twentieth Century* (Seattle: University of Washington Press, 2001), 4.
28 Ibid.
29 *Agreement between the Government of the Republic of India and the Government of the People's Republic of China on Border Defence Cooperation*, October 23, 2013, http://mea.gov.in/in-focus-article.htm?22366/Agreement+between+the+Governmen t+of+the+Republic+of+India+and+the+Government+of+the+Peoples+Republi c+of+China+on+Border+Defence+Cooperation (accessed 23 October 2013).
30 Rajeswari Pillai Rajagopalan and Rahul Prakash, *Sino-Indian Border Infrastructure: An Update*, Occasional Paper No. 42 (New Delhi: Observer Research Foundation, May 2013), www.orfonline.org/cms/sites/orfonline/modules/occasionalpaper/attachments/Occasional42_1369136836914.pdf (accessed 5 May 2014).
31 On the evolution of Chinese military doctrine, see Paul H.B. Godwin, 'Patterns of Doctrinal Change in the Chinese People's Liberation Army: From Threats to Contingencies to Capabilities', in N.S. Sisodia and S. Kalyanaraman, eds, *The Future of War and Peace in Asia* (New Delhi: Magnum Books, 2010), 111–135.
32 Sujit Dutta, 'Managing and Engaging Rising China', 131–133.
33 Pushpita Das, 'Evolution of the Road Network in Northeast India: Drivers and Breaks', *Strategic Analysis*, vol. 33, no. 1 (January 2009), 106–108.
34 Rajat Pandit, 'Army Gets Final Nod to Deploy 80,000 Troops along China Border', *Times of India*, 19 November 2013, http://timesofindia.indiatimes.com/india/Army-gets-final-nod-to-deploy-80000-troops-along-China-border/articleshow/26051747.cms (accessed 22 November 2013); Ajai Shukla, 'Preparing for a Two-front War', *Business Standard* (New Delhi), 10 September 2013, www.business-standard.com/article/current-affairs/preparing-for-a-two-front-war-113091001026_1.html (accessed 30 October 2013).
35 David Brewster, 'The India–Japan Security Relationship: An Enduring Security Partnership?', *Asian Security*, vol. 6, no. 2 (2010), 95–120; Frederic Grare, 'The India–Australia Strategic Relationship: Defining Realistic Expectations', *CEIP Paper*, 18 March 2014, http://carnegieendowment.org/2014/03/18/india-australia-strategic-relationship-defining-realistic-expectations/h4k3# (accessed 25 March 2014); Sumit Ganguly and Manjit S. Pardesi, 'The Evolving US–China–India Triangular Relationship', *CLAWS Journal* (summer 2010), 65–78; Ashok Kapur, 'Indo-Soviet Treaty and the Emerging Asian Balance', *Asian Survey*, vol. 12, no. 6 (June 1972), 463–474.
36 On what a static defence posture entails, see John J. Mearsheimer, *Conventional Deterrence* (Ithaca, NY: Cornell University Press, 1983), 48.
37 Since the 1960s, the Indian Army has not had 'a dedicated offensive formation against China', which is now being corrected. See Nitin Gokhale, 'India Military Eyes Combined Threat', *The Diplomat*, 17 January 2012, http://thediplomat.com/2012/01/india-military-eyes-combined-threat/?allpages=yes (accessed 14 May 2014).
38 K.C. Pant, 'Philosophy of Indian Defence', in Jasjit Singh, ed., *Non-provocative Defence: The Search for Equal Security* (New Delhi: Lancer International, 1989), 212–213. See also in the same volume Jasjit Singh, 'The Indian Experience', 220–224.

39 Carl von Clausewitz, *On War*, trans. and ed., Michael Howard and Peter Paret (Princeton, NJ: Princeton University Press, 1976), 370, 357.
40 Clausewitz, *On War*, 357, 370, 379.
41 This broad classification of two types of strategies was first postulated by the German historian Hans Delbrück, who derived it partly from military history and partly from Clausewitz's typology of the two types of war: total war in which the opposing military is completely defeated, and limited war in which a portion of territory is occupied either for outright annexation or for bargaining purposes. Gordon A. Craig, 'Delbrück: The Military Historian', in Peter Paret, ed., *Makers of Modern Strategy: From Machiavelli to the Nuclear Age* (Princeton, NJ: Princeton University Press, 1986), 341–342. For Clausewitz's identification of the two types of war, see *On War*, 69.
42 The strategy of exhaustion is also referred to as the strategy of attrition. The first of these formulations (exhaustion) has been used here mainly to distinguish it from the doctrine of attrition. For examples of studies that use the term 'strategy of attrition', see: Russell F. Weigley, *The American Way of War: A History of United States Military Strategy and Policy* (New York: Macmillan, 1973), xxii, 3–17; Efraim Inbar and Eitan Shamir, '"Mowing the Grass": Israel's Strategy for Protracted Intractable Conflict', *Journal of Strategic Studies*, vol. 37, no. 1 (February 2014), 65–90.
43 On the operational doctrine of attrition, see Edward N. Luttwak, 'The Operational Level of War', *International Security*, vol. 5, no. 3 (winter 1980/1981), 62–64; Mearsheimer, *Conventional Deterrence*, 33–35.
44 Jasjit Singh, *Defence from the Skies: 80 Years of the Indian Air Force* (New Delhi: KnowledgeWorld, 2nd edn, 2013), 114, 134.
45 Julian Stafford Corbett, *Some Principles of Maritime Strategy* (London: Brassey's, 1988 edn), 32, 115.
46 Ashley J. Tellis, 'Securing the Barrack: The Logic, Structure and Objectives of India's Naval Expansion', *Indian Defence Review* (July 1991), 137.
47 The terms *attrition* and *manoeuvre* are specifically employed in the context of land warfare. Since the ultimate decision in war is largely determined by events on land, albeit with the air and naval forces playing vital roles in shaping the battlefield and the overall context of the war, the army's doctrine has invariably had greater impact in terms of determining the contours of any military campaign. Hence the use here of the terms *attrition* and *manoeuvre* as the determining doctrinal features.
48 On the doctrine of manoeuvre, see Mearsheimer, *Conventional Deterrence*, 35–43; Luttwak, 'Operational Level of War', 64–76.
49 A tactical blockade is a blockade of ports, while a strategical blockade is aimed at the trade routes. Corbett, *Maritime Strategy*, 97.
50 On the objectives and methods of sea control, see Corbett, *Maritime Strategy*, 161–166.
51 The three-way classification of doctrines employed here draws upon Mearsheimer's typology of three distinct strategies: limited aims based on surprise, and unlimited aims or victory through attrition and manoeuvre. With respect to the role of surprise in a war involving limited aims, Mearsheimer notes that since the principal aim in such a war is the capture of a portion of the opponent's territory before it can mobilize its defences, an army, instead of adopting attrition or manoeuvre, is more likely to rely upon surprise in order to minimize contact with the opposing force. See Mearsheimer, *Conventional Deterrence*, 29–30. The imperative of avoiding contact is all the more important in a nuclear environment.
52 For a Pakistani record of this annexation bid through irregulars, see Akbar Khan, *Raiders in Kashmir* (Delhi: Army Publishers, n.d.).

53 Although Britain had declared that 'paramount power' will return to the princely states on Indian independence, it had also at the same time envisaged a 'Union of India embracing both British India and the States' in which the centre would be vested with the powers for conducting foreign affairs, defence and communications. Such an allocation of powers was, however, subject to an agreement between the princely state concerned and the Dominion (India or Pakistan) it chose to accede to. For a concise treatment of this issue, see S. Kalyanaraman, 'The Dawn of Independence and the Tribal Raid', in Virendra Gupta and Alok Bansal, eds, *Pakistan Occupied Kashmir: The Untold Story* (New Delhi: Manas Publications, 2007), 61–62.

54 For a record of the events surrounding Kashmir's accession to India, see Prem Shankar Jha, *The Origins of a Dispute: Kashmir 1947* (New Delhi: Oxford University Press, 2003).

55 For a diplomatic record of these developments, see Srinath Raghavan, *War and Peace in Modern India: A Strategic History of the Nehru Years* (Ranikhet: Permanent Black, 2010), 101–148; C. Dasgupta, *War and Diplomacy in Kashmir 1947–48* (New Delhi: Sage, 2002), 97–132.

56 On this aspect, see Dasgupta, *War and Diplomacy in Kashmir*, 45, 51, 92, 174.

57 For details of the army's operations see S.K. Sinha, *Operation Rescue: Military Operations in Jammu and Kashmir 1947–49* (New Delhi: Vision Books, 1977).

58 D.K. Palit, *The Lighting Campaign: The Indo-Pakistan War 1971* (New Delhi: Thomson Press, 1972), 100.

59 Jasjit Singh, *Defence from the Skies*, 54–65.

60 For the Indian military assessment in this regard, see Lorne J. Kavic, *India's Quest for Security: Defence Policies, 1947–1965* (Dehradun: EBG Publishing and Distributing Co, n.d.), 36–38. The quoted phrase may be found at 36.

61 Kavic, *India's Quest for Security*, 37.

62 The army was assigned the predominant role mainly because of the limited capabilities possessed by the air force and navy. For details, see Kavic, *India's Quest for Security*, 102, 118, 241.

63 Kavic, *India's Quest for Security*, 37.

64 On Operation Gibraltar, see B.C. Chakravorty, *History of the Indo-Pak War, 1965* (New Delhi: Government of India, 1992), ch. 4, 59–95, www.bharat-rakshak.com/LAND-FORCES/Army/History/1965War/PDF/1965Chapter04.pdf (accessed 16 April 2009).

65 Russell Brines, *The Indo-Pakistani Conflict* (London: Pall Mall Press, 1968), 318–321.

66 Harbakhsh Singh, *War Despatches: Indo-Pak Conflict 1965* (New Delhi: Lancer, 1991), 13–19; K.V. Krishna Rao, *Prepare or Perish: A Study of National Security* (New Delhi: Lancer, 1991), 128–129; Chakravorty, *History of the Indo-Pak War, 1965*, ch. 8, 228–244, www.bharat-rakshak.com/LAND-FORCES/Army/History/1965War/PDF/1965Chapter08.pdf. (accessed 16 April 2009).

67 Pakistan blew up some 70 bridges across this canal and turned the whole area into a moat. See Brines, *Indo-Pakistani Conflict*, 328, 336. See also Chakravorty, *History of the Indo-Pak War, 1965*, ch. 12, 328–336, www.bharat-rakshak.com/LAND-FORCES/Army/History/1965War/PDF/1965Chapter12.pdf (accessed 16 April 2009).

68 Brines, *Indo-Pakistani Conflict*, 340–341.

69 Calculated from data on air force sorties in Chakravorty, *History of the Indo-Pak War*, ch. 9, 269, 272, www.bharat-rakshak.com/LAND-FORCES/Army/History/1965War/PDF/1965Chapter09.pdf (accessed 16 April 2009). Counter-air sorties by fighter-bombers appear to have been classified under the category 'Others' in this official history of the war.

70 See the summary of the remarks made by Benjamin Lambeth in IDSA Round

Table on 'India's Transforming Air Posture: An Emerging 21st Century Heavyweight', 2 February 2011, http://idsa.in/event/IndiasTransformingAirPostureBenjamin_SLambeth (accessed 3 March 2014).
71 P.C. Lal, *My Years with the IAF*, ed. Ela Lal (New Delhi: Lancer, 2008 edn), 173–174.
72 A.S. Bahal, 'Strategic Roles of Air Power: Think, Plan, Equip and Train for it', *Air Power Journal* (spring 2007), 39. This role was assigned to the service in its early years because the Royal Air Force was then at hand to provide 'independent "strategic" capabilities'. See Jasjit Singh, *Defence from the Skies*, 216.
73 Chakravorty, *History of the Indo-Pak War*, ch. 9, 272.
74 Chakravorty, *History of the Indo-Pak War, 1965*, ch. 10, 283, 287–288, www.bharat-rakshak.com/LAND-FORCES/Army/History/1965War/PDF/1965Chapter10.pdf (accessed 16 April 2009).
75 Chakravorty, *History of the Indo-Pak War, 1965*, ch. 12, 328–338.
76 Luttwak, 'Operational Level of War', 63; Mearsheimer, *Conventional Deterrence*, 34.
77 In contrast, attrition has been successful during long-drawn-out conflicts such as the American Civil War and the two World Wars. See Michael Howard, 'The Forgotten Dimensions of Strategy', *Foreign Affairs*, vol. 57, no. 5 (summer 1979), 975–977.
78 Chakravorty, *History of the Indo-Pak War, 1965*, ch. 12, 330, 334.
79 Brines, *Indo-Pakistani Conflict*, 346.
80 Jasjit Singh, *Defence from the Skies*, 112.
81 For the gist of this assessment on the ceasefire provided to the then Defence Minister, see R.D. Pradhan, *Debacle to Resurgence: Y.B. Chavan – Defence Minister 1962–66* (New Delhi: Atlantic Publishers and Distributors, 2013), 285–286.
82 On China's military and diplomatic attempts to exert pressure on India as well as the international diplomatic efforts to bring about an early ceasefire, see Brines, *The Indo-Pakistani Conflict*, 353–381.
83 The quoted phrases are from a 'reported' conversation between the then Indian prime minister and the army chief. See Chakravorty, *History of the Indo-Pak War, 1965*, ch. 12, 333–334.
84 For additional details on this contingency plan, see Pran Chopra, *India's Second Liberation* (New Delhi: Vikas Publishing House, 1973), 103, 120.
85 Lal, *My Years with the IAF*, 174. See also *Official History of the 1971 War* (New Delhi: Government of India, 1992), ch. 10, 419–420, www.bharat-rakshak.com/LAND-FORCES/Army/History/1971War/PDF/ (accessed 10 November 2013).
86 Raju Thomas, *The Defence of India*, 204.
87 For a comprehensive account of the origins of the east–west divide in Pakistan, its denouement into a civil war and the impact on India, see Richard Sisson and Leo E. Rose, *War and Secession: Pakistan, India, and the Creation of Bangladesh* (Berkeley: University of California Press, 1990), *passim*.
88 P.N. Dhar, *Indira Gandhi, the 'Emergency', and Indian Democracy* (New Delhi: Oxford University Press, 2000), 158.
89 On US support for the Pakistan government during the civil war, see Gary J. Bass, *The Blood Telegram: India's Secret War in East Pakistan* (Noida: Random House India, 2013), *passim*.
90 The most prominent advocates of intervention were the Finance and Defence Ministers; the Foreign Minister opposed such a venture. See Chopra, *India's Second Liberation*, 77–78. The prime minister invited the army chief to a Cabinet meeting where he could voice his reservations and thus help silence the advocates. See Dhar, *Indira Gandhi, the Emergency, and Indian Democracy*, 157. It is not clear exactly when this meeting took place. In a 1996 interview,

91 Chopra, *India's Second Liberation*, 78–79.
 92 *Official History of the 1971 War* (New Delhi: Government of India, 1992), ch. 12, 498, www.bharat-rakshak.com/LAND-FORCES/Army/History/1971War/PDF/1971Chapter12.pdf (accessed 10 November 2013); Chopra, *India's Second Liberation*, 96.
 93 On what a forward defence posture involves and Pakistan adopting this approach, see Mearsheimer, *Conventional Deterrence*, 8–49, 206.
 94 For India's strategy and operations on the western front, see *Official History of the 1971 War*, chs 8, 9, 10 and 11.
 95 Palit, *Lightning Campaign*, 99–100.
 96 Because Pakistan was expected to automatically initiate a general war in the event of an Indian intervention in East Pakistan, the western front had to be well guarded. See Palit, *Lightning Campaign*, 100.
 97 Bass, *The Blood Telegram*, 282–286.
 98 Chopra, *India's Second Liberation*, 90–92.
 99 Palit, *Lightning Campaign*, 102–103.
100 Chopra, *Second Liberation*, 194.
101 *Official History of the 1971 War*, ch. 14, 592–595, www.bharat-rakshak.com/LAND-FORCES/Army/History/1971War/PDF/1971Chapter14.pdf (accessed 10 November 2013).
102 *Official History of the 1971 War*, 598, 602, 613–614.
103 Chopra, *Second Liberation*, 127; Palit, *Lightning Campaign*, 116.
104 *Official History of the 1971 War*, ch. 15, 645–649, www.bharat-rakshak.com/LAND-FORCES/Army/History/1971War/PDF/1971Chapter15.pdf (accessed 10 November 2013).
105 *Official History of the 1971 War*, 649–652.
106 See e.g. K. Subrahmanyam, *Our National Security* (New Delhi: Economic and Scientific Research Foundation, 1972), 58; Ravi Rikhye, 'Rethinking Mechanised Infantry Concepts', *Journal of the United Service Institution of India*, April–June 1972, 163–168; and Ravi Rikhye, 'A New Armoured Force for India', *Journal of the United Service Institution of India* (April–June 1973), 137–145.
107 The Committee considered that existing forces were sufficient to deter China. Amit Gupta, 'Building an Arsenal: The Indian Experience', in Gupta, ed., *Building an Arsenal: The Evolution of Regional Power Force Structures* (Westport, CT: Praeger, 1997), 49.
108 Raju G.C. Thomas, *Indian Security Policy* (Princeton, NJ: Princeton University Press, 1986), 165–166.
109 Gurmeet Kanwal, 'Strike Corps Offensive Operations: Imperatives for Success', *Indian Defence Review* (January 1988), 82.
110 Kanti Bajpai, P.R. Chari, Pervaiz Iqbal Cheema, Stephen P. Cohen and Sumit Ganguly, *Brasstacks and Beyond: Perception and Management of Crisis in South Asia* (New Delhi: Manohar, 1997), 25–30.
111 Gupta, 'Building an Arsenal', 59.
112 On the buildup of the air force's and navy's capabilities and their acquisitions during the 1980s, see Gupta, 'Building an Arsenal', 53–54, 60–63.
113 It is not clear when exactly Pakistan attained nuclear weapons capability. While President Zia-ul-Haq declared that Pakistan was a nuclear-capable country in 1987, the United States imposed sanctions on Pakistan under the Pressler Amendment in October 1990, an act that implicitly confirmed its possession of 'nuclear explosive capability'. For Zia's statement, see Vipin Narang, 'Posturing for Peace? Pakistan's Nuclear Postures and South Asian Stability',

(the then army chief mentioned that it may have been towards the end of April 1971. For a reproduction of this interview, see J.F.R. Jacob, *Surrender at Dacca: Birth of a Nation* (New Delhi: Manohar, 2001), Appendix 6, 181–183.)

International Security, vol. 34, no. 3 (winter 2009/2010), 48. For what US invocation of the Pressler Amendment implied, see *From Surprise to Reckoning: The Kargil Review Committee Report* (New Delhi: Sage, 2000), 66.
114 This view was articulated in the agreement Prime Minister A.B. Vajpayee signed with his Pakistani counterpart Nawaz Sharif. See the text of *Lahore Declaration*, 2 February 1999, http://mea.gov.in/in-focus-article.htm?18997/Laho re+Declaration+February+1999 (accessed 14 May 2014).
115 George Fernandes, 'Inaugural Address', *IDSA Seminar on The Challenges of Limited War: Parameters and Options*, 5 January 2000, www.idsa-india.org/defmin5–2000.html (accessed 17 May 2014). For the Indian debate on limited war, see Jasjit Singh, 'Dynamics of Limited War', *Strategic Analysis*, vol. 24, no. 7, October 2000, 1205–1220; V.R. Raghavan, 'Limited War and Nuclear Escalation in South Asia', *The Nonproliferation Review* (autumn/winter 2001), 1–18; Manpreet Sethi, 'Conventional War in the Presence of Nuclear Weapons', *Strategic Analysis*, vol. 33, no. 3 (May 2009), 415–425.
116 Shaukat Qadir, 'An Analysis of the Kargil Conflict 1999', *RUSI Journal*, vol. 147, no. 2 (April 2002), 26.
117 Qadir, 'An Analysis of the Kargil Conflict 1999', 26–27.
118 V.P. Malik, *Kargil: From Surprise to Victory* (New Delhi: HarperCollins, 2010 paperback edn), 139.
119 India's restraint was the principal factor in the United States not equating India and Pakistan during the conflict, with President Clinton even asserting that 'borders cannot be redrawn in blood'. Clinton's statement is cited in Sumit Ganguly, 'Will Kashmir Stop India's Rise?', *Foreign Affairs*, vol. 85, no. 4 (July/August 2006), 53. For the US' role during the conflict, see Bruce Riedel, 'American Diplomacy and the 1999 Kargil Summit at Blair House', in Peter R. Lavoy, ed., *Asymmetric Warfare in South Asia: The Causes and Consequences of the Kargil Conflict* (New Delhi: Cambridge University Press, 2009), 130–143; Kartik Bommakanti, 'Coercion and Control: Explaining India's Victory at Kargil', *India Review*, vol. 10, no. 3 (July–September 2011), 304–315.
120 In the Kargil sector alone, the army built up its force level to nearly a corps' worth, with one division serving as reserve. John H. Gill, 'Military Operations in the Kargil Conflict', in Lavoy, ed., *Asymmetric Warfare in South Asia*, 113, 128–129.
121 Malik, *Kargil*, 128–129.
122 A.Y. Tipnis, 'Operation Safed Sagar', *Force*, October 2006, 13.
123 Benjamin S. Lambeth, *Air Power at 18000': The Indian Air Force in the Kargil War* (Washington, DC: Carnegie Endowment of International Peace, 2012), 17.
124 Gill, 'Military Operations in the Kargil Conflict', 110.
125 For detailed treatments of these battles, see Y.M. Bammi, *Kargil 1999: The Impregnable Conquered* (Noida: Gorkha Publishers, 2002), 199–388; Malik, *Kargil*, 161–211.
126 Lambeth, *Air Power at 18000'*, 14, 20, 22.
127 Mathew Thomas, 'An Analysis of the Threat Perception and Strategy for India', *Indian Defence Review*, January 1990, 62.
128 For this interpretation of the Indian intent behind the mobilization, see S. Kalyanaraman, 'Operation Parakram: An Indian Exercise in Coercive Diplomacy', *Strategic Analysis*, vol. 26, no. 4 (October–December 2002), 478–492; Sumit Ganguly and Michael Kraig, 'The 2001–2002 Indo-Pakistani Crisis: Exposing the Limits of Coercive Diplomacy', *Security Studies*, vol. 14, no. 2 (April–June 2004), 290–324.
129 V.K. Sood and Pravin Sawhney, *Operation Parakram: The War Unfinished* (New Delhi: Sage, 2003), 73–76, 80–81.
130 Walter C. Ladwig, 'A Cold Start for Hot Wars? The Indian Army's New

Limited War Doctrine', *International Security*, vol. 32, no. 3 (winter 2007/2008), 162–163.
131 Gurmeet Kanwal, 'India's Cold Start Doctrine and Strategic Stability', *IDSA Comment*, 1 June 2010, www.idsa.in/idsacomments/IndiasColdStartDoctrine-andStrategicStability_gkanwal_010610 (accessed 2 June 2010); Nitin Gokhale, 'India's Doctrinal Shift?', *The Diplomat*, 25 January 2011, http://thediplomat.com/2011/01/25/india%E2%80%99s-doctrinal-shift/?all=true (accessed 29 November 2012).
132 Kanwal, 'India's Cold Start Doctrine'.
133 For an analysis of the first set of exercises in this regard, see Ladwig, 'Cold Start for Hot Wars', 178–181.
134 US Ambassador to India, Timothy Roemer, queried 'very high-level' officials of the Indian government including the National Security Adviser about their support for the doctrine and was assured that 'they have never endorsed, supported, or advocated for this doctrine'. See his cable to the US Secretary of State, 'Cold Start – A Mixture of Myth and Reality', 16 February 2010, www.wikileaks.nl/cable/2010/02/10NEWDELHI295.html (accessed 7 May 2011).
135 Manu Pubby, 'No "Cold Start" Doctrine: India tells US', *Indian Express*, 9 September 2010.
136 'Getting Leaner and Meaner? Army Practises Blitzkrieg to Strike Hard at Enemy', *Times of India*, 10 May 2011, http://timesofindia.indiatimes.com/india/Getting-leaner-and-meaner-Army-practises-blitzkrieg-to-strike-hard-at-enemy/articleshow/8212406.cms (accessed 14 May 2014); Jiby Kattakayam, '"Sudarshan Shakti" Aims to Transform Armed Forces', *The Hindu* (Chennai), 6 December 2011, www.thehindu.com/news/national/article2692709.ece (accessed 14 May 2014); 'India Conducts Wargames Near Pak Border', *The Nation* (Lahore), 23 December 2013, www.nation.com.pk/national/23-Dec-2013/india-conducts-wargames-near-pak-border (accessed 24 December 2013).
137 Shashank Joshi, 'India's Military Instrument: A Doctrine Stillborn', *Journal of Strategic Studies*, vol. 36, no. 4 (2013), 512–540.

7 The Indian Army adapting to change

The case of counter-insurgency

Vivek Chadha

India has faced both conventional intra-state as well as internal insurgency/terrorism threats for over six decades. The army's involvement in counter-insurgency (CI) operations is designated as its secondary responsibility, while its primary role remains the defence of the country against external threats.[1] However, the Indian Army has lost more soldiers in CI operations than conventional wars despite, since Independence, fighting three major wars and a limited conflict in Kargil in 1999.[2] In the case of Jammu and Kashmir (J&K), Pakistan's direct involvement adds an external factor, which transforms it into a hybrid threat with external and internal dimensions. The arc of insurgency and terrorism has expanded since the mid-1950s to include areas in north, east, northeast, central and south India. Violent groups within this large, affected land mass have threatened to rewrite the geographical and political boundaries of the country. This threat becomes all the more viable, given the direct involvement of Pakistan in some of these conflicts.[3] The police are the first line of defence against terrorism in India, while the Indian Army remains the instrument of last resort for the state against such threats and is only employed when all else fails to control violence. Despite this fundamental guideline for employment of the army for internal security duties, the army has continuously been involved in CI operations, and often in more than one region simultaneously.[4] This is a reflection of both the inadequate capacities of the state and central police organizations, as well as the gravity of the threat.

This chapter examines how the Indian Army has adapted to change in the face of challenges emerging from insurgencies and terrorism, with the aim of drawing lessons for armies and countries faced by similar challenges. The history of the past 60 years not only highlights the importance of CI operations for India; recent US experiences in Iraq and Afghanistan further reinforce the importance of CI operations for other countries and their armies. Furthermore, I employ India's CI experience to analyse military change and the ability of the army to adapt to evolving threats, political directives, technological advancements, and, in particular, the importance of bottom-up experiences. Unlike external threats, which require strategic shifts to bring about substantive change, change in a counter-insurgency environment can be brought about by operational adaptation. This chapter highlights a number of such changes undertaken by the army and, given past focus on external state threats, it contributes to a limited literature on such processes.[5] I analyse three major types of change: doctrinal, organizational and operational shifts, using case studies from different areas of operations of the army, including Northeast India,[6] Jammu and Kashmir (J&K), and Sri Lanka.[7]

The early years of India's experience

The Indian Army has many years of experience in counter-insurgency operations. Prior to 1947, in the pre-independence period, the British Indian Army operated against tribal warriors in the North West Frontier Province (NWFP), now in Pakistan, areas of Afghanistan, Nagaland and Mizoram (the latter two are now states in northeast India) in the form of expeditionary operations. Given the unique conditions of each of these areas, the army adapted to the varying complexion of threat posed in each theatre of conflict. The lessons on frontier warfare were incorporated into the Field Service Regulations, Part I Operations, 1909[8] and *Passing it On: Fighting the Pashtun on Afghanistan's Frontier* by General Sir Andrew Skeen. The British Indian Army followed the tactics of punitive expeditionary operations in the northeastern region of the country as well. The 1889 to 1890 expedition witnessed a severe reprisal by the British, as a result of repeated Lushai raids and killings. Subir Bhaumik quotes the report of the Chief Commissioner of Chittagong, W.S. Oldham: 'All villages, except Aitur and Malthuna in North Lushai Hills, were ruled by widows.'[9]

It was these sub-conventional operations that defined some of the pre-independence operations. However, the realities of post-independence required a change in the nature of operations undertaken by the army. The ability to cope with this change faced a series of challenges:

First, the political and senior military leadership had limited experience in managing the armed forces, exposure to higher direction of war and sub-conventional warfare.

Second, in the first decade after Independence, the army had to grapple with the contrast presented by pre and post-independence sub-conventional operations. This required a change in the policy of punitive military expeditions and strikes, while dealing with revolutionary uprisings. The army was no longer the sword arm of a colonial power, but the defensive shield of a democratic republic. It therefore needed clear doctrinal guidance for taking forward the political vision of the leaders of the country.

Third, the organization of the army, especially the battalions, was based on conventional threats, which saw a large number of men being tied down with heavier calibre weapons and logistic support infrastructure, which was not conducive for CI operations.

Fourth, there was a need to operationally transform a first-rate army, which had acquitted itself with distinction during the Second World War, to counter-insurgency operations. The training and experience gained during the war had its limitations against guerrilla fighters. This led to the induction of troops into difficult operational conditions with inadequate understanding of counter-insurgency operations. The regimental history of a Gurkha battalion highlights this very limitation. Referring to operations of an infantry battalion, it says that it 'had been inducted into this counter-insurgency area (Nagaland) without any specialised training'.[10]

Fifth, the army's forays into CI operations also brought forth the challenge of legal sanction for its deployment and operations. The existing laws gave these powers only to the police. However, the nature of responsibility of the army required a change in the laws to facilitate its deployment.

Adapting to change

In the following section I will analyse how the Indian Army has adapted to four different forms of change: doctrinal, organizational, operational and technological.

Doctrinal adaptation

The army adapted to doctrinal change both as a result of strategic direction from the highest political level in the country, as well as innovation within. The importance of political direction for the army stems from the fact that insurgencies are a violent expression of political problems.[11] Therefore, solutions also tend to pursue a political trajectory.[12] At the very outset of post-independence CI operations, this led to tempering the operational military approach with a clear political mandate. The political mandate prior to Independence in the country's periphery was aimed at suppressing violent tribes and maintaining a degree of order, even as the locals were allowed to undertake autonomous socio-economic activities.[13] The army ensured the security of British-controlled areas in the vicinity of such regions and treated some of these as buffer states. This led to occasional expeditionary operations to achieve these objectives.[14]

There was a need to reverse these politico-military objectives. At the heart of the army's involvement in counter-insurgency operations was Prime Minister Nehru's idea of balanced integration of the tribal people into the Indian mainstream, wherein their development was envisioned along 'the lines of their own genius'.[15] This socio-economic guideline was accompanied by a political directive to the army deployed to secure areas dominated by Naga tribals, the first ethnic group to take up arms. Nehru's biographer notes:

> But the military approach, while necessary, was not adequate; and Nehru insisted that soldiers and officials should always remember that the Nagas were fellow-countrymen, who were not merely to be suppressed but, at some stage had to be won over.[16]

This approach was a major departure from the erstwhile British policy of expeditionary military forays into the region and required the army to recalibrate its response. The foundation of ensuing doctrinal direction may be found in the Order of the Day issued by the then Chief of Army Staff to the troops being deployed for operations in Nagaland in 1956. It read:

> You are not there to fight the people in the area, but to protect them. You are fighting only those who threaten the people and who are a danger to the lives and properties of the people.[17]

It was these clear guidelines that laid the foundation for the army's military response. It is therefore not surprising that there was a distinct difference between the pre and post-independence period in terms of the nature of operations conducted.

The directions provided the necessary political and military objective; however, given the negligible experience up until then, they were expressed in the form of military derivatives of conventional operations. In the context of operations in Mizoram, Lt. Gen. Mathew Thomas says:

> We had to open all axes from south to north and had to clear all the major villages along the central route. In doing that, people were alienated, to a certain extent, due to the fact that we were not very sure about how to go about things. None of the battalions that were from 61 Mountain Brigade had any experience of counter insurgency.[18]

During the mid-1960s, the army's doctrinal orientation was characterized by 'search-and-clear' operations, without holding areas, which were frequented by the guerrillas. Bodies of troops were stationed at selective locations, with the direction to operate over long distances in substantial numbers, thereby reinforcing the writ of the state. The role of the army was focused towards such military operations, with limited time and resources invested for winning over the population and establishing a broad-based intelligence network.

The second phase of change was initially influenced by British operations in Malaya and other communist guerrilla movements during the period. This involved adoption of the rudimentary 'clear-and-hold' strategy, which in effect led to the adoption of a CI grid along with attempts at isolating the insurgents from the population. However, the means adopted were not conducive to winning over the population. This was borne out by the decision to move the people en masse to 'progressive villages' against their wishes. Brigadier (later Lt. Gen.) S.K. Sinha, writing in the *USI Journal* in 1970, gives an indication of the army's thought process during this period. He suggests measures to include: projection of an 'alternative ideal'; 'grouping of local population' in order to prevent local support; 'dilution of population' by outside settlements along with the Chinese example of Tibet; and 'sound development plans' for environmental improvement.[19] He goes on to reinforce the need for an integrated civil and military organization along the lines of the Malaya model. He also discards the notion of minimum force as applied in the case of aid to civil authorities when he writes:

> At the end of the scale is the mistaken notion that the Army should conduct counter insurgency operations as aid to the civil power. This is a totally unrealistic concept. The technique of dispersal of unlawful assemblies with minimum force cannot be applied to warfare of this type.[20]

Sinha correctly distinguishes between dispersing a mob in aid to the civil authorities and counter-insurgency; however, the importance of the principle of minimum force in CI operations remains a significant facet that cannot be discarded. The task of domination of the area was reinforced through a 'grid' system of deployment,[21] as was the emphasis on the conduct of security forces. Sinha writes, 'The conduct of Security Forces towards the locals must be exemplary at all times. Any attempt to harass, torture or otherwise maltreat the people must be ruthlessly stopped.'[22] Despite this clear understanding of the importance of winning over the population, options such as 'imposing of food rationing', as executed in Malaya, were reflective of an inadequate understanding of WHAM measures (winning hearts and minds).[23]

The limitations noticed during the initial period of adapting to change were corrected over a period of time by the army. This resulted in establishing an even balance between military operations and WHAM. The Doctrine for Sub Conventional Operations (DSCO) highlights this aspect in its overarching theme:

> Since the centre of gravity for such operations is the populace, operations have to be undertaken with full respect to Human Rights and in accordance with the laws of the land. This underscores the importance of people friendly operations which are conducted with a civil face.[24]

This balance is best described by the 'Iron Fist with Velvet Glove' strategy. General J.J. Singh, a former Chief of Army Staff, coined the term to illustrate the underlying theme of the doctrine. He says:

I have emphasised the concept of 'Iron Fist with Velvet Glove', which implies a humane approach towards the populace at large in the conflict zone. This also entails the use of overwhelming force only against foreign terrorists and other hardcore inimical elements, while affording full opportunity to indigenous misguided elements to shun violence and join the mainstream.[25]

The erstwhile limitation of being biased towards conventional operations was also addressed by highlighting the need to reorientate commanders and men towards CI operations. The DSCO highlights this aspect by emphasizing the protracted duration of operations, changed rules of engagement, requirement of multi-agency integrated operations, the challenge of fighting one's own people and the criticality of people-friendly operations.[26] It further emphasizes the need for specialized training to imbibe these facets prior to induction for training.

The grid system of deployment was formalized during this period and ensured security of population centres, lines of communication and critical infrastructure.[27] In addition, the threat of proxy war across the Line of Control (LoC) was simultaneously addressed through deployment aimed at controlling both infiltration and exfiltration, as was also the erection of the LoC fence.[28] This, in areas like J&K, continues to be a part of the three-tier deployment, with the army at the LoC, Rashtriya Rifles (RR) battalions in the hinterland and central police organizations along with local police in urban population centres. This strengthened local governance by involving them in day-to-day law and order duties, even as areas more prone to terrorism were controlled by RR.

There was also greater emphasis on clinical operations on the basis of specific intelligence in the DSCO, rather than operations being conducted on a prophylactic basis.[29] Small team operations received a fillip. It potentially increased the chances of contact with terrorists and the capability and flexibility of grouping and regrouping forces.

Perception management received special attention and emphasis in areas where the army was deployed for CI operations.[30] Existing headquarters were restructured by providing additional staff for these specialized functions. These organizational changes were accompanied by efforts to change the method of operations which aimed to limit the negative impact of collateral damage and prolonged presence of the army in populated areas.[31]

The trend of doctrinal progression represents a certain common framework, which was provided through political direction and military guidelines over a period of time.

Organizational adaptation

The army faced a dual challenge during the initial years in CI operations. The battalions were organized for conventional operations, wherein four rifle companies formed its cutting edge. In addition, a support company carried heavy weapons, such as three-inch mortars. It was soon realized that this organization, while suitable for conventional operations, was not the most efficient in terms of availability of manpower for the CI role. The terrain, nature of operations and need for accurate fire with least collateral damage limited the use of heavy weapons. This led to local pooling of manpower within battalions with the dual aim of neutralizing the organizational imbalance and, in the process, releasing additional manpower. This manpower was placed under an officer by creating an ad hoc rifle company. Thus, an infantry battalion which was designated to have four companies now had five.[32] This method of local improvisation addressed some of the existing limitations faced by battalions, thereby becoming an example of adjusting to the needs of the designated operational role more effectively.[33]

This local improvisation partially resolved existing organizational challenges at the battalion level. However, a number of limitations remained. The battalions were organized for the purpose of fighting conventional operations, and were equipped with heavy weapons like 81-mm mortars, anti-tank guided missiles and anti-material rifles. Local procurement and sector-specific weapons and stores were purchased to meet additional critical requirements such as generator sets, specialist vehicles, etc. This made the set-up ad hoc and unreliable, and often prone to delays and contractual breakdowns. The changeover of infantry battalions from an area after every two to three years disrupted the battalions' intelligence networks and neutralized the good work done in an area. A new battalion was forced to recommence the efforts of the battalion it replaced, often from scratch. The primary role of infantry battalions was the defence of the country from external threats. This forced field formations and battalions to simultaneously focus on this role even as they were involved in CI operations, thereby dividing their attention. Infantry battalions were self-contained in most respects and have the capability of carrying out most assigned tasks independently. The need for specialists in operations is met by brigade and divisional support elements. However, this arrangement is not ideal in CI operations, given the lack of homogeneity, a much-needed attribute for the successful conduct of operations. Further, it is a challenge for higher formations to spare the manpower ideally needed to support infantry battalions. The strength of infantry battalions dissipated due to support elements like mortars and missile detachments redeploying, and it forced local-level improvisations to re-create additional manpower by reassigning these troops from within for non-specialist duties. This proved to be effective in CI operations, but it impacted upon their training and battle worthiness, which was required to focus on their primary responsibility against external threats.

In order to address the limitations of infantry battalions, primarily for J&K, the Rashtriya Rifles (RR) Battalions were raised, drawing their men and officers from the Indian Army. After a modest beginning in 1990, RR has expanded to a total of 63 battalions.[34] The battalions are affiliated to specific regiments of the army, thereby ensuring regimental affiliations. Simultaneously, they also draw their troops from all arms and services to ensure a specialist cadre vis-à-vis infantry battalions. RR battalions focus solely on CI operations, thereby making them better suited for the role. The battalions have dispensed with heavy weapons to free up additional manpower and have six instead of four companies, thereby providing the much-needed additional boots on the ground. Each RR battalion is assigned approximately 1,200 soldiers, compared to 800 of an infantry battalion. The equipment profile of the battalions has been tailored to ensure that every requirement peculiar to CI operations can be catered for within its resources. Even as soldiers are rotated over a period of two years, the battalions continue to remain in CI operations, thereby ensuring continuity and better adaptation to local needs and challenges. The raising of these battalions was accompanied by headquarters, which are tailored to CI operations and remain stationed permanently in affected geographical areas. This helps them build an effective database on the area of operations and, in the process, improve institutional memory.

Adaptation in conduct of operations

The initial phase of operations following India's independence witnessed the lack of experience in counter-insurgency operations and largely addressed the issue through on-the-job training. Battalions deployed in the area adapted to the circumstances based on their experiences. A veteran of CI operations during the mid-1960s said: 'But we learnt on the way, we learnt very quickly I would say.'[35] On-the-job training remained the norm until 1 May 1970, when the first Counter Insurgency Jungle Warfare School was established at Veirangte, Mizoram. A similar situation came up when the army was deployed for CI operations in Sri Lanka, commencing from 1987. Troops were inducted without the requisite pre-induction training and perforce learned through on-the-job training instead. The initial induction for operations in Punjab, as well as in J&K, followed a similar trend, until the concept of Corps Battle Schools (CBS) came up and induction into any CI area was compulsorily preceded by structured training at various CBS.

Over time, the army evolved its operational approach in CI operations. In the case of Mizoram, where the insurgency broke out in 1966, the shock of Mizo National Front (MNF) cadres taking over Assam Rifles posts saw the government order stern military countermeasures. The series of actions witnessed the employment of air power in a punitive role, under what were perceived as exceptional circumstances. However, this adaptation for CI operations was an experiment, which was not repeated given the psychological impact it had on the population for years to come. It reflected a negative cost–benefit analysis as a result of the inherent inaccuracy of bombs, challenges of immediately exploiting the shock affect with limited troops and, most importantly, the resultant alienation of the population. These experiences are possibly the reason for punitive air strikes not being repeated following the Naga and Mizo insurgencies in India.[36] Subsequent attempts to employ even helicopters in offensive roles have been opposed by the army, despite heavy casualties in Naxal-hit areas of central India.[37]

Initially, the army conducted CI operations as an offshoot of its training for conventional operations, with large columns employed for 'search-and-clear' operations. However, this was subsequently replaced by 'clear-and-hold' operations.[38] This required greater numbers to establish the physical presence of the army. From purely a military perspective, the grouping of villages facilitated this process. In the early 1970s, this became the precursor to the existing grid system of deployment of the army. The grouping of villages was an attempt at adapting to the challenge of insurgents drawing their sustenance from the local population. To counter this, the army relied upon the experience of the British Army in Malaya for not only the military expertise gained, but also for the strategy adopted to defeat the insurgency. Lt. Col. S.P. Anand, writing in the *United Services Institution of India Journal* of 1971, reinforced this thinking. 'The pattern followed was the one proposed and practised by Gen Briggs in Malaya.' He describes these as follows:

a   Cut the insurgents off from the population which supports them.
b   Make the zones in which the underground move untenable.
c   Act simultaneously over a wide expanse for a long period of time, to wear out the insurgents.[39]

As part of this initiative, the grouping of villages was done to separate the insurgents from the population, since the villages were patrolled by the army, which was assisted by the 'I' and Central Reserve Police Battalions. The policy suffered from a lack of understanding of the local customs and culture, wherein the local people subsisted on shifting cultivation. By bringing them together in artificial colonies without the requisite job opportunities, the locals were potentially dependent on government support for subsistence.[40] They were also alienated from their traditional way of life, causing deep resentment over a period of time.[41] The fact that the government and the army did not resort to this practice after the Mizo counter-insurgency operations reinforces the inadequacy of the experience, despite having achieved eventual success in the counter-insurgency efforts.

There have been a number of examples of the use of local militia over the years. It began with the security forces facing stiff resistance from the guerrillas in northeast India. In one such operation, the Inspector General of Police, G.S. Arya, was killed in Mizoram. His successor, Brigadier G.S. Randhawa, applied the 'pseudo-gangs' experiment of Kenya to Mizoram.[42] He employed relatives of locals killed for reprisals against Mizo National Front (MNF) insurgents. The experiment was soon stopped, given the inherent lack of control over armed members of the population. The government attempted this experiment on subsequent occasions as well. In the short term, it facilitated military operations with the advantage of local knowledge and improved flow of information; however, almost every experiment has subsequently led to poor control and reprisal attacks, alienating the people in the long run. The cases of the surrendered United Liberation Front of Assam (ULFA) cadres, *Salwa Judum* in Chhattisgarh[43] and the *ikhwan* in Jammu and Kashmir, have provided similar lessons over the years.

The conduct of operations was also impacted by the size of force employed and their nature of deployment. The initial years witnessed large-sized troop movements for operational manoeuvres. This was justified on the basis of heavy fire power requirements against terrorist groups, which could only be generated from a strength of 70 to 80 soldiers.[44] These numbers persisted despite guerrillas deciding to move in smaller numbers to reduce their footprints and to maintain their ability to merge with the local population. The decision to operate in larger groups was based on a number of factors. First, the conventional bias of the army continued to influence their operations in CI operations. This led them to view companies and platoons as ideal fighting sub-units, despite a change in the operational context. Second, most operations were conducted with only generic intelligence. The absence of specific actionable intelligence precluded the ability to carry out clinical strikes with small teams. Third, this further forced the columns to maintain a certain degree of fire power which, besides larger calibre automatic weapons, could only come from a larger number of weapons which would require a bigger force.

This concept underwent a change when it was realized that a larger force was more cumbersome, inefficient and ineffective for CI operations. This was simultaneously accompanied by greater reliance upon more specific and reliable intelligence, which enabled precision operations. A mention of the same is made in the Indian Army Doctrine, released in 2004, which says that resources of the security forces will invariably be spread over a large area of responsibility. In such an environment, operations based on small teams backed by good or specific intelligence increase the chance of contact with and success against the insurgents.[45]

This led to battalions becoming more nimble, flexible and adaptable to the inherently fluid nature of CI operations.

A major change experienced in the conduct of operations was a shift from large-scale cordon and search operations based on generic intelligence to clinical operations on the basis of specific intelligence. These were increasingly based on both human and technical intelligence. There was a marked improvement in the precision achieved during operations, given steady penetration by the army's intelligence teams in local areas and the availability of sophisticated equipment for monitoring terrorist communications. The operations became swift, caused limited collateral damage and, as a result, did not alienate large sections of the population. The initial forays into this highly specialized field began in the mid-1990s and, over the years, in addition to the army, a number of intelligence and enforcement agencies, which function in close liaison with the army, have developed sophisticated capabilities.[46]

The existing laws related to law and order in India do not allow the army to search houses, to raid suspected areas or to arrest people. These powers remain within the police. This was realized soon after the deployment of the army for CI operations in the mid-1950s. It led to the framing of special laws which, when applied to a particular area, allowed the deployment of the army and simultaneously gave the powers to search, raid and arrest suspected insurgents and terrorists. The first major law in this regard was called the Armed Forces (Special) Powers Act, 1958.[47] This law was subsequently amended to enlarge its geographical applicability in northeast India. It was also enacted for the states of Punjab, and Jammu and Kashmir, with minor variations. Although the law was framed to facilitate the limited and occasional deployment of the army in CI operations, it has been employed almost continuously since then.

Technology-driven changes

Technology has remained a major driver of change in the conduct of warfare. However, unlike conventional threats, CI operations are not affected as much by technological changes, given the relative advantage held by security forces. However, this was not always the case for the Indian Army. There were instances during operations in Sri Lanka and in J&K where terrorists had technological advantage over the army in terms of basic weaponry and communication equipment.

The use of night vision devices in conventional state versus state conflicts has been a reality for a long time. This has forced a change in how night operations were planned and perceived, with the differentiation between day and night being bridged by increasingly sophisticated equipment. However, this change did not come about in a similar fashion in the case of CI operations. While the army made good use of night vision devices as part of its progressive modernization, insurgent groups and terrorists did not have access to it in the Indian environment. This provided a distinct advantage to the army and reversed the advantage terrorists enjoyed in case of night operations, given their ability to better exploit local terrain under most circumstances.

Surveillance devices offer a similar advantage. They enable the army, in conjunction with other supporting forces, to better monitor and, in a number of cases, locate terrorists. Operations on the LoC especially received a boost as a result of these inductions. Surveillance centres were now able to track and monitor terrorists approaching the fence. This enabled clinical operations timed to perfection in a bid to neutralize maximum terrorists.

CI operations require large-scale vehicular movement, both for operations and administrative support. However, a constant threat of remotely controlled IEDs caused a number of casualties over the years. An officer from the Electronics and Mechanical Engineers (EME) was instrumental in locally adapting to the challenge and creating a device which could jam the remote signals of an IED, thereby neutralizing the threat. Vehicles were equipped with this device and became an integral part of military convoys thereafter, likely saving innumerable lives. A more comprehensive solution was the procurement and deployment of Casspir mine protected vehicles (MPVs). These MPVs were purchased from South Africa and deployed in areas like J&K. A product of scientific research, the vehicles were successful in limiting the impact of underbody and lateral explosions.

The Indian Army's ability to cope with change

The previous section gave an overview of the challenges faced by the army while dealing with CI operations. It highlighted the manner in which the army adapted and changed in order to improve its response over the years. However, it is not merely change, but the quality, pace and resultant success that must define the ability to adapt to it.

The first aspect which merits assessment is whether change in the army came about as a result of internal introspection or was due to external circumstances and pressures. The army's shift from expeditionary operations to treating insurgents as misguided youth came about both as a result of political direction as well as guidance from the highest levels within the system. This approach went a long way, since it became one of the most important guidelines for the conduct of operations.

The initial attempts at innovation were based on lessons from other armies, as is evident from the adoption of the Briggs Plan and the creation of model villages in both Nagaland and Mizoram. Similarly, operations in Kenya possibly led to the use of the pseudo gang concept. The influence of experiences of other armies in CI operations became limited after a few of these experiments failed to deliver desirable results. Some of the experiences and lessons learned during the initial years in CI operations led the army to evolve its approach and to refine it according to the nature of the challenges faced. The most important change in this regard was the shift from a conventional approach to operations in counter-insurgency to those suited and tailored for it.

The army struggled during the initial years with the deeply ingrained conventional bias to operations. Over a period of time it was realized that the army needed to simultaneously prepare both for external threats and internal security challenges. The Indian Army Doctrine published in 2004 balances this dual responsibility and aims to ensure requisite preparedness for both. While this may not be the best option for an army, which is essentially required to fight external threats, under the prevalent circumstances it does provide the necessary balance. The raising of the RR further facilitated this approach by creating a more focused organization for CI operations.

There is little evidence to suggest that civil–military relations or political interference adversely impacted upon the conduct of CI operations by the army. In contrast, doctrinal changes were undertaken by the army as an in-house effort and received the support of the government of the day. While there have been differences of opinion between the state government and the army on issues like the Armed Forces Special Powers Act 1990 in J&K, however, this has not impinged upon the conduct of operations, given the backing for the army's point of view from the central government.

The evolution of the Iron Fist and Velvet Glove doctrine shaped the institutionalization of the army's attempts at ensuring people-friendly operations. However, there is little doubt that this was also influenced by the increasing pressure from the people at large, media and human rights groups to minimize collateral damage and civilian casualties. It forced greater accountability and transparency in the functioning of the army in CI operations.[48]

Second, the quality of change is an important aspect of change management. The army has successfully overcome the challenge of raising new organizations, as seen from the example of RR Battalions. In addition to this, a number of new departments have been established to improve the responsiveness of formations employed for CI operations. While these are the positives of change management, an equally pertinent question regarding the effectiveness of some of these organizations remains relevant. The setting up of specialist organizations can only be effective when officers and men with the requisite training and experience are posted to those organizations. The existing manpower policies have not been able to manage this requirement, with other conflicting conditions like maintaining regional posting profiles and inadequate profiling of officers in areas of expertise. This leads to a steep learning curve and inadequate contribution for highly specialized responsibilities. This has led in turn to the establishment of organizations with a suitable mandate, but with inadequate capability inputs to achieve it.[49] In contrast, operational innovation at the junior leader level remains exemplary and is the basis for successful operations, despite environmental disadvantages and challenging rules of engagement. Since junior leadership is the most critical facet of CI operations, their ability to adapt and innovate at the tactical level has made up repeatedly for systemic deficiencies.[50]

Third, the speed of adaption is important to ensure that an organization remains prepared for future challenges. Change management merely as a reaction to emerging threats and challenges keeps an army behind the curve of fast-moving events. There have been few cases wherein the army has been able to pre-empt situations by adapting to future challenges. An example, in almost every case and as highlighted in the previous section, is training institutions that were set up well after the induction of troops in the areas of CI operations. This led to avoidable casualties and loss of precious time in bringing troops up to the requisite professional standards. Changes were brought about after reverses and events forced them upon the army. Conversely, operational adaptation was more responsive, often despite limited resources. This reflects a weak institutional culture of innovation at the strategic level and an absence of structured training of a cadre of leaders, who are trained to think beyond immediate challenges, especially for senior positions, which demand vision and the ability to drive change.

Change management was not influenced very often by external threats faced by the army. However, a major influence was and remains the impact of Pakistan upon the ongoing proxy war in J&K. This has led the army to fine tune its overall strategy in the state to include threats from both external and internal factors. It has also had an impact upon the army's deployment and the conduct of operations. Further, technology has contributed towards creating a viable obstacle and surveillance system along the LoC.

Change management is influenced by institutional memory and the ready availability of declassified material.[51] Despite the winds of change having a major social impact on the army, certain policies in this regard have not changed. The declassification policy of documents and its accessibility to scholars remains buried in bureaucratic hurdles. The army has also failed to freely circulate assessments of past experiences in CI operations among every successive generation of leaders. This lack of critical assessment of past successes and failures could result in leaders relearning past lessons and, even worse, repeating past mistakes.

It is evident from the analysis of case studies examined that military change in CI operations in the Indian context took place in the strategic, organizational and operational spheres. It is also clear that this was driven essentially by operational threats and only marginally by environmental influences, technology and political directions. A number of changes at the operational level also contributed to the eventual strategic shifts, which came about more as part of incremental evolution, rather than revolutionary changes. This indicates the marginal impact of revolution in military affairs (RMA) with relation to military change in CI operations. Changes were instead governed more by structural shifts and evolutionary operational lessons. Similarly, in contrast to operations in Afghanistan, the influence of coalition politics and civil–military relations have also been limited.

Conclusion

This analysis finds three distinct stages of doctrinal shift that have taken place both in the face of threats and as a result of institutional introspection. These have impacted upon the fundamental guidelines which form the basis for the conduct of operations. In terms of organizational restructuring, major changes followed a top-down approach, as these required large-scale restructuring and government sanction, while local-level bottom-up adaptation was, and continues to be, carried out to achieve reasonable success. Unlike external threats, which require strategic shifts to bring about substantive change, change in a counter-insurgency environment can also be brought about by operational adaptation. This chapter has highlighted a number of such changes undertaken by the army in its endeavour to better respond to threats over a period of time. While the army's fundamental direction in CI operations came from political decision-makers and the top of the military hierarchy, it is evident that the bulk of adaptation was carried out by junior and intermediate-level commanders under battlefield conditions. Local adaptive steps taken by units and others in CI operations had a perceptible impact throughout the army.

Change management is a challenge both in war and peace. However, the parameters of its evaluation become different when it is carried out in respect of CI operations. For instance, while operational factors seem to be more important in driving military change in CI, they have limited relevance in conventional operations. It is evident that the study of military change in a CI environment is different from more conventional operational threats. This raises the need to carry out a separate analysis for such operations, even if the same army has the responsibility to undertake them.

Notes

1 Indian Army Doctrine, *Army Training Command*, Shimla, India (October 2004), 33, http://ids.nic.in/doctrine.htm (accessed 29 October 2013).
2 General (Retired) Deepak Kapoor, 'The Indian Army: An Insight', *India Strategic* (November 2013), www.indiastrategic.in/topstories3089_the_Indian_Army_an_insight.htm (accessed 22 April 2014).
3 It is widely acknowledged that Pakistan has been instrumental in spearheading the proxy war both in Punjab and after in Jammu and Kashmir. See Raymond Whitaker, 'Pakistan Warned it Could Be Declared a "Terrorist State": US Says Islamabad Must End its Support for Kashmiri Militants Pakistan Warned it Could be Declared a "Terrorist State"', *Independent*, 10 May 1993, www.independent.co.uk/news/world/pakistan-warned-it-could-be-declared-a-terrorist-state-us-says-islamabad-must-end-its-support-for-kashmiri-militants-pakistan-warned-it-could-be-declared-a-terrorist-state-2321994.html (accessed 22 April 2014).
4 The most challenging period for the army was its simultaneous deployment in Sri Lanka, northeast India, Punjab and J&K in the early 1990s.
5 One recent study in this field is *Military Adaptation in Afghanistan* edited by

130 V. Chadha

Theo Farrell, Frans Osinga and James A. Russell (Stanford, CA: Stanford University Press, 2013).
6 Northeast India has witnessed a number of insurgencies over the years. While it began with the Naga movement in the early 1950s, uprisings in Mizoram, Tripura, Manipur and Assam have followed since then.
7 The involvement of the Indian Army in Sri Lanka commenced as a peacekeeping operation in 1997. However, the failure of the negotiated settlement with the Liberation Tigers of Tamil Eelam (LTTE) soon saw it transform into a counter-terrorism operation. This is the only engagement of the Indian Army outside the country for counter-terrorism after India became independent in 1947.
8 See Field Service Regulations, Part I Operations, 1909, London 1912, 191–192, http://ia700300.us.archive.org/17/items/pt1fieldservicer00greauoft/pt1fieldservicer00greauoft.pdf (accessed 3 September 2013).
9 See Subir Bhaumik, *Insurgent Crossfire: Northeast India* (New Delhi: Lancers Publishers, 2008), 63.
10 Brigadier H.S. Sodhi and Brigadier Prem K. Gupta, *History of 4th Gurkha Rifles (Vol IV), 1947–1971*. Delhi, 1985, 150.
11 See *US Government Counterinsurgency Guide*, Bureau of Political–Military Affairs (January 2009), 6, www.state.gov/documents/organization/119629.pdf (accessed 20 November 2013).
12 Doctrine for Sub Conventional Operations. Integrated Headquarters of Ministry of Defence (Army), Headquarters Army Training Command, Shimla (December 2006), 20.
13 See B.G. Verghese, *India's Northeast Resurgent: Ethnicity, Insurgency, Governance, Development* (New Delhi: Konark Publishers, 2002), 16–19.
14 The British forces undertook a series of expeditions into then AGA areas commencing from 1866.
15 Jawaharlal Nehru, Foreword to the second edition of Verrier Elwin, *A Philosophy for NEFA*, 9 October 1958, www.arunachalpwd.org/pdf/Philosophy%20for%20NEFA.pdf (accessed 5 September 2013).
16 Sarvepalli Gopal, *Jawaharlal Nehru: A Biography*, vol. 2, 1947–1956 (Delhi: Oxford University Press), 212.
17 Quoted from Rajesh Rajagopalan, *Fighting Like a Guerrilla: The Indian Army and Counterinsurgency* (New Delhi: Routledge, 2008), 147.
18 Lt. Gen. Mathew Thomas in an interview with the author. Quoted from Lt. Col. Vivek Chadha, *Low Intensity Conflicts in India: An Analysis* (New Delhi: Sage, 2005), 342. The quote is in context with operations in Mizoram immediately after the induction of the Indian Army into the area.
19 Brig S.K. Sinha, 'Counter Insurgency Operations'. *USI Journal*, vol. C, no. 420 (July 1970), 262–263.
20 Ibid., 263.
21 Ibid., 266.
22 Ibid., 268.
23 WHAM (winning hearts and minds) has for long been part of the overall CI strategy. As part of this, while kinetic operations are conducted, measures are undertaken to alleviate the day-to-day problems of the local population. This can include steps like the provision of medical aid, transportation, opening schools, creating job opportunities, etc. This also helps in weaning away the population from insurgent influence back into the governmental fold.
24 Doctrine for Sub Conventional Operations, no. 15, 21.
25 Ibid., i–ii.
26 Ibid., 29.
27 Ibid., 30.

28 The LC fence went up in J&K, along the LoC, at a distance of approximately two to four kilometres from it. The fence, along with the associated dynamic deployment, has been instrumental in reducing the infiltration of terrorists from Pakistan-occupied Kashmir.
29 Doctrine for Sub Conventional Operations, no. 15, 32.
30 Perception management aims at positively influencing the thinking and behaviour of the intended audience in pursuit of organizational objectives.
31 Doctrine for Sub Conventional Operations, 35.
32 See Major R.D. Palsokar, M.C., 'Fighting the Guerrilla'. *USI Journal*, vol. 91, no. 385 (October 1961), 270.
33 Since infantry battalions continue to be employed in a secondary role in CI operations, even as their primary role and organization is suited for conventional operations, battalions carry out this local-level improvisation in a bid to optimize manpower utilization. The author, while commanding a battalion on the Line of Control between India and Pakistan in the J&K in 2007/2008, also resorted to this local adjustment to cover a larger area more effectively. He also witnessed its employment earlier during Operation Pawan against the LTTE in Sri Lanka in 1989 to 1990.
34 For a brief account of the evolution of RR see B. Bhattacharya, 'The Rashtriya Rifles', *Bharat Rakshak*, www.bharat-rakshak.com/LAND-FORCES/Units/Infantry/222-Rashtriya-Rifles.html (accessed 20 November 2013).
35 Lt. Gen. Mathew Thomas in an interview with the author, no. 21, 342.
36 See Vijendra Singh Jafa, 'Insurgencies in North-east India: Dimensions of Discord and Containment', in S.D. Muni (ed.) *Responding to Terrorism in South Asia*, (New Delhi: Manohar, 2006), 96.
37 Aloke Tikku, 'Army Nixes Govt Plan to Hit Nasals from Air', *The Hindustan Times*, 2 April 2013, www.hindustantimes.com/india-news/newdelhi/army-nixes-govt-plan-to-hit-naxals-from-air/article1–1036158.aspx (accessed 12 November 2013).
38 Vijendra Singh Jafa, no. 38, 96.
39 Lt. Col. S.P. Anand, no. 22, 153.
40 Ibid., 154–155.
41 Vijendra Singh Jafa, no. 38, 98.
42 Ibid.
43 The Salwa Judum, an experiment to employ tribal youth as local militia (special police officers), was banned by the Supreme Court of India. See J. Venkatesan, 'Salwa Judum is Illegal, Says Supreme Court', *The Hindu*, 5 July 2011, www.thehindu.com/news/national/salwa-judum-is-illegal-says-supreme-court/article2161246.ece (accessed November 12, 2013).
44 Major R.V. Jatar, 'Counter Insurgency Operations', *USI Journal*, vol. 98, no. 413 (October 1968), 415.
45 Indian Army Doctrine, no. 3.
46 An article by Pravin Swami documents this capability in the year 2000. Since then, there has been a further marked improvement in this capability. See Pravin Swami, 'Eyes and Ears Wide Open', *Frontline*, vol. 17, no. 9 (29 April to 12 May 2000), www.frontline.in/static/html/fl1709/17090230.htm (accessed 12 November 2013).
47 See http://indianarmy.nic.in/Site/RTI/rti/MML/MML_VOLUME_3/CHAPTER__03/457.htm (accessed 31 October 2013).
48 The inclusion of Dos and Don'ts as part of the Armed Forces Special Powers Act was the result of a judgment by the Honourable Supreme Court after a case was filed challenging the constitutional validity of the Act. Some inquiries into alleged human rights violations also followed pressure created by locals in the area of operations.

49 For a debate on the challenges of specialization in the army, see Anit Mukherjee, 'Facing Future Challenges: Defence Reforms in India', *RUSI Journal*, vol. 156, no. 5 (October/November 2011), 32–33.
50 Brigadier P.S. Mann, 'Conduct of Junior Leaders in Counterinsurgency Operations', *USI Journal*, vol. 141, no. 584 (April–June 2011), www.usiofindia.org/Article/?pub=Journal&pubno=584&ano=821 (accessed 20 November 2013).
51 For a detailed analysis of the problem see Anit Mukherjee, no. 51, 34.

Part II
Military change in Europe

8 Perspectives on military change and transformation in Europe

Sven Bernhard Gareis

Over the past two decades military change and transformation in Europe has been characterized by three major trends. First, in the absence of an immediate threat since the end of East–West antagonism, most European countries wanted to realize their 'peace dividend' by downsizing their armed forces and shrinking their defence budgets. Second, the central mission of the military shifted from territorial defence to participation in international crisis management – often out of the European area. Finally, in order to maintain crucial capabilities that were no longer available on the national level, European countries started to build multinational military structures and units.

At the same time, military contributions to common security remained a neglected topic on the political agenda of the EU. The EU relied on its strengths in the traditional areas of treaty-based multilateral arrangements and close economic cooperation with its neighbours more than it relied on military power. The incentives of an association agreement with, or the prospect of eventual access to, the most prosperous community of states were strong enough for the partners to peacefully settle their internal and external disputes. The EU thus succeeded in establishing an extraordinarily stable zone of interconnectivity and interdependence encompassing practically all European nations on different levels of integration, ranging from partnerships to association agreements to full EU membership.

In the wake of the Kosovo war of 1999, the Common Security and Defence Policy (CSDP; the original name was the European Security and Defence Policy (ESDP) until it was renamed in the Lisbon Treaty in 2009) started out as an ambitious project. In practice, however, the EU's military capabilities are still confined to small, low-intensity missions. Most notably, the CSDP has remained a strictly intergovernmental policy field that by no means served as a framework for the coordination of military change and transformation of the member states' militaries. Military and defence issues largely remained in the national realms. Even the European Council held in Brussels on 19 and 20 December 2013, which had long been heralded as a kind of European Security Summit, was in the end dominated by economic topics, such as the negotiations on the European Banking

Union. Any hopes for a new roadmap for intensified European security and defence efforts were dashed. In their conclusions, the Heads of State and Government called for enhanced cooperation among the various actors in the CSDP context without, however, mentioning substantial steps towards more concerted military reforms, not to speak of more integration in the field of European armed forces.[1]

This security setting, however, is being challenged by three major developments. First, in 2011, US President Barack Obama proclaimed the rebalancing of his country's interests to the Asia-Pacific region.[2] Although Europe and NATO are still considered to be America's most important partners, as a result of this 'pivot' the US will have to decrease the number of its soldiers in Europe as well as its overall commitment to European security. Second, with its forcible annexation of the Crimea, its hybrid warfare in eastern Ukraine, and its blatant threats to former Warsaw Pact allies and Soviet republics that joined NATO and the EU or seek a close relationship with the two Western organizations, Russia has clearly expressed its dissatisfaction with the post-Cold War order in Europe.[3] As a result, collective defence against new forms of military threats has returned to the European agenda. Third, the global financial crisis that started in 2008 still strongly affects the EU and its members with heavy debt burdens and financial constraints, not least with regard to their defence budgets.

The EU and its member states will therefore have to reconsider their Common Security and Defence Policy. This chapter argues that more and closer European integration will be needed in the security and defence policy field. In order to offer a 'big picture' of European security policy, it will examine the larger trends of military change and transformation in Europe, discuss the achievements reached thus far as well as the persisting shortfalls and challenges for the CSDP, and propose new approaches to more integrated European armed forces. Before doing so, however, a brief analysis of the current security environment will highlight the political backdrop against which military change and transformation in Europe is taking place.

Europe's changing security environment

The end of the bipolar world order brought paradigmatic changes to the European security setting. The immediate threat by the Soviet Union and its allies fell away. As early as in June 1990, at its meeting in Turnberry, UK, the North Atlantic Council offered to 'extend to the Soviet Union and to all other European countries the hand of friendship and cooperation'.[4] In its London Declaration, NATO member states committed themselves to strive for a cooperative order in Europe based on confidence and force reduction.[5] The former adversaries in the Warsaw Pact and – following the dissolution of the Soviet Union at the end of 1991 – even former Soviet Republics turned to the West and sought a secure position among democratic nations.

For Europe, the historic turning point offered the opportunity to continue an integration process that in the West had proved to be the most successful concept for promoting peace in the continent's history.

The improvement in the regional and global security environment, however, could not prohibit the emergence of new risks and challenges in Europe and on its periphery. The dissolution of Yugoslavia at the beginning of the 1990s brought war back to Europe, affecting the entire continent and bringing with it mass atrocities and massive streams of refugees, and destabilized Southeast Europe. Although a new regional order could be established in the Western Balkans, the EU and NATO remain involved in an enduring stabilization process providing a broad array of assistance to Bosnia and Herzegovina and Kosovo in fields including governance, security, the economy and rule of law.

Globalization

On the global level, the end of the East–West conflict unleashed enormous change-producing forces. Within one generation, the world witnessed two major transitions in the international system: the end of the decade-long bipolar world and the subsequent collapse of the Soviet Union left the US as the sole global power, with unprecedented opportunities and capacities to shape international relations. This unipolarity, however, would turn out to be a brief period, a 'unipolar moment',[6] rather than a lasting global order. With the rise of newer powers like China, this unipolarity came to an end and a new international system began to emerge.

Increasingly borderless travel, the free flow of goods, services and finance, and sky-rocketing trade and commerce led to tremendous economic gains, and contributed significantly to the reduction of poverty and famine in many parts of the world. On the other hand, the same channels needed to harvest the fruits of globalization turned out to be the gateways for malign actors or negative appearances. Interconnectivity and interdependency thus not only improved cooperation among states as well as regional and global stability but also increased the vulnerabilities of (not only) European nations and societies: conflicts spilled over borders (even in remote regions) as concerns rose about the proliferation of weapons of mass destruction, terrorism and organized crime, and the spread of infectious diseases. The world has indeed turned into a global village, one which faces all of these risks as well as opportunities.

New tasks for Europe

Against the backdrop of the globalization process, the EU had to broaden its scope from the mainly economic orientation of its precursors to a more comprehensive participation in world politics. The violent collapse of Yugoslavia in the 1990s displayed the Europeans' inability to react to major crises in their immediate vicinity. Under the aegis of its Common Foreign and Security Policy (CFSP), the EU started to create a number of political and military instruments designed to enable the Union to contribute to the international management of the many crises that occurred after the end of the bloc confrontation. As early as 1992, the members of the Western European Union (WEU) adopted the Petersberg Declaration, in which they offered 'to make available military units from the whole spectrum of their conventional armed forces for military tasks',[7] such as humanitarian assistance, peacekeeping or peace enforcement. Step by step, those 'Petersberg Tasks' became integral elements of the Treaty on European Union that in 2000 (Nice Treaty) actually absorbed the WEU and its functions in the field of European collective defence.

The Kosovo crisis and the subsequent war which US-led NATO fought in 1999 taught the European nations that in the transatlantic relationship 'there is not too much U.S. but too little Europe'.[8] In the wake of the Kosovo war, the EU launched its European Security and Defence Policy (ESDP, known since the 2009 Treaty of Lisbon as the Common Security and Defence Policy (CSDP)), which quickly established an institutional arrangement that includes common bodies like a Political and Security Committee (PSC), an EU Military Committee (EUMC), an EU Military Staff (EUMS), and the nucleus of an operational headquarters.[9]

In 2003, the EU took over its first military mission from NATO, in which European forces monitored the observation of the Ohrid Framework Agreement in Macedonia (EUFOR Concordia). Since 2003, the EU has also been conducting a number of military missions at a low-intensity level, including EUFOR ALTEA in Bosnia and Herzegovina or the naval operation ATALANTA off the shores of Somalia and the Horn of Africa, as well as support missions to the UN and (sub-) regional organizations in Africa (AU, ECOWAS), including missions in the DR Congo (2003, 2006), Chad and the Central African Republic (2008/2009), in Mali (started in 2013), and again in the Central African Republic (2014).

Both the CFSP and CSDP, however, remained relatively weak institutions. Other than in the field of the Common Market, EU member states refused to transfer substantive sovereignty rights to the Union. External action of the EU has never been subject to intergovernmental decision-making procedures, in particular with regard to military deployments. Although the EU increasingly took over the role of a player in global security policy, the prevalence of national interest again and again resulted in disharmonies and even political splits. The disputes over the recognition of the successor states of the former Yugoslavia in the early 1990s; the disruption over support to the US invasion of Iraq in 2003; the lack of EU consultation in the Libya crisis in 2011; and the hesitant responses to Russian measures against Ukraine again and again disclosed the weaknesses of the EU, both with regard to reaching a common situation awareness and to taking determined joint action.

Furthermore, the CSDP never became a framework for a coordinated process of the reduction of national military forces or the maintenance of complementing capacities.[10] Decisions on force structures, armament procurement, capabilities and doctrines remained mainly in the national realms – even if some multinational arrangements among a number of cooperation-willing nations could be established both in the field of procurement (e.g. aircraft like the Airbus A400M) and force structures (e.g. the close cooperation between Germany and the Netherlands).

In response to the new tasks and challenges (and to heal the wounds of the European split over Iraq), the EU adopted the European Security Strategy (ESS) in December 2003 as its first and thus far only orientation paper for its external action. It noted that 'as a union of 25 states with over 450 million people producing a quarter of the world's Gross National Product (GNP), and with a wide range of instruments at its disposal, the European Union is inevitably a global player'[11] – a statement that did not express an abundant determination to actually take on that role. However, in its 14 pages, the ESS named the major challenges (global underdevelopment and the continent's energy security) and the key threats to its security (terrorism; proliferation of weapons of mass destruction; regional conflicts; state failure; and organized crime), and defined its own objectives in tackling those challenges. Here again, the EU emphasizes its strong reliance on international cooperation:

> In a world of global threats, global markets and global media, our security and prosperity increasingly depend on an effective multilateral system. The development of a stronger international society, well-functioning international institutions and a rule-based international order is our objective.[12]

The ESS concludes with a chapter on the policy implications for the Union. Here it stresses the necessity to 'develop a strategic culture that fosters early, rapid, and when necessary, robust intervention'.[13] Although the EU launched a brief report on the implementation of the ESS in 2008, in which cyber security and climate change were added to the list of common threats,[14] the ESS remains a 'stand-alone document'[15] that urgently needs to be updated.

New strategic requirements

In 2015, the EU once again finds itself in a changing strategic environment. With the Russian aggression directed against Ukraine and the threats to Eastern European member states, the issue of collective defence has returned to the European agenda. Russia's 'hybrid warfare' in Ukraine, which includes open and covert, direct and indirect forms of use of force against a sovereign state, poses new challenges to the cohesion of

EU and NATO. Both organizations have to reassure their Eastern European members of the security guarantees they gained through their membership while simultaneously seeking common grounds with Russia for a stable political order in Europe. Differing perceptions and assessments of Russia's actions have to be discussed and harmonized among the partners as well as the choice of the appropriate instruments to cope with the new threats. With regard to the military, the Europeans have to display their cohesion without delivering to Russia pretexts for further escalation.

The cohesion among the Europeans – both in the EU and NATO – becomes even more significant at a time when the US is refocusing its strategic centre of gravity to Asia in order to balance the growing influence of China. By visiting Asia in the spring of 2014 during the culmination of tensions over Ukraine, President Barrack Obama made it clear that the US shift of interest would continue. In the Libya crisis of 2011, the US already showed a relatively low military profile and it continues to do so with regard to the reassurance of its allies in Eastern Europe.

> 18 fighter jets deployed to Poland and Lithuania, an extended Black Sea cruise for a Navy destroyer and about 300,000 prepackaged field rations (but no weapons) for the Ukrainian armed forces. U.S. support for Ukraine 'has not pulled any assets away' from Asia or any other region, Hagel noted.[16]

Indeed, any hopes of a US re-rebalancing to Europe are not very realistic. The US, after two unwinnable wars in Iraq and Afghanistan and itself under enormous budgetary pressure, will have fewer capacities left for the defence of its European partners from an increasingly assertive Russia.[17]

The increasing challenges posed by Russia and the drawdown of US engagement in Europe hit the continent amid its struggle to recover from the financial crisis that started in 2008/2009 and still constrains state expenditures including the defence budgets. As Friedberg notes, by 2010 the European NATO partners collectively had reduced their military expenditures to an average of 1.7 per cent of their GDP compared to 3.1 per cent in the late 1980s (see also following section).[18] Under the auspices of austerity, defence budgets are not very likely to rise, so that new forms of defence cooperation are being discussed under the headlines of 'smart defence' in NATO and 'pooling and sharing' in the EU.[19] Both concepts propose ways to combine specialized national capabilities within the framework of common security and defence efforts.

The new geopolitical power shifts will require from Europe ever-increasing efforts to look after its security interests and to develop appropriate strategies and instruments to pursue them successfully. That is not only true for its original region: Europe is a global economic and trading power that is closely interconnected with the Asia-Pacific region and its rapidly growing markets. Europe therefore has a vital interest in security and stability in Asia – without, however, possessing major leverage to influence political developments there. Europe will have to complement its economic interests (and weight) through increased unity as well as a stronger engagement in its own region and on the global level. Otherwise, the continent will run the risk of being reduced to the role of an observer of a global order dominated by powers like the US and China. In this context, the armed forces will not be playing a decisive role but an important one nonetheless. Therefore, the following section will present some major trends in European military cooperation and integration that will then be discussed with regard to their effective contribution to strengthening Europe's role as a security actor.

Cooperation and integration: multinational forces arrangements complement national capabilities

The rapid and fundamental improvement of the European security environment after the end of the Cold War allowed all countries to realize major peace dividends by massively reducing their armed forces as well as their defence budgets.[20] All across the EU member states, armed forces were downsized by approximately two-thirds compared to 1989 (see Figure 8.1).

European defence budgets shrank or stagnated nominally at ever-shrinking percentages of the growing GDPs (See Figure 8.2). A similar

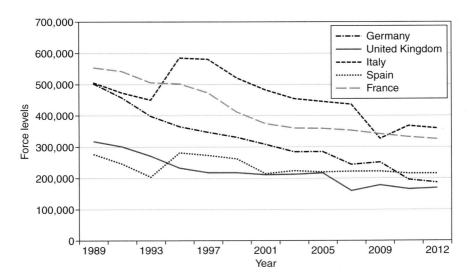

Figure 8.1 Decreasing force levels of important EU member states (source: Author's own compilation, data from SIPRI 2014).

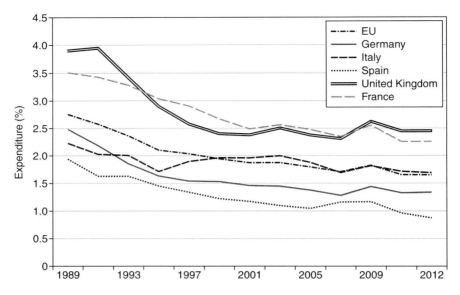

Figure 8.2 Military expenditures in per cent/GDP of important EU countries and EU average (source: Author's own compilation, data from SIPRI 2014).

trend may be observed in the national defence budgets. In Germany, for example, defence expenditures went down from around 17 per cent of the overall annual federal budget in 1989 to 10.7 per cent in 2013.

At the same time, however, the emergence of new risks and challenges required the maintenance of an operational set of military capabilities. For this purpose, European states started to rely on multinational headquarters and units that were considered to be capable of achieving a number of objectives.

Maintain capabilities

Already in the Alliance's New Strategic Concept (1991), NATO member states pointed out that:

> Multinational forces demonstrate the Alliance's resolve to maintain a credible collective defence; enhance Alliance cohesion; reinforce the transatlantic partnership and strengthen the European pillar. Multinational forces, and in particular reaction forces, reinforce solidarity. They can also provide a way of deploying more capable formations than might be available purely nationally, thus helping to make more efficient use of scarce defence resources. This may include a highly integrated, multinational approach to specific tasks and functions.[21]

Table 8.1 Important multinational military structures in Europe

Unit	Founded	Contributing nations
German-French Brigade (Müllheim/Germany)	1989	Germany, France + Belgium, Luxembourg, Netherlands
Allied Command Europe Rapid Reaction Corps (ARRC) (Gloucester/UK)	1991	Belgium, Canada, Czech Republic, Denmark, France, Germany, Greece, Netherlands, Norway, Poland, Portugal, Spain, Turkey, United Kingdom, United States
Eurocorps (Strasburg/France)	1992	Belgium, France, Germany, Luxembourg, Spain + six more nations
German/Netherlands Corps (Münster/Germany)	1995	Germany, Netherlands + ten more nations
Eurofor (Florence/Italy)	1995	France, Italy, Portugal, Spain
Multinational Land Force (MLF, Udine/Italy)	1998	Hungary, Italy, Slovenia
Multinational Peace Force, Souteastern Europe (rotating)	1998	Albania, Bulgaria, Greece, Italy, Macedonia, Romania, Turkey
Multinationales Korps Nordost (Szczecin/Poland)	1999	Denmark, Germany, Poland + eight more nations
Lithuanian-Polish-Ukrainian Brigade (in preparation)	2011–2014	Lithuania, Poland, Ukraine

Source: author's own compilation.

As a consequence, national headquarters on the operational level (e.g. army corps) and national units (divisions, brigades) were step by step replaced by multinational headquarters, which meant that not only were the smaller European countries no longer left with any operational capabilities of their own, but also that Germany had completely transferred this level of command to multinational structures (See Table 8.1).

Contribute to stability and integration

Germany and France soon recognized the importance of close cooperation between the armed forces, first as a symbolic act, and later as a practical step towards the consolidation of the European integration process. By creating the German-French Brigade in 1988[22] and the Eurocorps in 1991, both states paved the way for more multinational cooperation as a pacemaker for increased integration. NATO's 'Partnership for Peace' programme was based on the same idea: to grant former adversaries from Warsaw Pact countries a role in Euro-Atlantic structures and to enhance the transfer of stability towards Eastern and Central Europe. The new partners in Eastern and Central Europe started to appreciate the technical and organizational as well as the stability-related advantages of joint, sustainable and experienced military units, and modelled the new bi- and multinational formations on the German-French examples.[23] The extent to which these multinational cooperation efforts among the armed forces contributed to overall political integration becomes evident if we bear in mind that most of the participating states have in the meantime joined the EU and/or NATO.

Peace operations

In the meantime, the multinationality concept has evolved into much more than a structural basis for military units; it has become the guiding principle of European peace missions. Since 2003, the EU has completed four military missions under the ESDP (Concordia in Macedonia 2003; Artemis in the Democratic Republic (DR) of Congo 2003; EUFOR in the DR of Congo 2006; and EUFOR Chad/Central African Republic 2008/2009) and is currently running five missions: the European Union peace force EUFOR 'Althea' in Bosnia and Herzegovina, the training missions EUTM Mali and EUTM Somalia, as well as the maritime anti-piracy operation 'Atalanta' off the Horn of Africa and EUFOR RCA in the Central African Republic (See Table 8.2). In order to enhance its quick-response capabilities, the EU in 2007 started to keep two battle groups on stand-by and ready for deployment within ten days after a EU Council decision over an area within a radius of 6,000 kilometres around Brussels. The battle groups usually consist of an infantry battalion plus combat support elements with a strength of around 1,500 soldiers from several nations. For deployment a battle group will be associated with a mobile force headquarters and enablers like airlift and logistics. After preparatory training a battle group remains on stand-by for six months and is then followed by another one. The EU has developed a multi-year planning calendar indicating nations' commitments to contribute forces or capabilities to a battle group. Although since the initiation of the concept the EU has launched a number of military operations, no battle group has been activated thus far.[24]

These multinational structures were meant to guarantee, through joint structures supported by several nations, the maintenance of those military capabilities and command levels that could not be retained at a national level due to security-related and economic reasons. At the same time, however, the contributions to these joint structures were to remain a matter of national sovereignty and to take into account a nation's military traditions, identity and leadership culture. Multinational staffs and units therefore reflect the attitude of those intergovernmental bodies and mechanisms in Europe that do allow for collective action in security matters while simultaneously restricting that action by unanimity voting requirements and national caveats that limit the scope of mandates and the duration of missions.

Table 8.2 Military operations of the EU (as of March 2015)

Mission	Deployed to	Purpose	Duration
Concordia	Macedonia	Secure implementation of the Ohrid Framework Agreement	2003
Artemis	Democratic Republic of the Congo	Securing the situation in Bunia	2003
EUFOR ALTHEA	Bosnia and Herzegovina	Securing stabilization process	since 2004
EUFOR DR Congo	Democratic Republic of the Congo	Assistance to UN mission MONUC to secure the election process	2006
EUNAVFOR ATALANTA	Somalia/Horn of Africa	Countering piracy	since 2008
EUFOR Chad/RCA	Chad/Central African Republic	Provide safe environment for the buildup of UN mission MINURCAT	
EUTM Somalia	Somalia	Training of Somali armed forces	since 2010
EUTM Mali	Mali	Training of Malian armed forces	since 2013
EUFOR RCA	Central African Republic	Provide safe and secure environment in the Bangui area	since 2014

Source: author's own compilation.

This contradiction between collective requirements and national caveats has made multinational cooperation among Europe's armed forces problematic in spite of the undeniable achievements in their daily work and during missions. A closer examination of the challenges to a multinational European military force inevitably raises the question of whether it may be a good idea to equip European armed forces with a legal basis of their own and to develop a common military culture as an improvement over the current form of cooperation and the side-by-side existence of national contingents and contributions.

Challenges to multinational military structures

On the one hand, there are bright prospects and opportunities for multinational military cooperation in Europe; on the other hand, there are obstacles and limits. Those are usually due to the fact that in the tradition of the Western world – at least since the days of the American Revolutionary War and the French Revolution – the nation and the military have always been two sides of the same coin. The military symbolizes the nation state, and a well-functioning army symbolizes and at the same time guarantees the nation state's sovereignty. The size, the structure, the equipment and the mission of an army reflect the strategic culture of a nation and its society.[25] The continued existence of the nation – in the sense of a nation being the community of fate of a people – represents a core value for which soldiers are willing to risk their health or, if necessary, even their lives. More than anything else, this willingness to sacrifice everything requires a good cause: a rationale which is defined by law and, more importantly, has a strong emotional impact. As a consequence, nation states are reluctant to give up this important instrument of power along with the responsibility for their soldiers and are unwilling to pass it on to some intergovernmental body, while the soldiers themselves – at least initially – see and assess multinationality through the lens of their national backgrounds.

This limitation of the nation state's power to make use of its military is one of the characteristics of increasing multinational integration,[26] but states tend to keep multinational interaction under control through rules based on respect for their sovereign rights. At the same time, soldiers in combat units are expected to feel just as safe under a foreign commander and with unfamiliar comrades while risking their lives as they would under their national command. For everyone involved, reaching such a degree of trust requires a long learning process, a lot of effort, a lot of time and a lot of attention. This makes integrated military multinationality a difficult endeavour that is often prone to cause irritation among the participating nations. The nation states' desire to maintain as much control as possible over their soldiers during multinational missions leads to conflicts because of their different national legal and political rules and regulations, and also because of the different military and cultural backgrounds of their soldiers in day-to-day routines and rights in the midst of an operation. If there is one conclusive finding that German and international research on military multinationality has produced, it is the limitations and complications of cooperation resulting from national caveats.[27]

In general, the joint task of a multinational unit is laid down in its foundation documents. But at the practical level of daily cooperation, it turns out that national operational principles, military traditions and military identity largely determine how this joint task is interpreted. Is it derived from an offensive use of force, from a combat mission, or rather from a defensive concept such as peace consolidation and stabilization? These questions need answering before a military operation can begin. In order to ensure that their soldiers will not participate in any operation inconsistent with their national concepts, many states set up formal limitations on what their soldiers are allowed to do (national caveats), which then act as a brake on the whole mission.[28] As a result of these numerous limitations to the mandates of multinational forces, European missions such as the one in Africa in 2006 (DR Congo) and 2008/2009 (Chad/Central African Republic) were of a certain symbolic value, but militarily not particularly efficient.

Multinational military structures are based on agreements between states that regulate the rights and duties, the decision-making processes, procedures, principles of operation, etc. of all parties concerned. They tend to emphasize the limitations imposed by national legislation. As a result, multinational structures turn into highly complex bureaucratic constructs, which are nevertheless unable to bridge the gap between joint action and the nationally imposed 'red lines' that may not be crossed. This leads not only to frustration among soldiers, but also to less procedural efficiency.

In spite of the soldiers' professionalism, their goodwill and their readiness to work in a multinational context, experience shows that as soon as a situation becomes difficult, confusing or dangerous, soldiers will resort to their national structures and behave and act like they used to in a national environment. This undermines the common cause and leads to partial disintegration of cohesion or to the duplication of multinational structures by national networks. Soeters *et al.* discovered that withdrawal into national networks and, consequently, a lower degree of cooperation occurs more frequently when a small number of equally strong military cultures is involved.[29] Abel was able to prove that this also applies to the German-French Brigade.[30]

In contrast to this, multinational procedures work much more smoothly if one national military culture is dominant and therefore shaping the behaviour of the members of the other armed forces, or if there is a colourful mix of many diverse military cultures. In the latter case, soldiers need to be pragmatic when working out joint procedures and cannot withdraw into their national corners. This point seems to be relevant for the future of military multinationality.

A European army?

In view of the dominant role played by national regulations and the limited scope of action in a multinational environment, a stronger harmonization of those regulations may be helpful. But then again, considering how important armed forces are to their national governments, such aspirations are unlikely to be crowned with success. The establishment of common European Armed Forces – a recurring subject of political and academic debate[31] – may be a way out of this dilemma. Such a step would not only be of high symbolic value for the integration of Europe. It might also spare the participating countries many of the problems discussed above, help reduce the military spending required for the upkeep of 28 national armies, and eventually improve the EU's capability to act in the field of regional and global security policy. Given the fact that 22 out of 28 EU member states are also members of NATO, enhanced European military capabilities could also be beneficial to the Alliance.

Although the idea of European Armed Forces is not yet very popular all across the continent, this option is presented in the final report submitted by the 11 Ministers of Foreign Affairs of the 'Future of Europe Group' on 17 September 2012:

> We believe that once the Euro crisis has been overcome, we must also improve the overall functioning of the European Union. In particular, the EU must take decisive steps to strengthen its act on the world stage. This should be tackled beyond and separately from EMU reform.[32]

The measures to improve the European Union's performance in regional and global security policy therefore should also aim 'for a European Defence Policy with joint efforts regarding the defence industry (e.g. the creation of a single market for armament projects); for some members of the Group this could eventually involve a European army.'[33] It seems worthwhile to consider in the next steps under what conditions such a project would be feasible.

Framework conditions

Such armed forces would require a common European framework, not to mention the numerous political preconditions and the strong public support needed for the development of their internal structures. But since the relationship between the soldier and the nation state has always been so special, this would be an extremely difficult task. One of the most vital preconditions for success would be an increase in commonalities and a decrease in differences. It will not be possible to create European Armed Forces by mixing national contingents. They will have to be built from within, based on new formations. Any decision in favour of establishing European Armed Forces will inevitably require further decisions on common (command) structures, on career tracks, training facilities, and a unified service law and remuneration system. These are the aspects that refer to the external framework of European Armed Forces and, from a legal point of view, are not too difficult to implement, since there are already models to emulate (such as the EU bureaucratic system in the Commission and other supranational bodies, or the EU External Action Service).

The much more difficult issue is the legitimacy of European Armed Forces with respect to the soldiers, as well as with respect to the societies of European states. Even though military personnel would be recruited exclusively on a voluntary basis, it remains unclear which authority would be entitled to demand that soldiers risk life and limb, and for what purpose such a sacrifice could be demanded. In addition, the European societies would need to know by whom and in whose name their armed forces could be deployed. There is indeed a lot of scepticism in this respect in Europe. The decision-making processes within the CSDP, which is based on the Charter of the United Nations, do offer a high degree of legitimacy, but they lack a parliamentary dimension that could be remedied by including the European Parliament in the process. This type of formal legitimacy would at least prevent any abusive, improper, indiscriminate or wilful use of European Armed Forces, so this aspect does not present an insurmountable obstacle either.

The real challenge lies in the development of the 'internal legitimacy' of a European military culture, the raison d'être for the armed forces as a military organization, and the reason why soldiers join it, identify with it and take action. This is all the more difficult, since 'Europe' – as opposed to the 'nation'– is perceived as something abstract, something not related to the mind-set of the individual. What it takes to build this legitimacy is shared experiences and a long time spent together so that the trust in one's own abilities, in the political and military leadership, and in one's comrades can grow. Educational and training facilities run by the European Armed Forces for the development of military action and doctrine could accelerate such a process. This is not a hopeless undertaking, because national debates on the subject are also confronted with the challenge of making soldiers and societies understand the legitimacy and the reason for multinational missions abroad in the context of NATO, the EU or the UN.[34] Changes in the security environment in Central Europe have led to a different perception of the military, and national defence is no longer seen as their only task. National contingents have gone on military missions abroad to defend European or transatlantic interests that are rather abstract in nature. Some countries, such as Luxembourg or Belgium, have invited EU citizens to join their armed forces, Spanish and French forces also accept citizens from other countries into their ranks, and at the beginning of 2011, Germany started considering opening up the Bundeswehr to citizens from other nations.[35] As one can see, the problems of the EU and those of the nation states are more or less the same, only on a different scale.

Approaches to realization

So what are the options for the creation of European Armed Forces? The first steps in such a process would be taken in a structure that is complementary to the national armed forces. Because even in countries open-minded enough to welcome foreign nationals into their armies, a complete abolition of national armed forces would be out of the question. To have European armed forces as a supplement, however, is considered reasonable.[36] During the initial stage and probably well beyond, national armed forces would be the most important recruitment pool for European structures. Basic military skills, as well as career-oriented training for officers, non-commissioned officers and specialists would first be offered at the national level. Later, soldiers would be able to apply for service in the European Armed Forces. This model allows for the selection of soldiers from different ranks and with a variety of special skills according to a procedure established beforehand. But since it is difficult enough to attract highly qualified volunteers into armed service, many national armed forces would be reluctant to let the best and the brightest leave to join European forces.

All in all, it seems that the creation of European Armed Forces on the basis of an agreement among the 28 member states is very unlikely. Taking into account the EU's experience with common decision-making, it is much more probable that a group of vanguard states will volunteer and give it a try. The 'permanent structured cooperation' laid down in the Lisbon treaty could serve as the basis for the establishment and implementation of common military structures. Just as with any other multinational body, a treaty under international law would be required to regulate all the complex issues involved, and their legal, political and social implications. The treaty itself would have to be very comprehensive, because apart from the military arrangements, new administrative structures would have to be created as well. The Eurocorps might serve as the core of European Armed Forces, all the more since it has proved to be very capable and ready to act in various missions in the Balkans and Afghanistan.[37] Its five 'framework' nations (Belgium, Germany, France, Luxembourg and Spain) could form the vanguard group. The associated nations represented in Strasbourg, particularly Poland, but also Italy, Austria, Romania, Turkey and the US, have already worked out flexible mechanisms of cooperation with the corps.

In summary, the establishment of common European Armed Forces appears as a project that faces many political, cultural and administrative difficulties but that is feasible eventually.

Prospects for a stronger Europe

Europe has to find answers to the security challenges it faces. It remains to be seen in which direction Vladimir Putin will steer Russia after openly displaying his geo-strategic power policy course since the beginning of 2014 in Crimea and in Ukraine. In any case, Moscow's confrontational and expansionist policy should be a wake-up call for Europe's capitals and make them realize that traditional security risks still exist and that common strategies are required to manage those risks. It is also difficult to say to what extent the US will stay engaged as a European security provider after its pivot to Asia. In any case the Europeans will have to shoulder more burdens for their own security, as a strong European pillar in a renewed transatlantic alliance or for the sake of its own capacity to political action. The warning that then US Secretary of Defense Robert Gates gave to the allies in 2011 still resonates in the European capitals:

> The non-U.S. NATO members collectively spend more than $300 billion U.S. dollars on defense annually which, if allocated wisely and strategically, could buy a significant amount of usable military capability. Instead, the results are significantly less than the sum of the parts.[38]

In these days of financial constraints, however, it seems very unlikely that the European governments are inclined to significantly increase their national defence budgets. A more efficient allocation of fewer financial resources will be a must for almost all European states; governments will be forced to keep on pooling capabilities and forces, and to share these resources among themselves (*pooling and sharing*). The nation state's exclusive right to make use of its military forces will erode, as will the principle of national defence as the most important source of legitimacy for military service and military action.[39] Against this backdrop the establishment of European Armed Forces may be a policy option that is difficult to realize but by no means a utopian endeavour.

At the same time, the path ahead is full of trials and tribulations. The European Union will have to convince its citizens that it is indeed a capable institution able to offer more than national structures can. But confidence in the EU has been dwindling for years. Will EU citizens be ready to entrust their militaries and their security to the European Union after having already given up national symbols such as their currencies? That remains to be seen.

Still, the first steps have been taken and there is more to come. The multinational arrangements mentioned above have proved to be significant, although they are far from perfect and, apart from being more suited to low-intensity conflicts to begin with, these units will find it difficult to fulfil military missions in a consistent and coherent manner.

Europe's role as a security actor will remain limited if the old cooperation models persist. European Armed Forces, once they have been tried and tested, could help fill the capability gap in Europe's toolbox.

Europe has been trying to establish common security structures for more than 60 years but, unlike in 1954, when the European Defence community failed, there is no attempt being made now to turn a vision into a political fact. This time, an idea is taking shape and being developed gradually, always open to changes and always allowing for a change in pace. So this time it will be easier to make the states and societies of Europe familiar with a new type of common armed forces and perhaps increase willingness to promote military and security cooperation to new levels and in new structures in order to face the new challenges confronting Europe today.

Notes

1. European Council, 'Conclusions of the European Council' (19/20 December 2013), Part I, para. 1–22, www.consilium.europa.eu/uedocs/cms_Data/docs/pressdata/en/ec/140245.pdf (accessed 25 October 2014).
2. The White House, 'Remarks by President Obama to the Australian Parliament', 17 November 2011, www.whitehouse.gov/the-press-office/2011/11/17/remarks-president-obama-australian-parliament (accessed 25 October 2014).
3. See Katarzyna Zysk (Chapter 9, this volume).
4. NATO, Message from Turnberry. North Atlantic Council Ministerial Meeting 7–8 June 1990, www.nato.int/docu/comm/49–95/c900608b.htm (accessed 25 October 2014).
5. NATO, London Declaration On A Transformed North Atlantic Alliance. Issued by the Heads of State and Government participating in the meeting of the North Atlantic Council, 1990, www.nato.int/docu/comm/49–95/c900706a.htm (accessed 25 October 2014).
6. Charles Krauthammer, 'The Unipolar Moment', *Foreign Affairs*, vol. 70, no. 1, 1990/1991, 23–33.
7. Western European Union, Petersberg Declaration, II, 2, 1992, www.weu.int/documents/920619peten.pdf (accessed 25 October 2014).
8. Gerhard Schröder, 'Rede von Bundeskanzler Schröder auf der Conférence de Montréal am', 25 June 2002, www.deutsche-aussenpolitik.de/daparchive/dateien/2002/07024.html (accessed 25 October 2014).
9. Gunter Hauser, 'The Common Security and Defense Policy (CSDP) – Challenges after "Lisbon"', in Sven Bernhard Gareis, Gunter Hauser and Franz Kernic (eds), *The European Union – A Global Actor?* (Berlin and Toronto: Opladen, 2013), 32–51; Lisa Karlborg, 'EU Military Operations: Structures, Capabilities, Shortfalls', in Sven Bernhard Gareis, Gunter Hauser and Franz Kernic (eds), *The European Union – A Global Actor?* (Berlin and Toronto: Opladen, 2013), 88–107.
10. Bastian Giegerich, 'Military Transition in the CSDP', in Sven Bernhard Gareis, Gunter Hauser and Franz Kernic (eds), *The European Union – A Global Actor?* (Berlin and Toronto: Opladen, 2013), 75–87, at 75.
11. European Council, 'A Secure Europe in a Better World, European Security Strategy', Brussels, 2003, www.consilium.europa.eu/uedocs/cmsUpload/78367.pdf (accessed 25 October 2014).
12. European Council 2003, 9.

13 Ibid., 8.
14 European Council, 'Report on the Implementation of the European Security Strategy. Providing Security in a Changing World', Brussels, 2008, www.consilium.europa.eu/ueDocs/cms_Data/docs/pressdata/EN/reports/104630.pdf (accessed 25 October 2014).
15 Gustav Lindström, 'The European Security Strategy – The Way Ahead', in Sven Bernhard Gareis, Gunter Hauser and Franz Kernic (eds), *The European Union – A Global Actor?* (Berlin and Toronto: Opladen, 2013), 52–64, at 55.
16 Doyle McManus, 'Chuck Hagel: The Asia Pivot is Still On', *Los Angeles Times*, 30 March 2014, http://articles.latimes.com/2014/mar/30/opinion/la-oe-mcmanus-column-hagel-asia-20140330 (accessed 25 October 2014).
17 Oystein Tunsjø, 'Europe's Favorable Isolation', *Survival*, vol. 55, no. 6, 2013, 91–106, at 94. See also Austin Long (Chapter 13, this volume).
18 Aaron L.Friedberg, 'The Euro Crisis and US Strategy', *Survival*, vol. 54, no. 6, 2012, 7–28, at 17.
19 Anders Fogh Rasmussen, 'Building Security in an Age of Austerity. Keynote Speech by NATO Secretary General Anders Fogh Rasmussen', 2011 Munich Security Conference, 4 February 2011, www.nato.int/cps/en/natolive/opinions_70400.htm (accessed 25 October 2014); Christian Mölling, 'Pooling and Sharing in the EU and NATO', in Ina Wiesner (ed.) *German Defense Politics* (Baden-Baden: Nomos Verl.-Ges, 2013), 359–369.
20 Jolyon Howorth, *Security and Defense Policy in the European Union* (Basingstoke and New York: Palgrave Macmillan, 2007), 100.
21 NATO, 'The Alliance's New Strategic Concept', 1991, section 53, www.nato.int/cps/en/natolive/official_texts_23847.htm (accessed 25 October 2014).
22 Heike Abel and Marc-Randolph Richter, 'Militärkooperation im deutsch-französischen Alltag – Einflussfaktoren und Probleme aus Sicht der beteiligten Akteure', in Nina Leonhard and Sven Bernhard Gareis (eds) *Vereint marschieren – marcher uni. Die deutsch- französische Streitkräftekooperation als Paradigma europäischer Streitkräfte?* (Wiesbaden: VS Verlag für Sozialwissenschaften, 2008), 137–182.
23 Eva Feldmann and Sven Bernhard Gareis, 'Polens Rolle in der NATO. Zur Bedeutung externer Hilfen bei der Stabilisierung Osteuropas', *Zeitschrift für Politikwissenschaft*, vol. 3, 1998, 983–1005; Marybeth Peterson Ulrich, 'Militärische Multinationalität und die Streikräftereformen in Mittelosteuropa', in Sven Bernhard Gareis and Paul Klein (eds) *Handbuch Militär und Sozialwissenschaft*, 2nd edn (Wiesbaden: VS Verlag für Sozialwissenschaften, 2006), 424–434.
24 Karlborg, 'EU Military Operations', 88.
25 Heiko Biehl, Bastian Giegerich and Alexandra Jonas (eds), *Strategic Cultures in Europe. Security and Defense Policies Across the Continent* (Wiesbaden: Springer Fachmedien, 2013), 7.
26 Sven Bernhard Gareis and Kathrin Nolte, 'Zur Legitimation bewaffneter Auslandseinsätze der Bundeswehr – politische und rechtliche Dimensionen', in Sabine Jaberg, Heiko Biehl, Günter Mohrmann and Maren Tomforde (eds) *Auslandseinsätze der Bundeswehr – Sozialwissenschaftliche Analysen, Diagnosen und Perspektiven* (Berlin??, 2009), 27–50, at 40.
27 Sven Bernhard Gareis, Ulrich vom Hagen, Per Bach and Adam Kolodzieczyk, *Conditions of Military Multinationality. The Multinational Corps Northeast in Szczecin* (Strausberg: SOWI, 2003), 63; Abel and Richter, 'Militärkooperation im deutsch-französischen Alltag', 137–182; Ulrich vom Hagen, Paul Klein, René Moelker and Joseph Soeters, *True Love. A Study in Integrated Multinationality within 1st (German-Netherlands) Corps* (Strausberg: SOWI Forum International No. 25, 2003); Joseph L. Soeters, Delphine Resteigne, René Moelker and

Philippe Manigart, 'Smooth and Strained International Military Co-operation', in Ulrich von Hagen, René Moelker and Joseph Soeters (eds) *Cultural Interoperability. Ten Years of Research into Co-operation in the First German-Netherlands Corps* (Strausberg: SOWI Forum International No. 27, 2006), 131–162, at 131.
28 Robert Bergmann, 'Multinationale Einsatzführung in Peace Support Operations', in Sven Bernhard Gareis and Paul Klein (eds) *Handbuch Militär und Sozialwissenschaft*, 2nd edn (Wiesbaden: VS Verlag für Sozialwissenschaften, 2006), 374–379.
29 Soeters *et al.*, 'Smooth and Strained International Military Co-operation', 131.
30 Abel and Richter, 'Militärkooperation im deutsch-französischen Alltag', 183.
31 Jürgen Groß and Andreas Weigel, 'Fernziel: Europäische Armee', *Sicherheit und Frieden*, vol. 1, 2009, 60–62, at 60.
32 Future of Europe Group, 'Final Report of the Future of Europe Group of the Foreign Ministers of Austria, Belgium, Denmark, France, Italy, Germany, Luxembourg, the Netherlands, Poland, Portugal and Spain', 2012, 3, www.auswaertiges-amt.de/cae/servlet/contentblob/626322/publicationFile/171783/120918-AbschlussberichtZukunftsgruppe.pdf (accessed 25 October 2014).
33 Future of Europe Group 2012, 7.
34 Gerhard Kümmel, 'Militärische Aufträge und die Legitimation der Streitkräfte', in Sven Bernhard Gareis and Paul Klein (eds) *Handbuch Militär und Sozialwissenschaft*, 2nd edn (Wiesbaden: VS Verlag für Sozialwissenschaften, 2006), 104–111, at 104.
35 Die Welt, 'Guttenberg holt Ausländer in die Bundeswehr', *Die Welt*, 13 February 2011, www.welt.de/politik/deutschland/article12530529/Guttenberg-holt-Auslaender-in-die-Bundeswehr.html (accessed 25 October 2014).
36 Sven Bernhard Gareis, Paul Klein, Giulia Aubry and Barbara Jankowsky, 'Opinion publique et défense européenne en Allemagne, en France et en Italie', Paris, 2005.
37 Paul Klein, 'Das Eurokorps', in Sven Bernhard Gareis and Paul Klein (eds) *Handbuch Militär und Sozialwissenschaft*, 2nd edn (Wiesbaden: VS Verlag für Sozialwissenschaften, 2006), 416–423.
38 Robert M. Gates, 'The Security and Defense Agenda (Future of NATO)', 2001, www.defense.gov/speeches/speech.aspx?speechid=1581 (accessed 25 October 2014).
39 Mölling, 'Pooling and Sharing in the EU and NATO', 359f.

9 Managing military change in Russia

Katarzyna Zysk

Whilst numerous Western governments are reducing their defence spending and are searching for further reductions, Russia, in contrast, has been systematically increasing its defence investments. Since Vladimir Putin came to power, and especially since his second presidential term (2004–2008), there has been a pronounced effort to restore Russia's great power status and bring the outdated Soviet era military up to modern standards. The top political leadership has demonstrated the will and financial commitment to improve the state of the armed forces through what has become one of the most comprehensive military modernization programmes in Russia's history.

The Russian authorities attempted to reform the armed forces several times following the collapse of the Soviet Union. The military reform under Boris Yeltsin's presidency was, however, focused merely on troop reductions and changes in the set-up and number of services. This minimalist approach continued during Vladimir Putin's first years as President in the early 2000s.[1]

It has become imminently clear, however, that the military organizational structure, operational doctrines and means of Soviet provenance, oriented towards a traditional large-scale war, were not only ill suited to the radically changed security environment; they were also unsustainable in terms of economic, material and human resources. Therefore, in his second term as President, Putin initiated a sweeping military reform and large-scale modernization aimed at creating, in essence, a new army. Although the Russian Armed Forces are still far from achieving many of their ambitious goals and are facing structural problems, the political leadership has succeeded in introducing a number of changes with a positive impact on the overall state of the defence sector.

This chapter examines how the process of military change has been managed in Russia. The analytical framework is based on two levels of investigation: national strategic level changes and institutional level changes. Analysis is provided in two parts: the first identifies motives, drivers and enablers of military change, while the second examines processes, outcomes and barriers to military change. In particular, this chapter identifies Russian intellectual debates regarding military innovation and military organization, perceptions and emulation of novel concepts, ideas and weapons technologies.

Strategic level changes

Deterioration and urgent need for modernization

The Russian authorities, military leadership and experts alike have called for urgent military reform and modernization for a long period of time. Among major reasons was the rapid and dramatic deterioration in the armed forces. The military top brass warned repeatedly that capabilities were sinking rapidly and that the military was unable to carry out many of its basic missions. The problems concerned the entire system, which was underfunded, underdeveloped and rapidly ageing due to a lack of investment and political interest. For instance, existing ships, on average 20 years old, were expected to be capable of very little in the next five to ten years. In addition, the nuclear arsenal was ageing rapidly. The Chief of the General Staff, Nikolai Makarov (2008–2012), noted that in 2008 over 50 per cent of the military's equipment was faulty.[2]

The defence industry, which remained unreformed after the fall of the Soviet Union, also faced a significant problem as it was unable to deliver quantity and quality on time, and for a reasonable price. The Russian leadership was aware that the industry significantly lagged behind the West in technological development. One example of this was the T-90 main battle tank, which was considered obsolete already at the time of its production, and significantly inferior in survivability, fire power and other features to its Western equivalents.[3]

In addition, the Russian authorities have struggled with the endemic corruption that plagues the armed forces, undermining the organization both materially and morally. Following an audit conducted by the Minister of Defence Sergei Ivanov in 2006 to 2007, Putin was informed that 40 per cent of the military budget was being stolen by generals and officers, making the MoD the most corrupt department.[4] This development aggravated the already undermined reputation of the military service, which was troubled by multiple problems including hazing, a high rate of combat deaths and suicides, low salaries and social benefits, and poor living conditions.

The 2008 war in Georgia was broadly presented by the Russian authorities and many experts as an external shock that revealed the serious level of decay and inadequacy in Russian defence.[5] The army still employed Soviet 'tank-centric' rather than 'network-centric' operational concepts, with an overwhelming use of ground forces and heavy artillery that compensated for shortages in high-tech, sophisticated weapons and technology. Among the lessons learned were poor interoperability, military planning and field operations, ineffective communication, command and control systems, as well as obsolete weaponry and equipment, including a lack of or inability to use guided munitions. The conclusion was that Russia was poorly prepared to fight a modern war, even against a far weaker opponent.

The modernization of the Russian Armed Forces had been ongoing since the middle of the first decade of the 2000s. It had already brought some positive results during the Georgian war as compared with Russia's war-fighting capability (i.e. number of forces employed, effectiveness of operations, duration, casualties) during the protracted second war in Chechnya (1999–2000[6]). However, there is no doubt that the way the Russian political and military leadership presented the experiences from the Georgian war helped them justify further, more radical and painful reforms they deemed necessary to implement.

Changes in civil–military relations

The determination of key Russian decision-makers and a consensus that military change was necessary have been central ingredients in the process. Likewise, the state's governmental system and political environment, including legal and regulatory frameworks, give the top political leadership broad room to manoeuvre and pursue desired goals.

Moreover, the top political authorities have broken the organizational rigidity in the defence sector by removing the most influential opposition and by opening up the military to outside influence. Introducing Anatolii Serdyukov as a defence minister (2007–2012) was of key importance. Contrary to his predecessors, Serdyukov managed to implement major change to the military in response to policy priorities. His task was facilitated by the fact that as a civilian from outside of the military organization, he did not have organizational stakes in existing military practices, equipment and structures.[7]

Having the support of the central decision-makers, Serdyukov won a number of key battles with the most conservative parts of the military brass. Within two years of his ministership, the vast majority of central opponents of the reform, including Chief of the General Staff (2004–2008) General Yurii Baluyevskii, had retired or were moved to other positions.[8] The backing of the top political leadership was key to overriding institutional conservatism, to increase the responsiveness of the military to civilian policy and the ability of the latter to effect military innovation. Moreover, Serdyukov expanded civilian control over the armed forces and made an attempt to divide the civilian (responsible for financial, economic and supply questions) and military branches of the establishment, which helped him to keep a better grip over the armed forces.[9]

Economic component of the military modernization

Economic growth in Russia, embedded in massive income from energy sales and high oil prices that have sky-rocketed since 2003, has been a major enabler of military change, as well as an incentive for the expansion of foreign policy ambitions. It has allowed a sharp and systematic increase in defence spending since the early 2000s. For instance, in 2008 the defence budget reached 3.3 per cent of GDP, and in recent years 3.9 per cent in 2012 and 4.1 per cent in 2013; in 2000, Russia spent US$9.2 billion on defence, while in 2013, spending increased to US$87.8 billion.[10] Until the early 2000s, state armament programmes (GPV), which planned for arms procurement, research and development, modernization and repair of military technology, were either never realized (in the 1990s), or were implemented to a minimal degree (1996–2005).[11]

The situation changed however, with the GPV of 2006 to 2015, which planned on spending US$169 billion on armament.[12] Despite economic problems caused by the 2008 global financial crisis, the Russian authorities have continued implementation of the programme. The defence spending on the following state armament programme for the period up to 2020 has further increased to more than US$670 billion (20 trillion roubles).[13] In addition, Russia plans to commit US$85 billion (3 trillion roubles) on modernization of the defence industry by 2020.[14] These numbers make Russia the third largest defence spender in the world.[15] Despite economic difficulties, the authorities promise to increase the defence budget by US$19 billion in 2015.[16] The likelihood of this trend continuing in the near future is, however, questionable. As it will be argued in the second part of this chapter, the economic challenges are likely to pose a significant impediment to further military innovation.

Political philosophy and threat perceptions

The importance given to the armed forces in Russia has deep roots in Russian perceptions of the world order. Since Putin has come to power, attitudes towards the armed forces have undergone significant changes. Among major reasons were conclusions drawn from developments on the international stage. According to Russian decision-makers, the role of military force in international relations has not decreased following the end of the Cold War. Quite the contrary; its importance, especially in US and NATO's policies, has grown systematically. They point to NATO's eastward enlargement, the use of force in Bosnia, Kosovo, Afghanistan, Iraq, Libya and Syria, as well as the increasing Western influence in the post-Soviet sphere as evidence and an argument for a military buildup.

Russian decision-makers make a direct connection between influence on world affairs and military strength. The military is seen as a key foreign policy tool in peacetime, in addition to its traditional wartime role.[17]

Rebuilding Russia's influence and great power status is therefore closely connected to strengthening the armed forces. The dominance of hard power in the Russian view of the international system is a foundation upon which military change and modernization have been built.

The picture of Russian threat perception is that of a long list without clear priorities, which, as discussed below, has an impact on the direction of the modernization programme. Official Russian threat perceptions are similar to Western assessments. Security is seen as complex and multidimensional, with a blurred line separating internal and external security. The definition includes a broad spectrum of societal and human dimensions, and is viewed in the context of economic growth, education, demography, health care, ecology, living standards, and even in the context of cultural development.[18] Russia also recognizes that many postmodern security problems such as international terrorism, cyber, biological and organized crimes have a transnational character and therefore require international collaboration and often non-military approaches.

On the other hand, key Russian security policy documents, as well as statements by the political and military leadership, simultaneously expose distinct patterns of continuity in strategic thinking. They view the security system as being predominantly about political–military power playing a determining role in world affairs today. A representative example was the draft of the European Security Treaty, presented by Dmitrii Medvedev in 2009.[19] It aimed at rearranging the existing security architecture that emerged after the Cold War, a system which Russia has never accepted. Russia views it as disproportionately weighted in favour of the US and NATO, and as one that failed to integrate Russia, offering the country little beyond a formalistic recognition.

The Russian authorities, including Foreign Minister Sergei Lavrov, have long postulated 'an acute need' for a new, flexible form of collective leadership by great powers, to 'steer the global boat into untroubled waters'.[20] The Russian leadership's view of the world order reflects a traditional realist culture. It is a variation of a system based on the balance-of-power principle and non-confrontational mutual containment, which was established during the Congress of Vienna in 1815. In an address in September 2007, Lavrov called for a new 'concert of the powers of the 21st century'. He underlined continuity in Russia's foreign policy and argued that 'Russia has now borne a considerable share of the burden of equilibrium maintenance in European and world politics for 300 years'. As he put it, in periods when Russia gave up this responsibility, it 'tended to lead the continent to catastrophe'.[21]

A central pillar in Russian political philosophy is the idea of 'spheres of privileged interests'.[22] Russia supports the classical Westphalian order, with its traditionally defined state sovereignty, core principle of territoriality, clear distinction between domestic and foreign affairs, and principle of non-interference in internal affairs. Western military interventions, unauthorized by the UN Security Council, appear as hegemonic attempts and irresponsible destabilizations of the very foundation of the international system, with potentially unpredictable consequences.[23]

In the Russian understanding, sovereignty is, however, not so much a right as it is a capacity for economic independence, military strength and cultural identity. According to this definition, only great powers can be fully sovereign. While the post-Soviet states can be independent and self-governed, their independence derives from relations they develop with great powers, not from their own strengths. Russia's attempts at reintegrating the post-Soviet space, by creating the Collective Security Treaty Organization, the Custom Union and the Eurasian Economic Union, is at root driven by the notion that if Russia does not fill the space, others will.

A natural sequel to this reasoning is that the Russian foreign policy and military doctrine allows for use of force in case of encroachment upon its sphere of influence, and reserves the right to protect Russian citizens beyond the borders of the Russian Federation, what the military interventions in Georgia and Ukraine demonstrated in practice.[24]

Russia has traditionally taken into account a wide range of potential threats along its borders. The Russian elites continue to remain preoccupied with the West and its intentions. The concern has been increasingly pronounced since Putin's second presidential term. The key Russian post-Soviet foreign policy concept – the vision of the multipolar world presented by Minister Yevgenii Primakov in 1996 – aims, *inter alia*, at undermining the Western dominance.[25]

In a lecture given in November 2011, General Makarov argued that nine of the 12 main military threats to Russia were expected to be coming from the US, NATO and the EU.[26] The list includes dangers posed by strengthened foreign military presence in the proximity of Russian borders and adjacent areas, including NATO enlargement; Western military-technological superiority, in particular the development of high-precision and strategic conventional weapons, ballistic missile defence, and the miniaturization of combat and reconnaissance systems.[27]

Russia also faces a spectrum of real and immediate security threats along its southern border. Conflicts smoulder in Chechnya, Ingushetia and Dagestan, as well as in Abkhazia and South Ossetia, and Nagorno-Karabakh. Widespread corruption, high unemployment, organized crime and drug trafficking, in addition to the expansion of radical Islamism and terrorism, contribute to the destabilization of the Caucasus and Central Asia.

In the 2013 foreign policy concept, the Russian government acknowledged the shift of global power towards the Asia-Pacific region and announced pivot to Asia by turning to eastern markets, developing Siberia and the Far East, and by building a 'strategic partnership' with China. The development reflects an evolution in the perception by the Russian elites, which remain nevertheless focused on the West as a counterbalance to US dominance.[28]

As for China, the country was barely mentioned in key Russian security documents, and then only in the context of international cooperation.²⁹ Yet maintaining 'a reliable military force able to counter neighbouring China's numerical supremacy, rapidly growing economy, military strength and political influence has been a major goal in Russian military modernization programmes. Russian military leaders and experts argued that Russia had to prepare to fight a multi-million-man army in the east.³⁰ They see non-strategic nuclear weapons as a 'nuclear equalizer', compensating for conventional supremacy of potential opponents such as NATO and China.³¹ Similar views were expressed by former Commander of the Siberian Military District Lieutenant General Vladimir Chirkin. Referring to China's military strength, he argued that friendship is possible only among strong nations, those that are able to 'quiet down a friend with a conventional or nuclear club'.³²

Russia has been regularly rehearsing various security scenarios in the Asia-Pacific region by staging high-profile war-games, such as the 'Vostok' in July 2008, 2010 and 2013 and in September 2014. The potential opponent in the region is also the US and its allies. As Deputy Defence Minister Vladimir Popovkin argued in April 2010, one of the main missions for the *Mistral* ships which Russia planned to buy from France would be to protect the Kuril Islands against Japanese demands.³³ Russia has also shown a determination to maintain nuclear deterrence capability in the Pacific Fleet, although with mixed results thus far.³⁴ In the early 2000s, Russia concluded that limited nuclear strikes against a conventionally superior enemy might be effective only if combined with well-equipped, combat-ready general purpose forces, which served as yet another argument for military modernization.³⁵

The Arctic and Russia's northern borders are seen in Moscow as being among the most peaceful regions. Nevertheless, in response to the growing international interest in the Arctic, the activity and presence of Russian Armed Forces in the region have expanded systematically, since 2007 in particular.³⁶ The region continues to play a central role in Russia's traditional nuclear deterrence and naval strategy. In addition, as the Arctic Ocean opens up for increased human activity, the political and military authorities have called for improved control, surveillance and ability to exercise sovereignty and deterrence by enhancing the state's military presence.

A long-standing element in the Russian security thinking relates to the expected increase in global competition for dwindling energy reserves worldwide, in particular in the Middle East, the Caspian Sea, Central Asia, as well as in the Barents Sea and in the Arctic.³⁷ Russia expects that some actors may use military force in that rivalry, and that Russia could drown in such a conflict. The General Staff has argued for years that Russia, with its abundant natural resources, may become an object of a large-scale expansion.³⁸ In the opinion of the Chief of the General Staff, Valerii Gerasimov (2012–), the likelihood of the threat may increase by 2030.³⁹

In recent years, the Russian authorities have been increasingly concerned about the impact of the promotion of democratic institutions and the norm of humanitarian intervention on security of Russia and its 'near abroad'. The Russian leadership views the 'Western technology of colour revolutions', as they are called in the Kremlin, as a result of direct or indirect Western involvement aimed at overthrowing inconvenient political regimes. According to the Chief of The Main Operational Directorate in the Russian General Staff, General Vladimir Zarudnitskii, these interests include corroding the attacked country's economic and military potential, thus facilitating its conquest in the first place by decomposition from within rather than conquering it in the traditional way.[40]

The Russian military and political leaders argue that the problem not only concerns uprisings that swept away several regimes in Russia's 'near abroad', such as the 2003 Revolution of Roses in Georgia, the 2004/2005 Orange Revolution in Ukraine, and the 2005 Tulip Revolution in Kyrgyzstan. This is seen as a means of destabilizing states that are not aligned with the US in the Middle East, North Africa, Central Asia and South Asia. Since many of the Western interventions failed, they are viewed as a central source of terrorism, extremism, transborder crime and other problems that destabilize the international security environment.[41]

The Ukrainian crisis that erupted in 2013/2014 was as much about Russia's foreign and security policy goals and geopolitical vision, as about its domestic interests. The Maidan protests were also feared because of the possible spill-over effect on 'hearts and minds' in Russia and its 'sphere of influence'. The ultimate nightmare scenario for the current regime is a 'colour revolution' on Moscow's streets. As expressed in April 2014 by parliamentarian Andrei Lugovoi, 'Russia's enemies are constantly looking for weak spots in our ranks', and the government's critics are seeking to disrupt normal life in Russia, 'possibly supported or guided by the West'.[42]

While such views are often peddled by Russia as propaganda both at home and abroad, the obsession with the aims of Western power are deep-seated within the ruling elite. An example is the 2006 assessment by Vladislav Surkov, an influential ideologist and a close associate of Putin, in which he named 'colour revolutions' as one of the four key threats to Russia's sovereignty, along with international terrorism, lack of economic competitiveness and a direct military confrontation.[43] The massive pro-democracy, anti-Putin protests that erupted in Russia in 2011, in response to what were widely perceived as rigged elections, served as a reminder of such a scenario.

Russian assessment of modern warfare

Apart from Russia's own experiences gained in conflicts in the post-Soviet sphere, the intellectual debates on military innovation, military organization and perceptions of new concepts and weapons technologies, have often drawn lessons from Western experiences resulting in the partial emulation of some of their concepts and ideas.

In the early 2000s, the Russian MoD concluded that the armed forces were not prepared to fight in a contemporary warfare environment, which had changed fundamentally. This realization has served as an argument for pursuing military change and innovation.

The Russian strategic planners view conventional, large-scale state-on-state conflict as unlikely. The focus is now on asymmetric, non-contact conflicts, where air superiority, precision munitions, special operations forces (*spetsnaz*) and strategic mobility are key assets for victory.[44] Overwhelming manpower is no longer viewed as a prerequisite to create battlefield superiority, since modern warfare requires highly skilled personnel. The General Staff argues that the centre of wartime activities is moving from the traditional three-dimensional format (land–air–sea) to the fourth dimension, i.e. the cyber sphere and space.[45] It emphasizes the complexity of modern warfare, where the dividing lines between peace and war, and offensive and defensive actions are less distinguishable. A crucial consequence is that not only the military but also the whole population and social infrastructure become targets and participants. A natural follow-on to that reasoning is a broad spectrum of tools to be used in the conflict: military, economic, political-diplomatic, criminal structures, psychological and information warfare. In the assessment of the General Staff, rather than conquering a state through direct force, the use of these tools is aimed at weakening the enemy internally, facilitating its decomposition from inside.[46]

The General Staff views the 'colour revolutions' as a new type of Western warfare, called *An adaptive approach to the use of military force*. It includes such elements and stages as: (1) the search for or creation of a pretext for military intervention, such as the protection of civilians and foreign citizens, or accusations of the use of weapons of mass destruction – both of which allow the aggressor to maintain a positive image in the international community; (2) the use of military force to pressure governments and inflict chaos; military training of rebels by foreign instructors; the supply of weapons and resources to anti-government forces; the reinforcement of opposition units with foreign fighters supported by foreign states; the application of special operations forces and private contractors, disguised as rebels, making the identity of the aggressor unclear; (3) operations by a coalition of countries to get rid of government forces and assist armed opposition to seize power.[47]

Despite this critique, Russia has seemingly incorporated some of the elements of 'Western warfare' as perceived in Moscow into its own concept of operations. The 'lessons learned' bear a rather striking resemblance to the type of unorthodox, non-contact covert operations effected by the Russian high-readiness forces (special operations forces, airborne troops and naval infantry) in Crimea in March 2014. Many aspects of the employed tactics and doctrine have been old. An example of this was the old Soviet *maskirovka* (deception), aimed at creating ambiguity with regard to the presence and dislocation of forces and objectives of the operations, and providing the authorities with time and a plausible deniability while weakening the opponent from within, both physically and morally. Simultaneously, Russia combined several 'civilian' tools, such as security services, paramilitary, civilian 'self-defence forces', with political and diplomatic measures, economic pressure and informational warfare, in a more comprehensive approach.

Institutional level changes

New command and force structure

One of the lessons learned by the Russian MoD from past conflicts was that unified command and control, with more autonomous individual services concentrated in one pair of hands, as well as joint operations, have become increasingly important in contemporary warfare.[48] As a result, in 2010 six military districts were reorganized into four military districts/Joint Strategic Commands (JSC): West, East, Centre and South. JSC combine various types of services in the armed forces, as well as other components that did not belong to the MoD, such as the Federal Security Service (FSB) and Interior Troops. All services, including the Navy, are subordinate to the new commands within their respective geographical areas of responsibility, with the exception of Strategic Missile Forces, the Airborne Forces and Aerospace Defence Forces which are directly subordinated to the General Staff.

Likewise, the Soviet four-tier system (military district–army–division–regiment) was replaced by a Western-inspired three-echelon chain of command (military district/strategic command–operative command–brigade). Military leadership has signalled a strong interest in transitioning to an information-based and technologically equipped military with modern Command, Control, Communications, Computers, Intelligence, Surveillance and Reconnaissance (C4IRS).[49]

A central aspect of military change has been the reorganization of the force structure, especially in the army, which had remained almost intact since the Soviet period. As with almost all Western countries, Russia decided to move towards a smaller, more flexible, more lethal military, with a primary emphasis on brigades rather than on heavy divisions and armies. The force is to be fully manned, permanently combat-ready, aimed primarily at small- and medium-scale conflicts.

Initially, light, medium and heavy brigades were to be the basic military unit. At the end of 2012, there were 100 brigades.[50] Only the airborne forces and the strategic missile forces were to maintain the division structure. However, by May 2013, Minister Shoigu had re-established some of the divisions in the Western Military District: the famous Tamanskaya and Kantemirovskaya divisions, although the staffing level will probably be lower than those of a division. This move may be seen as populist given that both divisions have long and heroic, symbolically important histories, and their reorganization was unpopular among conservatives. It remains uncertain whether other divisions will be re-established, especially given limited human resources.[51] The MoD is contemplating other reversals of previous reforms, however, including returning to six military districts, potentially splitting the Western and Eastern Military Districts reportedly owing to command difficulties exposed during military drills.[52]

The Russian authorities have directed strong attention towards improving military training and exercises, including surprise inspections and drills, which have increased sharply and systematically in recent years. Interoperability across various types of armed forces and security services, high readiness and strategic mobility are seen as core force multipliers.

For instance, the exercise 'Vostok' conducted in July 2013 in the Military Districts East and Centre included, according to official sources, 160,000 soldiers, 1,000 military vehicles, 130 planes and 70 ships from the Pacific Fleet.[53] The following year, the drills involved most of the operational and tactical units of all troops and divisions of the Central and Eastern Military Districts and the Pacific Fleet, reaching 155,000 men.[54] Simultaneously, exercises were conducted in the Arctic and the European part of Russia, totalling approximately 200,000 soldiers, training on 14 exercise fields simultaneously, with several thousand armoured combat vehicles, several hundred planes and helicopters, and nearly 100 warships and auxiliary units, which made it comparable to large-scale Soviet-era drills.[55]

Changes in personnel structure

Another key aspect of military change in Russia is the professionalization of the armed forces. It has become increasingly difficult to rely on uncommitted recruits in building a modern military, given mounting technical demands and a growing need for skilled personnel able to operate the high-tech weapons that Russia is introducing. At first, conscription was to be abandoned altogether, but the authorities ultimately settled for a mixed system. In January 2008, conscription was reduced from two years to one year of service with the goal to recruit 425,000 contract soldiers by 2017.[56] As argued below however, this number appears overly optimistic.

Military modernization implies radical changes in the personnel structure. In line with a presidential decision, Russia's armed forces should comprise 1 million men (cf. in 2002: 2.3 million; 2008: 1.134 million), and up to 700,000 in a mobilization force.[57] The General Staff identified the need to reduce the number of senior and warrant officers who constituted 80 per cent of the armed forces.[58] The rank which the General Staff wanted to increase in terms of numbers was that of younger officers.[59] The aim was to create professional non-commissioned officers, and increase the contract-based corps of sergeants, junior leaders and specialists to approximately 400,000.

Serdyukov, referring to the 'global standards' of 7 to 20 per cent of staffing levels,[60] decided to make a bold move and cut the officer corps from 355,000 to 150,000 by 2012.[61] Radical changes have also affected the positions in the MoD and the General Staff, which was reduced from 22,000 to 8,500.[62] Serdyukov also eliminated the positions of many staff officers, such as lawyers and physicians.[63] He disbanded the warrant officers corps of 140,000 in 2009,[64] although Shoigu re-established it in February 2013, albeit in a limited form (55,000).[65] The numbers of units, bases, military cities and other objects were reduced accordingly. Financial means freed from the reductions were to be redirected to other central parts of the reform, such as procurement, and research and development.[66]

Makarov announced in December 2009 that the radical reductions were completed.[67] The leadership concluded, however, that the officer corps should be enlarged again, which resulted in the recall of 70,000 senior officers.[68] Such actions appeared inconsistent and had a negative impact on the perceptions of the reforms in the officer corps, giving an impression that Serdyukov did not know exactly what he was doing.

Radical changes in the force structure required an adequate military education system that could prepare the forces to apply new concepts and new technologies. Serdyukov reduced 65 various military education units to 16: three military educational centres, 11 military academies and two military universities. The idea was that an officer would gain a basic education before receiving the rank of a lieutenant, and then complete short-term courses later during service.[69]

Shoigu, however, reversed the changes.[70] While this has been met with a positive reception among the military, it is an open question as to what extent the system will be able to satisfy the needs of the modernized and reduced organization.

Defence procurement and the military-industrial complex

A fundamental part of the reform is the modernization of weapons, equipment and infrastructure, on which Russia plans to spend approximately US$670 billion (21 trillion roubles) by 2020. Seventy per cent of weapons and military technology is to be modernized by 2020, which is a very ambitious goal considering that Russia must increase the renewal of weapons from the current 2 per cent to 10.7 per cent a year. The plans appear more realistic if the word 'modern' also includes modernized older weapons and equipment.[71]

The Russian strategists' emphasis on air superiority in modern warfare is consistent with the high priority given to air power in the modernization programmes.⁷² The Air Force is to be modernized by '70 per cent or more' by 2020.⁷³ Like other military branches, the Air Force was severely worn down after the fall of the Soviet Union. The decay was made eminently clear especially during operations in Georgia, which demonstrated short range and a limited ability to coordinate with troops on the ground.

According to Russian sources, 170 new aircraft were delivered in 2012, 207 in 2013 and 300 in 2014.⁷⁴ These included Su-34 and Su-30SM attack aircraft, Su-35 fighters, Yak-130 jet-aircraft, attack aircraft trainers and fifth-generation fighter aircraft based on the prototype T-50. Purchasing plans also provide for 1,000 new military helicopters (Ka-52/Ka-52K, Mi-35, Mi-28N) and 200 new military transport planes. Along with modernized older aircraft, Russia will have approximately 2,500 aircraft and helicopters by 2020.⁷⁵ There is also a growing focus on unmanned aircraft. The General Staff promises Russian-made drones by 2020, but the domestic industry has been struggling with producing indigenous drones despite almost US$900 million (5 billion roubles) spent for that purpose.⁷⁶

By 2020, Russia also plans to complete the Aerospace Defence Forces (ADF); created in 2011, they are responsible for, among other tasks, air and missile defence and the operation of military satellites. Russia has also intensified the development of precision munitions, such as Kalibr and Club. The procurement plan stipulates that all ground forces missile brigades will receive Iskander-M ballistic missile systems by 2020, and the ADF brigades are to receive S-500 and S-400 anti-air missile systems.⁷⁷

Since 2007 in particular, Russia has devoted a considerable share of attention to modernizing the Navy. A fleet with global power projection capability has been regarded as a condition for Russia's recognition as a first-rate international power.⁷⁸ According to Putin, the GPV-2020 allocates five of the 20 trillion roubles for the Navy (*c.* US$177 billion), which is a threefold increase from the previous GPV–2015.⁷⁹

Russia is continuing a surface combat ship modernization programme, giving priority to capacities in the littoral zone, i.e. corvettes and missile ships of various classes, anti-submarine ships and minesweepers. Construction of the main ship in that class, the new *Steregushchii* corvette, continues with designers developing modernized versions. Russia is also building ships for long-range operations with the main asset being frigates of the *Admiral Gorshkov* class. Of the 20 planned by 2020, however, one is undergoing sea trials, and three other are being built. Since the construction is slow and expensive, Russia has also ordered several cheaper frigates of the *Admiral Grigorovich* class, a modified version of the *Talwar* class built for the Indian Navy. The plans for the aircraft carriers that were broadly advertised in 2008/2009 seem to have been abandoned for the time being given the realities of limited resources, while work on a new destroyer has not yet left the conceptual phase.⁸⁰

Although military modernization focuses on upgrading conventional weapons, this does not challenge the top priority given to nuclear capabilities, which continue to play a central role in Russian strategic thinking.[81] Russia plans to spend approximately US$48 billion (nearly two trillion roubles) by 2020 on upgrading the nuclear triad.[82] A significant portion of the budget is consumed by the construction of the fourth-generation strategic submarines (SSBN) of the *Borei* class, the future foundation of Russia's sea-based strategic forces. Two ships joined the Navy and three others of the eight to be built by 2020 are under various stages of construction.

Two of the *Borei* submarines are to replace the old *Delta III* SSBNs deployed with the Pacific Fleet in 2015. The construction of the base infrastructure in the settlement Rybachii on the Kamchatka Peninsula has been delayed, which angered Minister Shoigu, who issued a stern warning to those responsible, underlining the vital importance of the project.[83]

The defence procurement programme also gives high priority to the construction of smaller, fourth-generation nuclear-powered attack submarines (SSNs) of the *Yasen'* (or *Severodvinsk*) class, of which 'not less than eight' are to be built by 2020. Construction has also been delayed, as only one has been completed and three others are under construction. Russia is also building diesel submarines and modernizing existing non-strategic submarines, including *Oscar* class nuclear-powered cruise missile submarines and *Akula* SSN.

Russia's modernization programme includes upgrading land- and air-based strategic missile forces: new RS-24 Yars and Topol-M ICBM. According to government sources, Russia is in the process of developing a new type of ICBM. Development of a new strategic bomber, PAK-DA, which will replace the current Tu-95MS and Tu-160, is underway, but is not expected to be ready before 2025 to 2030. Meanwhile, Russia is upgrading older aircraft, including three Tu-160s.[84]

Nuclear weapons will remain central in the Russian military doctrine in the foreseeable future as the ultimate guarantee of the country's security and great power status. Still, a successful modernization of conventional capabilities, as well as an increasing focus on strategic conventional weapons and precision weapons in Russia and abroad, may in the long term challenge the position of the nuclear deterrent.

Newer elements in Russian strategic thinking are related to the development of network-centric defence and new technologies of communication, surveillance and reconnaissance, as well as modern command-and-control systems to improve tactics and operations.[85] The MoD has expressed interest in establishing a cyber-command with both defensive and offensive roles, including defending one's own and disrupting the opponent's communication and management systems, and influencing public opinion through the internet, TV and other mass media.[86]

Many of the official figures are difficult to confirm given the secrecy and limited transparency surrounding defence spending and procurement. Nevertheless, the government acquisition time frames appear in many cases unrealistic, despite significant funding. One of the main problems is that the modernization depends on the domestic defence industry. It will take considerable time to reverse the tendency towards low production capacity, insufficient quality control, exorbitant prices, deteriorating expertise and the mismanagement of resources that plagues the Russian military–industrial complex.

Under Serdyukov, the MoD showed increasingly less understanding for the state of affairs and ended in open conflict with the defence industry in 2011, which earned Serdyukov even more powerful enemies. The MoD decided to purchase some central components abroad, such as the *Mistral* amphibious helicopter carriers from France, drones from Israel, infantry vehicles and small arms from Italy, and combat training centres for brigades and smaller units in Germany. These purchases were not only important to rapidly satisfy the needs of the armed forces, but also to facilitate the modernization by giving the defence industry access to the latest technology and forcing it to become more competitive.

However, Shoigu announced the reversal of this trend and promised to limit purchases abroad and favour indigenous industry, through assisting in its modernization with such incentives as tax reliefs, low interest loans, and help with obtaining key components for major weapons systems. In addition, President Putin connects the development of the military-industrial complex to Russia's general economic future and position on the global market. The massive defence order is meant to become a source of a national technological innovation, fuelling economic growth by creating new jobs and new, high-tech products, and later by switching military production to new commercial products for domestic and foreign markets.[87]

A radical and unpopular reform of the defence industry under current unstable economic and political conditions seems unlikely. As Putin noted, about two million people work in the sector; together with their families, that number comes to almost seven million, which is a force the Russian authorities have to reckon with.[88]

Socio-economic impediments

The fact that military reforms and modernization are not made public fuels uncertainty, scepticism and anxiety in the military about the outcomes. Large parts of the defence policy, military budget, armament programmes and purchases are kept secret. Excessive secrecy limits scrutiny by the Parliament and public debate, leaving the process of shaping defence policy, budget and priorities to the defence bureaucracy. Russia's political and legal system, and regulatory frameworks help hide lobby groups' activities, which limit the effective use of resources and constitute a fertile ground for corruption. The situation has apparently not improved in recent years. In May 2011, Russia's chief military prosecutor acknowledged that as much as one-fifth of defence expenditure was stolen.[89]

As the experience over the past few years has demonstrated, the human resource situation in Russia is another obstacle to reaching modernization goals. Russia has struggled to attract sufficient numbers of contract soldiers, which in December 2013 were reported to total 220,000 (cf. the goal: 425,000 in 2017).[90] There are several reasons for this. A dramatic demographic decline in Russia since the late 1980s means that there are steadily fewer eligible for military service. In addition, health challenges, drug or alcohol abuse and criminal records further reduce the available pool of recruits. A report from 2012 noted that 60 per cent of teenagers are not subject to military service for medical reasons, and 50 per cent of those called up have health restrictions.[91] In addition, draft dodging is widespread: each year some 190,000 do not report to enlistment offices; in 2013 the number increased reportedly to 250,000.[92] The MoD also has to compete for recruits with other armed and security services (e.g. the FSB, Interior Troops, Federal Penitentiary Service).[93] Military service conditions, even if they are improved as an important part of the modernization programme, continue to scare away potential recruits.[94] In 2012, the shortfall from the 1 million goal was at least 200,000. Several reports show that the lack of personnel in some brigades has led to understaffing of 30 to 50 per cent.[95] The continued decline challenges the professionalization plans and possible further reductions in the number of units in the future force structure.

Institutional conservatism is another impediment to the modernization efforts. The military remains one of the most conservative institutions in Russia, resenting civilian interference. Many civilians brought to the MoD by Serdyukov (the so-called 'battalion in skirts') have been to a large degree replaced by military in uniforms.[96] A weakened civilian control in the MoD is likely to make it harder to implement changes that are unpopular among the military.

The condition of the Russian economy constitutes another factor of uncertainty. The national economy is on the verge of stagnation: the GDP growth is expected to be close to zero (0.8 to 0.2 per cent). The IMF and the World Bank forecast that investment in Russia will remain weaker for longer. The prospects are characterized by significant risks given low oil prices, several rounds of the Western sanctions and Russian counter-sanctions, as well as the weakening value of the rouble, lowered domestic demand and high inflation. The World Bank's baseline scenario for Russia is one of stagnation with an expected 0.3 to 0.4 per cent growth in 2015/2016.[97] A low-level recession should not be excluded.

In order to boost economic growth, Russia would have to implement major structural reforms, i.e. improve economic institutions to ensure stable public finances; improve education and infrastructure to increase productivity and competition; and create a predictable investment environment to attract private capital and generate change in consumer sentiment.[98] Since such reforms are not on the agenda to a sufficient degree, defence spending is likely to be negatively affected, unless the government takes resources from, for example, the health budget, education and social spending. The result of the disproportionate ratio of warfare versus welfare may be, however, political instability in Russia.[99]

Finally, in October 2014, Finance Minister Anton Siluanov maintained that the economic situation had worsened sufficiently as to require a reduction in the next armament programme for the period up until 2025. Siluanov argued that it should be smaller because its design was based on earlier optimistic macro-economic forecasts, and Russia 'simply cannot afford this now'.[100]

Conclusions

The Russian Armed Forces and military thinking are undergoing a potentially historic transition, which has accelerated significantly since 2008. Military change has occurred in overlapping processes including speculation, experimentation and implementation, and spanning incremental modernization and discontinuous transformation.

Driving such change is an understanding among the ruling elite that traditional great power politics remains at the heart of the international system. Thus, maintaining a strong military is a prerequisite to ensuring Russian prestige and international recognition, great power status and corresponding influence on world affairs. Military weakness has been identified as a major reason for Russia's repeated setbacks on the international stage following the fall of the Soviet Union.

Western and particularly American power and perceived hegemonic ambitions has provided further impetus for military change. Likewise, Russia has acknowledged the need to radically transform the armed forces, in line with the fundamentally changed nature of contemporary warfare. Sustained by consistent economic growth, defence spending has sharply increased, accelerating the military modernization process.

The role of the individual level factor for military change has been decisive. The top political leadership has demonstrated the capacity to mobilize resources, converge institutional norms and foster the conditions needed to pursue innovation. It has limited conflicts between various competing security agencies, strengthened civilian control over the military and opened it up to outside influence. The leadership's ability to control and coordinate security and defence has been strengthened by making the Security Council the key institution for supervising security affairs and strategic planning, thus undermining the previously dominant position of the General Staff.[101] Another step in that direction was creating a National Management Centre in 2014 aimed at improving coordination of the numerous agencies engaged in security and defence policy-making.

Military change and modernization has resulted in strengthening the Russian Armed Forces, which have become a more effective and flexible foreign policy tool, despite a number of remaining shortcomings. The overall capability has improved. New military formations have been established, together with doctrinal revisions to accommodate new ways of war (e.g. joint operations, strategic mobility, permanent readiness and rapid reaction), and massive acquisition of new weapons and technology. Training and exercises have increased sharply in scope, quantity and quality, and have brought positive results.

As the 2014 operations in Crimea have demonstrated, lessons learned from observation of modern conflicts and Russia's own experiences, as well as the effect of the military modernization, have resulted in a new approach to warfare and the better coordination of a spectrum of tools aimed at achieving one strategic goal. The military component – high-readiness and special operation forces – has been more professional, disciplined, clearly better trained and equipped with modern weapons and military technology. The comprehensive approach to the use of a variety of military and civilian tools has been tested successfully and is likely to influence further doctrinal changes and military modernization goals.

The future of Russian military change and modernization remains, however, uncertain. The state's political and legal system, and regulatory frameworks constitute a breeding ground for endemic corruption that undermines military modernization efforts across a variety of areas. Excessive secrecy and limited scrutiny exacerbate the problems, while Russia's military-industrial complex is proving highly resistant to modernization. Furthermore, institutional rigidity and demographic problems make military change a particularly complex endeavour.

Numerous time frames and the scope of many of the giant procurement programmes are unrealistic, given the limited funding and production ability of the Russian defence industry. Economically and militarily, Western sanctions following the conflict in Ukraine have created new impediments to modernization efforts, limiting opportunities for the Russian Armed Forces to engage in – and learn from – exercises with Western militaries. The Russian government has had to revise partly the current armament programme and postpone some of the expenditures for the defence order beyond 2016.[102] Even if Russia's centralized power structure allows for greater freedom in decision-making when compared with democracies in Europe, there is awareness among some members of the government that there is a limit to what can be committed to defence spending without challenging popular support and economic stability.

Their views, however, have not garnered much traction thus far. The top decision-makers promise further increases in defence spending and have adopted a range of new laws and policies aimed at restricting public rallies, cracking down on opposition, NGOs and remaining independent media pockets, in particular since Putin's return as President in 2012. In a pre-emptive move to face potential broader public unrest, in May 2014 Putin appointed his loyal ally, General Viktor Zolotov, as Commander of the Internal Troops, which would be used in such a scenario. In addition, the MoD has also been taking contingency measures. In November 2011, Makarov argued that Russia's army has to prepare for a possible 'colour revolution' by, among other measures, increasing the number of snipers in military units, as these are the most relevant forces for clashes in cities.[103]

The process of military change in Russia is characterized by strategic incoherence: problems with connecting strategy, policy, procurement, implementation, as well as available economic and human resources in a coherent manner. The debates on military innovation, perceptions of novel concepts, strategic and operational rationale, and utility relative to the security conditions continue to evolve. Persistent tendencies towards conservatism in strategic thinking in terms of old (Soviet) concepts, remains of organizational structures, practices and equipment, merge in theory and practice with modernizing, mostly Western-inspired, approaches. Inconsistencies in military thought translate into various contradictions in the orientation of the military. Shoigu's reversals and adjustments to several of the already implemented reforms by the former team seem to confirm what Makarov acknowledged in 2011, namely that the debate on what kind of armed forces Russia should have is still going on and is yet to be decided.

Notes

1 Marcel de Haas, *Russia's Military Reforms: Victory after Twenty Years of Failure*, Netherlands Institute of International Relations, Clingendael Papers no. 5, The Hague, 2011, 5; Carolina Vendil Pallin, *Russian Military Reform. A Failed Exercise in Defence Decision Making* (Abingdon: Routledge, 2009), 64–117.
2 Nikolai Makarov, lecture at the Military Scientific Academy in Moscow, published partly in 'Reforma dayet nuzhnye rezultaty', *Nezavisimoe voennoe obozrenie*, 17 February 2012.
3 Mikhail M. Rastopshin, 'U rossiiskoi armii net sovremennykh tankov', *Nezavisimoe voennoe obozrenie*, no. 43, 5–11 December 2008.
4 Stephen Blank, 'Civil–Military Relations in Contemporary Russia', in Stephen Cimbala, *Civil–Military Relations in Perspective: Strategy, Structure and Policy* (Farnham: Ashgate, 2012), 58; Dale Herspring, 'Anatoly Serdyukov and the Russian Military. An Exercise in Confusion', *Problems of Post-communism*, November to December 2013, 43. See also Tor Bukkvoll, *Russian Military Corruption – Scale and Causes*, FFI Report (Kjeller, 2005).
5 See *Russia's New Army*, ed. Mikhail Barabanov, David Glanz and Center for Analysis of Strategies and Technologies (Moscow, 2011); Carolina Vendil Pallin and Fredrik Westerlund, 'Russia's War in Georgia: Lessons and Consequences', *Small Wars and Insurgencies*, vol. 20, no. 2 (2009), 400–424; Keir Giles, 'Russian Operations in Georgia: Lessons Identified versus Lessons Learned', in Roger N. McDermott, Bertil Nygren and Carolina Vendil Pallin, *The Russian Armed Forces in Transition. Economic, Geopolitical and Institutional Uncertainties* (Abingdon: Routledge, 2012), 9–28.
6 The period of regular military campaign.
7 *A Transformation Gap? American Innovations and European Military Change*, ed. Terry Terriff and Frans Osinga (Palo Alto, CA: Stanford University Press, 2010), 7–8.

8 Baluyevskii was moved to the Security Council of the Russian Federation.
9 Aleksander Golts, 'One General Too Many', *Moscow Times*, 17 December 2012.
10 *The SIPRI Military Expenditure Database*. Stockholm International Peace Research Institute, www.sipri.org.
11 Susanne Oxenstierna and Fredrik Westerlund, 'Arms Procurement and the Russian Defense Industry: Challenges Up to 2020', *Journal of Slavic Military Studies*, vol. 26 (2013), 4; Haas, *Russia's Military Reforms*, 13.
12 Oxenstierna and Westerlund, 'Arms Procurement and the Russian Defense Industry', 4.
13 Haas, *Russia's Military Reforms*, 22.
14 'Russia to Inject 3 Trln Roubles into Defence Industry – Putin', *Itar-Tass*, 7 October 2014.
15 *Jane's Defence Budgets*, 2 October 2014.
16 'Oboronnyi byudzhet Rossii vyrastaet na chetvert', *Voenno-promyshlennyi kurer*, no. 39 (22 October 2014).
17 Lecture by Chief of the General Staff Valerii Gerasimov, in Oleg Falichev, 'Budushchee zakladyvaetsya segodnya', *Voenno-promyshlennyi kurer*, 13 March 2013.
18 *Strategiya natsional'noi bezopasnosti Rossiiskoi Federatsii na period do 2020 goda*, 12 May 2009, Security Council of the Russian Federation, www.scrf.gov.ru.
19 *The Draft of the European Security Treaty*, 29 November 2009, President of Russia, http://eng.kremlin.ru.
20 Sergei Lavrov, speech at MGIMO University, 3 September 2007, available at the homepage of the School of Russian and Asian Studies, www.sras.org.
21 Lavrov, speech at the MGIMO University.
22 Ivan Krastev, Mark Leonard, Dimitar Bechev, Jana Kobzova and Andrew Wilson, *The Spectre of a Multipolar Europe* (London: European Council on Foreign Relations, 2010), 40.
23 Gerasimov, speech at the Conference 'Military Security of Russia: 21st Century', 14 February 2013, *WPS Analysis, Defense & Security*, no. 458 (4 March 2013).
24 *Voennaya doktrina Rossiiskoi Federatsii*, Ministry of Defence, 5 February 2010 and of 26 December 2014, President of Russia, www.kremlin.ru.
25 Yevgenii Primakov, 'Na gorizonte – mnogopolyusnyi mir', *Nezavisimaya gazeta*, 22 October 1996.
26 'Genshtab gotovitsya k voine', *Kommersant*, 18 November 2011; 'Real'nye ugrozy dlya Rossii, SShA i ES – v nyneshnei nestabil'nosti', *Kommersant*, 12 January 2012.
27 *Voennaya doktrina Rossiiskoi Federatsii*.
28 *Kontseptsiya vneshnei politiki Rossiiskoi Federatsii*, Moscow, 12 February 2013, www.mid.ru; Bobo Lo, *Russia's Eastern Direction – Distinguishing the Real from the Virtual*, IFRI Reports, no. 17 (January 2014), 10.
29 Cf. *Strategiya natsional'noi bezopasnosti*.
30 According to Chief of the Ground Forces Staff Lieutenant General Sergei Skokov, quoted in Aleksandr Khramchikhin, 'Starye osnovy novoi doktriny', *Voenno-promyshlennyi kurier*, 17 February 2010.
31 Aleksei Arbatov, 'Foreword', in Anatoli Diakov, Eugene Miasnikov and Timur Kadyshev (eds), *Non-strategic Nuclear Weapons: Problems of Control and Reduction* (Dolgoprudny: Center for Arms Control, Energy and Environmental Studies, Institute of Physics and Technology, 2004), 3; Dmitrii Trenin, *Russia's Nuclear Policy in the 21st Century Environment*, IFRI, Proliferation Papers (autumn 2005), 11–12.
32 Chirkin, quoted in 'Ukrepchenie granitsy s Kitayem', *Argumenty nedeli*, 10 March 2010.

33 Interview with Popovkin in "Mistral" – otvet Yaponii na vechnyi kuril'skii vopros', *Kommersant*, 8 April 2010.
34 The Ministry of Defence of the Russian Federation, press release, 23 September 2014.
35 *Aktual'nye zadachi razvitiya vooruzhennykh sil Rossiiskoi Federatsii*, The Ministry of Defence of the Russian Federation, Moscow, 2 October 2003, published in *Krasnaya zvezda*, 11 October 2003.
36 For an analysis of Russia's military activity in the Arctic, see Katarzyna Zysk, 'Military Aspects of Russia's Arctic Policies: Hard Power and Natural Resources', in J. Kraska (ed.), *Arctic Circumpolar Security in an Age of Climate Change* (Cambridge: Cambridge University Press, 2011), 85–106.
37 *Strategiya natsional'noi bezopasnosti*.
38 *Strategiya natsional'noi bezopasnosti*; Gerasimov, quoted in Yurii Gavrilov, 'Genshtab otsenil ugrozy', *Rossiiskaya gazeta*, 15 February 2013; Deputy Chief of the General Staff Anatolii Nogovitsin, quoted in Olga Kolesnichenko, 'Arktika – prioritet rossiiskoi vneshnei politiki', *Voenno-promyshlennyi kurer*, 26 August 2009; Dmitrii Medvedev, *Vystuplenie na zasedanii Soveta Bezopasnosti po voprosam razvitiya sudostroeniya*, 9 June 2010, President of Russia, www.kremlin.ru.
39 'Genshtab dolozhil o real'nykh rezul'tatakh voennoi reformy', *Odnako.org*, 14 February 2013.
40 Gerasimov and Zarudnitskii, Presentations at the Moscow Conference on International Security; Vladimir Mukhin, 'Rossiiske voiska gotovy k provedeniyu mirotvorcheskoi operatsii v Ukraine', *Nezavisimaya gazeta*, 26 May 2014.
41 Mukhin, 'Rossiiske voiska gotovy k provedeniyu mirotvorcheskoi operatsii v Ukraine'.
42 Alexey Eremenko, 'Russia Targets "Traitorous" Dual Citizenship Holders', *Moscow Times*, 29 May 2014.
43 Vladislav Surkov, *Suverenitet – eto politicheskii sinonim konkurentosposobnosti*, remarks at the Education and Training Centre of the United Russia Party, 7 February 2006, www.rosbalt.ru.
44 *Aktual'nye zadachi razvitiya vooruzhennykh sil*.
45 Gerasimov, 'Tsennost' nauki v predvidenii', *Voenno-promyshlennii kurer*, no. 8, 27 February 2013; *Aktual'nye zadachi razvitiya vooruzhennykh sil*; Gavrilov, 'Bulava k kontsu goda', *Rossiiskaya gazeta*, 25 February 2011; Marianna Yevtodeva, 'Strategichteskaya stabilnost': Ugrozy dlya Rossii', *Nezavisimoe voennoe obozrenie*, 20 July 2012; Georgii Dvali and Yelena Chernenko, 'Evolyutsiya ugroz', *Kommersant*, 26 December 2011.
46 Gerasimov, 'Tsennost' nauki v predvidenii'; *Aktual'nye zadachi razvitiya vooruzhennykh sil*; General Vladimir Zarudnitskii, Presentations at the Moscow Conference on International Security; Mukhin, 'Rossiiske voiska gotovy k provedeniyu mirotvorcheskoi operatsii v Ukraine'; Alexey Eremenko, 'Russia Targets "Traitorous" Dual Citizenship Holders', *Moscow Times*, 29 May 2014.
47 Gerasimov, Zarudnitskii, Presentations at the Moscow Conference on International Security.
48 *Aktual'nye zadachi razvitiya vooruzhennykh sil*.
49 Makarov and Gerasimov, quoted in *Rossiiskaya gazeta*; *Krasnaya zvezda*, 15 February 2013; Gavrilov, 'Genshtab otsenil ugrozy'.
50 *The Military Balance*, 202.
51 Oleg Vladykin, 'Nedelya v armii: Vozvrat k pervichnomu zamyslu voennoi reformy', *Nezavisimaya gazeta*, 13 May 2013; 'Divizii ne dlya parada', *Nezavisimoe voennoe obozrenie*, 17 May 2013.
52 Aleksei Nikolskii, 'Vozvrat k shesti okrugam', *Vedomosti*, 20 December 2013.

53 Transcript from Putin's meeting with Shoigu, 'Rabochaya vstrecha s Ministrom oborony Sergeyem Shoigu', 23 July 2013, President of Russia, http://kremlin.ru.
54 Andrzej Wilk, *Is Russia Making Preparations for a Great War?*, 24 September 2014, Centre for Eastern Studies, Warsaw, www.osw.waw.pl.
55 Transcript from Putin's meeting with Shoigu.
56 Makarov, lecture at the Military Scientific Academy in Moscow; Aleksei Arbatov and Vladimir Dvorkin, 'Voennaya reforma: proschety i puti ikh ispravleniya', *Voenno-promyshlennyi kurer*, 27 June 2012.
57 Dmitrii Medvedev, *Ukaz Prezidenta Rossiiskoi Federatsii 'O nekotorykh voprosakh Vooruzhennykh Sil Rossiiskoi Federatsii'*, Presidential Decree of 29 December 2008, www.kremlin.ru; interview with Viktor Zavarzin, Chair of the Defence Committee in the State Duma, 'Voina nikomu ne nuzhna, no Rossiya gotova', *Utro.ru*, 18 November 2008.
58 Transcript of an interview with Nikolai Makarov in 'To Think and Work in a New Fashion', *Russian Military Review*, no. 10 (October 2008); Lev Makedonov, 'Genshtab pritormozil zvezdy', *Gazeta.ru*, 17 December 2008.
59 Interview with Viktor Zavarzin.
60 Serdyukov, quoted in 'Russian Military to be Fully Rearmed by 2020', *Moscow News*, 21 November 2008.
61 Gavrilov, 'Generalskoe sokrashchenie', *Rossiiskaya gazeta*, 15 October 2008.
62 Mariya Ivanova, 'Armiya smykaet ryady', *Vzglyad*, 14 October 2008.
63 'Sut' voennykh reform v Rossii – plan Serdyukova', *Ria novosti*, 18 December 2008.
64 Only 20,000 were assigned to different positions; the rest had to go: 'Genshtab VS zakonchil perekhod'.
65 'Proporshchik vozvrashchaetsya', *Gazeta.ru*, 26 February 2013.
66 Haas, *Russia's Military Reforms*, 20.
67 'Genshtab VS zakonchil perekhod k novomu obliku armii', *Vesti.ru*, 21 December 2009.
68 Roger N. McDermott, *Russia's Conventional Military Weakness and Substrategic Nuclear Policy* (The Foreign Military Studies Office at Fort Leavenworth, Kansas, 2011), 15.
69 Golts, 'One General Too Many'.
70 'Advantages and Disadvantages of Post-Serdyukov Reforms', WPS Analysis, *WPS Agency Bulletin, Defense and Security*, no. 552 (Moscow, 21 October 2013); Golts, 'One General Too Many'.
71 Susanne Oxenstierna and Bengt-Göran Bergstrand, 'Defence Economics', in Carolina Vendil Pallin (ed.), *Russian Military Capability in a Ten Year Perspective*, FOI Report (August 2012), 48.
72 *Aktual'nye zadachi razvitiya vooruzhennykh sil.*
73 *Arms-Tass*, 24 April 2013.
74 *Arms-Tass*, 24 April 2013.
75 Gavrilov, 'Bulava k kontsu goda'; Genshtab dolozhil o real'nykh rezul'tatakh'.
76 Vladimir Popovkin, quoted in Mikhail Sergeyev, 'Nebesnyi dolgostroi', *Rossiiskaya gazeta*, 25 February 2013; 'Genshtab dolozhil o real'nykh rezul'tatakh'.
77 'Genshtab dolozhil o real'nykh rezul'tatakh'.
78 Zysk, 'Russia's Naval Ambitions. Drivng Forces and Constraints', in Robert Ross, Peter Dutton and Øystein Tunsjø (eds), *Emerging Naval Powers in the 21st Century: Cooperation and Conflict at Sea* (Abingdon: Routledge, 2012), 112–135.
79 'Rossiya vtroe uvelichit finansirovanie VMF', *Lenta.ru*, 22 April 2011; Vladimir Shcherbakov, 'Nash flot prevzoidet amerikanskii. No tol'ko po tipazhu', *Nezavisimoe voennoe obozrenie*, 25–31 March 2011.
80 For details on the shipbuilding programmes, cf. Zysk, 'Russia's Naval Ambitions'.

81 *Voennaya doktrina Rossiiskoi Federatsii*.
82 Gavrilov, 'Bulava k kontsu goda'.
83 The Ministry of Defence of the Russian Federation, press release, 23 September 2014.
84 *Ria novosti*, 26 July 2013; see also Hans M. Kristensen, *Trimming Nuclear Excess: Options for Further Reductions of U.S. and Russian Nuclear Forces*, Federation of American Scientists, Special Report no. 5, December 2012.
85 Makarov, interview, 2008; Gavrilov, 'Genshtab otsenil ugrozy'; Gerasimov, quoted in *Rossiiskaya gazeta*; *Krasnaya zvezda*, 15 February 2013.
86 Oleg Falichev, 'Zhdem pomoshchi ot voennoi nauki i "oboronki"', *Voenno-promyshlennyi kurer*, 8 February 2012; Makarov, interview, 2008.
87 See for instance Vladimir Putin, Poslanie Prezidenta Federal'nomu Sobraniyu, Presidential Address to the Federal Assembly, The Kremlin, Moscow, 12 December 2013 and 12 December 2012, President of Russia, available at http://eng.kremlin.ru.
88 Putin, *Address to the Federal Assembly*, 2013.
89 Richard Weitz, 'Global Insights: Modernization Leaves Russia's Military Improved but Limited', *World Politics Review*, 15 April 2014.
90 Makarov, interview, 2008; Arbatov and Dvorkin, 'Voennaya reforma: proschety i puti ikh ispravleniya'.
91 Roger McDermott, 'Serdyukov Confirms Long-term Dependence on Dwindling Military Conscription', *Eurasia Daily Monitor*, 23 October 2012; *Russian Military Capability in a Ten-year Perspective*, 16–17.
92 Aleksander Golts, 'Putin's Paper Army', *Moscow Times*, 21 January 2014; according to General Staff, quoted in Pavel Felgenhauer, 'Shoigu to Build Office and Command Center Separate from General Staff', *Eurasia Daily Monitor*, 31 October 2013.
93 *Russian Military Capability in a Ten-year Perspective*.
94 Arbatov and Dvorkin, 'Voennaya reforma: proschety i puti ikh ispravleniya'.
95 *The Military Balance*, 202.
96 Golts, 'One General Too Many'.
97 The World Bank, *Russia Economic Report*, no. 32 (September 2014); *Country Report: The Russian Federation*, 8 October 2014, www.worldbank.org; *IMF World Economic Outlook (WEO) Update*, July 2014, 1–3; *Emerging Europe: Geopolitical Tensions Taking a Toll*, IMF Survey, 10 October 2014, www.imf.org.
98 The World Bank, *Russia Economic Report*.
99 Numerous statements by Kudrin are available in the media; see e.g. an interview with him in *Der Spiegel*, 21 January 2013.
100 'Oboronnyi byudzhet Rossii vyrastaet na chetvert', *Voenno-promyshlennyi kurer*, no. 39 (22 October 2014).
101 Haas, *Russia's Military Reforms*, 11.
102 'Novaya gosprogramma vooruzhenii budet byudzhetnee staroi', *Kommersant*, 8 October 2014.
103 'Genshtab gotovitsya k voinie'.

10 Military change in Britain and Germany in a time of austerity

Meeting the challenge of cross-national pooling and sharing

Tom Dyson

'Balanced forces' and the imperative of European defence cooperation

At the beginning of the 1990s advances in C4ISTAR (Command, Control, Communications, Computer, Intelligence, Surveillance, Targeting Acquisition and Reconnaissance) were viewed by key figures within the US defence establishment as embodying a 'Revolution in Military Affairs' (RMA).[1] The doctrines and capabilities of this new RMA would, it was claimed, deliver the US and its allies the ability to engage in conflict against near-peer competitors without the necessity to deploy large numbers of ground forces and at little cost in terms of civilian casualties.[2] As the post-Cold War era progressed, infantry-led stabilization and counter-insurgency operations in the Balkans, Iraq and Afghanistan highlighted the erroneous nature of the assumption that C4ISTAR could fundamentally transform the nature of conflict. In such 'wars among the people', technology emerged as very much secondary to the ability to deliver improvements in governance and economic development.[3] While the RMA's vision of future warfare proved exaggerated, other operations, such as NATO's air campaign in Libya, Operation Unified Protector (March to October 2011), highlight the continued important role which stand-off precision-strike technology has in modern warfare. Hence, the lessons of the operational experiences of the post-Cold War era demonstrate the importance of the development by the West European great powers (the UK, France and Germany) of balanced forces which can undertake high-intensity warfare, as well as prepare for land-based stabilization operations of rapidly varying intensity.[4]

Following operations in Afghanistan, the Balkans and Iraq, Germany and the UK have developed a strong level of expertise in counter-insurgency and stabilization. Yet, the capacity of these states to deploy land forces in future crisis-management operations is becoming increasingly limited.[5] The 2010 Strategic Defence and Security Review (SDSR) initiated cuts to the British Army of one-fifth of total manpower, from 102,000 to 82,000. These cuts were accompanied by a real-term defence budget cut of

1.9 per cent.[6] The SDSR's planning assumptions anticipate that the UK will be able to deploy a one-off intervention force of 30,000, or to simultaneously undertake an enduring (more than six months) stabilization operation of 6,500, a non-enduring (less than six months) complex intervention of up to 2,000 personnel and a non-enduring simple intervention of up to 1,000 troops.[7] When compared with the force of 45,000 that was mobilized by the British military in support of the initial stages of the Iraq War in 2003, the SDSR represents a clear reduction in the UK's strategic ambition.

The Bundeswehr has also undergone far-reaching defence reform during the CDU/CSU/FDP coalition of 2009 to 2013. The reforms initiated by Defence Minister Karl-Theodor zu Guttenberg (2009–2010) and implemented by Thomas de Maziere (2010–2013) have instigated several changes which will have a positive impact on the deployability of the Bundeswehr and its capacity to burden-share within NATO and CSDP. The first major step has been to begin to remedy the gap between Germany's strategic ambitions and military means. The 2006 Defence White Paper contained the ambition to be capable of deploying 14,000 troops overseas at any one time, from a total force of 245,000 troops. Yet, in 2010 the Bundeswehr was unable to meet this target and found itself overstretched, deploying the 8,300 troops who participated in expeditionary operations that year.[8]

Zu Guttenberg's reform has gone some way towards enhancing the deployability of the Bundeswehr by suspending conscription and replacing it with a voluntary civil service allowing young people to undertake military service or community work.[9] These changes have freed up funds for investment in deployability and have increased the number of troops available for deployment to 10,000.[10] However, Germany continues to punch below its weight in defence.[11] These improvements in deployability have been realized through reforms – such as the abolition of conscription – which should have been enacted in the late 1990s. While Germany is Europe's pre-eminent economic power, in 2012 it committed only 1.3 per cent of its GDP to defence in contrast to the UK (2.5%) and France (2.5%).[12]

Moreover, even before austerity-driven defence cuts had begun to fully manifest themselves in a reduction of operational capability, the UK and German militaries had only a limited ability to undertake higher intensity precision-strike operations within their geopolitical neighbourhood. While Germany did not participate in Operation Unified Protector, its involvement would have done little to remedy the continued dependence of European nations upon the US for key strategic enablers that characterized the operation – notably ISTAR and logistical support, as well as deficits in precision-strike munitions.[13] The EU's 2010 Foreign Affairs Council instigated 11 projects which focused on tackling deficits in a number of enablers. There has been progress in areas such as air-to-air refuelling, helicopter training, the development of a medical field hospital and satellite procurement.[14] However, further progress is needed in areas such as drones, smart munitions and military satellites. In addition, as Biscop notes, existing projects like air-to-air refuelling require an expansion in terms of the depth and breadth of EU member state involvement.[15]

The pivot towards Asia in US defence policy and plans to reduce the US defence budget by US$450 billion during the next decade have important implications for European security. The decline of Europe's major military powers will be accentuated by the growing capacity of rising economic powers, such as China and India, to translate their economic growth into military power. As Nick Witney, former Head of the European Defence Agency (EDA), highlights, 'the real challenge to the security and prosperity of Europe's peoples is to continue to count – to avoid being marginalised in a world where newer and more hard-nosed powers make the rules and assert their interests and values while Europe retreats into retirement'.[16] Furthermore, the ability of European states to respond to these challenges on an individual basis is undermined by cuts in defence budgets. In short, while it is very unlikely that the US commitment to NATO's Article 5 functions will be weakened, European states are going to have to pick up a greater share of the security burden in dealing with challenges within their geopolitical neighbourhood – in Africa, the Balkans, Caucuses and the Middle East.[17]

There are two key possibilities for European states to meet the challenge of the Asia Pivot. The first possible avenue for European cooperation is CSDP's Ghent Framework which forms the most logical institutional framework for European pooling and sharing. The Ghent Framework was agreed by the European Council in December 2011, following a German-Swedish initiative to try to spur Europe towards more concrete action to address its military capability deficits. It asks EU members to consider how to increase the interoperability of national capabilities; explore where there are possibilities for pooling capabilities, and examine the opportunities for role and task sharing in capabilities and support structures.

The second institutional forum for cooperation is NATO's Smart Defence Initiative launched in February 2011 and which has a similar set of goals to the Ghent Framework. The Smart Defence is an initiative with the potential to act in a complementary manner to the Ghent Framework. However, the initiative has been met with accusations from leading commentators that it is a means for the US to prompt its Alliance partners to purchase US capabilities and it has also been highlighted that the unwillingness of the US to provide financial backing for Smart Defence projects also undermines the credibility of the initiative.[18]

Given these problems and the reduced role for the US within NATO, it is the Ghent Framework that should take priority for European states, as this would allow CSDP to emerge as a more substantial 'European Pillar' of the Atlantic Alliance by providing capabilities that could not only be used in the event of autonomous European missions, but also in operations involving the US. Furthermore, in CSDP Europe has existing institutional architecture – not least the European Defence Agency (EDA) – to coordinate the procurement of precision-strike capabilities and strategic enablers. The US and the nations of the Weimar Five (France, Germany, Italy, Poland and Spain) are broadly supportive of European NATO members routing defence cooperation more intensively through CSDP, aware that CSDP and the Atlantic Alliance are largely complementary organizations.[19]

In short, the strategic imperative for Britain, Germany and other European nations is clear: to begin to fill CSDP's institutional architecture with substantive pooling and sharing initiatives or face military decline and growing insecurity around Europe's borders. However, Germany and the UK have provided only very limited leadership on behalf of the pooling and sharing of troops and capabilities within CSDP.[20] While there is recognition among defence policy commentators, think-tank experts and academics within both countries that the 'Asia pivot' will necessitate greater European cooperation, this has not been elucidated in official policy documents. The strategic perceptions of policy-makers in Britain and Germany appear to be shaped by the erroneous assumption that the US will continue to be willing to bail out Europe's capability deficits when European states need to tackle challenges in their geopolitical neighbourhood. In short, the UK and Federal Republic are continuing to bandwagon too heavily on US power.

This chapter will now examine the three key factors which are constraining the ability of Britain and Germany to meet the imperative of greater cooperation through CSDP. The first of these factors derives from the pressure of the international system. The final two issues are domestic in nature and include ideological path dependency and the perceived implications of pooling and sharing for the national defence technological and industrial base (DTIB) in the UK and Germany. Following this analysis the chapter will examine the theoretical implications of the empirical findings.

The alliance security dilemma

The first major 'systemic' impediment to cooperation through CSDP is the 'alliance security dilemma'.[21] The choice of alliance partners by European states involves a calculation of the risks of abandonment or entrapment.[22] Entrapment in the policy decisions of alliance partners presents a risk for European states, notably in their relationship with the US which has a high level of leverage within NATO. However, the intergovernmental characteristics of defence and security cooperation within NATO and CSDP permit a substantial level of freedom of action for the West European great powers, minimizing the problem of entrapment. Instead it is abandonment by alliance partners – be it de-alignment, re-alignment with an opponent, or failing to deliver on key commitments – that holds the greatest fear for European states.[23] These various forms of defection from alliances have occurred on a number of occasions during 20th-century European history. The lessons of history highlight the importance of being wary about the potential threat of losses in relative power which can derive from defection and the need for caution in relinquishing sovereignty in defence policy.[24] The scope and depth of defence cooperation under CSDP will, therefore, be inherently limited.

However, in the context of the Asia pivot, European states are too heavily focused on the dangers of alliance partner defection. Pooling and sharing that can deliver significant savings may be achieved without fundamentally jeopardizing the capacity for the West European great powers to deliver a broad spectrum of capabilities and forces on a unilateral basis if necessary.[25] In addition to the alliance security dilemma, a further 'systemic' factor undermines Europe's capacity to pool and share force and capabilities. Variance in the external vulnerability of European states also creates significant difficulties for European states in crafting common policy positions on key foreign and security policy issues.

Variance in external vulnerability

Given their relatively common geographical position, size and relative power, the UK and Germany face pressing incentives to cooperate with other European states to deal with a range of security threats. These threats include the challenges of failed and failing states in Europe's geopolitical neighbourhood; international terrorism; managing the implications of rising powers for Europe, particularly Russia and Iran, and ensuring secure access to sea lanes for international trade and energy security.

However, while Europe's response to these challenges has been characterized by a notable level of isomorphism, significant differences persist in the keenness by which these threats are felt and the response that is deemed appropriate. This differentiation was manifest, for example, in the 2011 Libya crisis, where the UK provided enthusiastic leadership on behalf of military action, while Germany refused to participate, undermining the ability of CSDP to play a significant role in the conflict. Some commentators have emphasized the importance of the German 'culture of restraint' in undermining military action;[26] however, a focus on variance in energy dependency also provides significant purchase in understanding Germany's reticence in becoming involved in the conflict.

The UK, France and Germany have very different energy security concerns. While the price of oil is set by the international oil market, states have the leeway to choose their trading partners in order to ensure security of supply.[27] During the 2000s all three states began to plan for the inevitable decline of North Sea oil (and gas). While Germany has focused on a 'strategic ellipse' for new gas and oil sources, which spans the Caucuses and includes Russia and a number of Soviet successor states, the UK and France have looked, among other states, to Libya to replace North Sea oil.[28] In addition, both countries have strong economic interests in oil exploration and production. As a consequence the UK and France had a strong incentive to topple the Gaddafi regime once it appeared that Libya could be heading for a protracted conflict. Germany, on the other hand, had little incentive to risk its forces in the operation.

Hence, while Germany's reticence to deploy high-intensity force (explored below) may have been a factor informing the decision by Guido Westerwelle (the German Foreign Minister from 2009 to 2013) to abstain from the conflict, variance in German external vulnerability also played an important role.[29] The impact of Germany's high level of dependence on Russia for oil and gas is highlighted by the disruptions to Germany's oil supplies caused by Russia's January 2007 closure of the Druzhba pipeline in an energy price dispute with Belarus. The pipeline supplies around 20 per cent of German oil and the consequent oil shortfalls could not be compensated for by tanker deliveries.[30] This dependence on Russian oil and gas also appears to have been a factor in Germany's unwillingness to support suspending Russia's membership of the G8 in the context of Russian occupation of Crimea. As the foreign policy commentator Simon Tisdall notes:

> Germany does not do wars any more, only peacekeeping, and that reluctantly. Berlin's interest is in trade, which means keeping Russia, whose energy supplies keeps the lights on and the factories running, sweet. It is Germany, no longer Britain, that is Europe's nation of shopkeepers.[31]

In summary, while European states are subject to some common pressures from the international security environment, this convergence is far from complete. Hence, in addition to the Alliance Security Dilemma, variance in external vulnerability further weakens the ability of the UK and Germany to fill the institutional structures of CSDP with meaningful action in pooling and sharing. Instead, this context incentivizes bilateral cooperation, such as that which has taken place between the UK and France in the 2010 Lancaster House Treaty and 2014 Brize Norton Summit. While any efforts to improve European military capabilities are welcome, as Biscop highlights, more extensive, multilateral cooperation among European nations is necessary to comprehensively respond to the challenges of austerity and the US 'Asia pivot'.[32]

The ideological legacies of the Cold War and low executive autonomy

The 'hangover' effect of the ideological frameworks of Cold War defence and security policies also has a pronounced effect on the ability of the UK and Germany to alter the strategic focus of their defence postures. However, culture and ideology are not primary variables. Instead, as the following section demonstrates, the capacity of policy leaders to mobilize culture, ideology and nationalism in an instrumental manner on behalf of strategic challenges is constrained by the level of autonomy enjoyed by the core executive in defence policy.

Britain: the special relationship and national strategic autonomy

For post-colonial Britain, retaining power and influence at a global level during the Cold War – and influence in key regional contexts where British had particularly pronounced strategic interests, such as the Middle East – necessitated a close relationship with the US. This led to a close defence and security relationship with the US, including a high level of dependency on the US for military technology transfer and intelligence sharing, and a consequent commitment to NATO rather than efforts to attain European military autonomy.[33] A public narrative of the 'special relationship' between the US and UK in defence and security was developed to frame and legitimate this dependency. At the same time, a parallel, equally powerful – and sometimes contradictory – narrative of national strategic autonomy has also remained from the colonial era. However, both ideologies are beginning to undermine a clear view of the UK's strategic interests in a world of shifting power vectors and fiscal austerity.

The UK has taken the lead on a number of central CSDP initiatives, including the 1998 St Malo Summit and 2004 Battlegroup Initiative. Furthermore, 'bottom-up' cooperation in pooling and sharing has emerged in recent years between the UK and select European counterparts, most notably with France through the 2010 Franco-British Lancaster House Treaty and the 2014 Brize Norton Summit. However, Britain's overall role within CSDP under the Conservative–Liberal Democrat coalition has been less constructive. Prime Minister David Cameron has displayed a highly reactive approach to the Eurosceptic backbench of the Conservative Party, who have exerted an excessive level of influence on Conservative European policy. It is vital that Cameron begins to challenge this Euroscepticism and deeply embedded Atlanticist ideology within his own party. Even US President Barack Obama has conceded that if the UK is to remain a useful alliance partner to the US, it can only do so through closer cooperation with its European neighbours.[34] In the era of austerity and the Asia pivot, the UK's global influence may be best secured by leading rather than hindering the development of an autonomous European military capability through the EU, notably through helping to create greater synergies between established and more recently initiated 'islands' of subregional defence cooperation within CSDP.[35]

British leadership on CSDP would not single-handedly resolve Europe's military deficiencies. Only a handful of the 26 European NATO members spend more than the NATO target of 2 per cent of GDP on defence, with Europe's economic powerhouse, Germany, being a particular laggard.[36] As already outlined, Germany is performing poorly in the translation of its economic power to military effectiveness, spending only 1.3 per cent of its GDP on defence in 2011.[37] Nevertheless, should British leadership on CSDP emerge it would provide an opportunity for German policy-makers to frame pooling and sharing as part of Germany's historic responsibility to pursue European integration.[38] This narrative would dovetail more comfortably with the German public's deeply entrenched anti-militarism than initiatives to frame pooling and sharing within NATO which is perceived as a more militaristic organization. The presence of the German Social Democrats as part of the Grand Coalition in the legislative period 2014 to 2017 also provides an excellent opportunity for stronger British-German leadership on CSDP. The electoral programme of the SPD places strong emphasis on the importance of pushing ahead with pooling and sharing under the auspices of CSDP.[39]

It is, however, unlikely that the UK will commit to stronger initiatives within CSDP during the current Parliament. The capacity of the core executive to provide leadership on EU defence cooperation is limited by the low autonomy of the core executive in defence. The UK's 'first past the post' electoral system usually delivers a significant majority for the victorious political party. Combined with the low salience of UK local elections, the UK government usually enjoys a high measure of autonomy in setting and implementing defence policy, particularly in the years immediately following a general election, when a substantial window of opportunity opens up to enact unpopular changes (such as military downsizing or changes to the ideological framework of defence).

However, the Hung Parliament of the 2010 general election and consequent Conservative–Liberal Democrat coalition has reduced the autonomy of the core executive in defence. The predominantly pro-European Liberal Democrats have a marginal level of influence over defence policy, as the UK Defence Ministry and Foreign Office are occupied by Conservative ministers. Instead, the coalition has acted as a constraint on meeting the CSDP imperative due to the way in which it sensitizes the Conservative Party leadership to Eurosceptic backbenchers in the party and to the electoral threat posed by the UK Independence Party. The potential loss of political capital associated with the bold leadership necessary to reorientate UK defence policy to the imperative of CSDP will not be forthcoming within this context of low executive autonomy. The lack of a large parliamentary majority for the Conservatives has also reduced the ability of the core executive to obtain the consent of Parliament for the deployment of military force, as illustrated by the defeat of Cameron over Syria. Yet the UK is not alone in suffering from ideological path dependency and low executive autonomy – it is a problem that has been manifest at the strategic level of German defence policy for the entirety of the post-Cold War era, as the following section will demonstrate.

Germany: the culture of restraint

Germany's institutionally and societally embedded ideological commitment to the use of force only in self-defence, or what a number of scholars have identified as a German 'strategic culture', is a key factor in reducing the ability of Germany to participate fully in pooling and sharing initiatives.[40] After the Second World War, Germany was committed to two overriding foreign policy objectives: gaining regional and international support for reunification, and returning the Federal Republic to the Western community of nations.[41] These imperatives necessitate the establishment of a strategic narrative that highlighted Germany's peaceful intentions emphasizing the key principles of multilateralism and the use of force only in self-defence. Over the Cold War period these principles took an ideological embeddedness within the key institutions of German security policy and broader German society.

Following the end of the Cold War, new security challenges requiring expeditionary crisis-management operations necessitated significant alterations to these principles, particularly German reticence to use force as a tool of foreign policy.[42] However, the frequent regional elections of the German federal state narrow the opportunity to refashion the strategic narratives used to frame German defence and security policy, thereby strengthening the rigidity of ideological path dependency in German policy.[43]

The reticence to deploy high-intensity force is also manifested in the German Constitution that restricts the ability of the core executive to initiate military operations by requiring the Bundestag to sanction overseas troop deployments. This institutional mechanism reduces Germany's capacity to burden-share by slowing the pace at which troops may be deployed as part of the EU battle groups or NATO Response Force. Crucially, it also raises the fear among Germany's Alliance partners that Germany may abandon its Alliance partners in the event that the core executive is unable to gain parliamentary approval for multinational operations, thereby sharpening the Alliance Security Dilemma.

However, recent changes in German defence policy should enable Germany to play a more proactive role in pooling and sharing through the European Defence Agency. After Germany's backseat role in the Libya crisis, the Merkel administration came under pressure from the US to reform the role played by the Bundestag in the deployment of forces overseas.⁴⁴ The Chancellor's Office worked closely with two key defence policy experts in the CDU/CSU Parliamentary Party, Roderich Kiesewetter and Andreas Schockenhoff, to instigate proposals to reform the parliamentary approval process for troop deployment.⁴⁵ On 30 May 2012 Kiesewetter and Schockenhoff released a paper entitled 'Strengthening Europe's Ability to Act in Security Policy: It is High Time'. The document proposed permitting the Bundestag to give approval in advance to the core executive for multilateral operations under the auspices of CSDP.⁴⁶

The proposal was met with vigorous opposition from the SPD and other political parties.⁴⁷ However, a public debate did not emerge, due mainly to the poor timing of the paper's release. The NATO Chicago Summit had taken place only ten days prior and there was a sense of saturation with defence and security issues at this point in time, leading the major newspapers to place the story a long way down the news agenda. Nevertheless, these proposals resurfaced in the CDU/CSU September 2013 federal election programme and, despite SPD opposition, there is no commitment in the coalition agreement to protect the rights of Parliament in respect of mandating troop deployment. Hence reforming the Bundestag's role in multilateral troop deployment has emerged once more as a theme in the 2013 to 2017 legislative period with concrete proposals to be developed by a cross-party Commission led by former CDU/CSU Defence Minister Volker Ruehe.⁴⁸

In summary, the legacy of colonialism and the Cold War have created deeply entrenched ideological path dependencies which have become embedded in the wider societies and core institutions of defence and security policy in Germany and the UK. The UK's Atlanticist ideology and Germany's high level of restraint in the use of force as a tool of foreign policy have slowed the speed at which these states have been able to meet the challenge of US disengagement from Europe.

Yet Europe faces the serious prospect of significant military decline, a scenario that will necessitate leadership from the West European great powers. After the last shift in power vectors following the fall of communism, the Franco-German motor drove the adoption of the Euro and EU enlargement. Europe now requires similarly bold leadership in the military sphere. Given Germany's low executive autonomy, the Federal Republic is not likely to provide this leadership. Although the May 2015 UK general election is likely to yield a Hung Parliament, should a Labour minority government supported by the Scottish National Party, or a Labour–Liberal Democrat coalition emerge, it will provide an opportunity for more proactive British leadership on European defence.

The high level of Euroscepticism amongst the UK electorate would appear to militate against such leadership. However, the example of France's return to NATO's integrated command structures under former French President Nicolas Sarkozy (2007–2012) in 2009 provides an instructive example of how high levels of executive autonomy can facilitate bold policy leadership on behalf of strategic reorientation, despite strong domestic opposition. Reintegration into NATO command structures contravened a central element of Gaullist ideology that had formed the ideological contours of French defence policy since the late 1960s. It met with opposition not only from French public opinion, but also from politicians from across the political spectrum.[49]

The capacity of President Sarkozy to counter opposition within his own party and the broader defence and security policy system was enhanced by the constitutionally enshrined powers of the President in the field of defence and the prolonged window of opportunity to enact radical change. This facilitated a malleable and somewhat paradoxical approach to Gaullism. President Sarkozy argued that reintegration to NATO's Command Structures was formed of a 'break with method, not principles' and was a means to achieve French 'independence'.[50] Playing a more active role in NATO's military command structures would, Sarkozy argued, facilitate further European integration in defence by assuring France's alliance partners that France was committed to complementarity between CSDP and NATO. Yet, at the same time, reintegration would also bolster France's power influence over the US: 'a state alone, a solitary nation, is no nation with influence.'[51]

Hence this example demonstrates how, under conditions of high executive autonomy, with elections far on the horizon, ideology may be employed on a selective basis to justify convergence with the imperatives of the international security environment. Despite Gaullism's emphasis on the centrality of national strategic autonomy, it was not only used by Sarkozy to justify the Europeanization of French defence policy, but also to frame the growing Atlanticization of French defence. Given the increasing imperative of cooperation under CSDP, an opportunity for similarly bold policy leadership may also emerge for the UK government in 2015 despite the high level of Euroscepticism prevalent within the British political system and public opinion.

Pooling and sharing, and the national defence technological and industrial base

Germany's ability to take advantage of the EU's institutional architecture in the field of pooling and sharing – the EDA – and pursue a sober and rational approach to multinational cooperation in armaments is weakened by its domestic armaments industry that does not wish to lose its market share.[52] The influence of the German armaments industry on policy-makers is particularly pronounced due to the federal political system, whose regular regional elections give the German regions (Laender) a high level of power and create very influential regional politicians within the CDU/CSU and SPD. Several post-Cold War German defence ministers (Volker Ruehe (1992–1998), Rudolf Scharping (1998–2002) and Karl-Theodor zu Guttenberg (2009–2011)) had the ambition of reaching the position of Chancellor.⁵³ It was not, therefore, in the interests of these ministers to initiate collaborative European armaments procurement projects which could have negative repercussions for industry at a regional level.⁵⁴ The Laender have also been able to block the participation of Germany in cross-national European armaments projects through the Bundestag's Budgetary Committee which is responsible for approving procurement projects costing over 25 million euros. On a number of occasions politicians on the Budgetary Committee, motivated by the concerns of the German defence industry, have blocked efforts by the Defence Ministry to participate in European projects.⁵⁵

Like Germany, the UK is only interested in pursuing EDA projects which will help strengthen its domestic DTIB. The core executive is also sensitive to the demands of the UK defence industry which supplies 10 per cent of the UK's manufacturing jobs.⁵⁶ However, in contrast to Germany, the UK's unitary political system does not grant such a disproportionate level of influence to the national defence lobby. This permits a more pragmatic approach to European armaments cooperation that seeks to marry the opportunities for savings with the need to sustain a broad DTIB and protect UK jobs. As Mawdsley notes,

> the British national tradition on European armaments cooperation is one of pragmatism. They will only involve themselves in projects that make economic sense and will allow them to maintain British capabilities that they otherwise could not maintain.⁵⁷

Instead, the UK's reticence in using the EDA derives from factors rooted in ideology and low executive autonomy explored earlier in this chapter. Furthermore, concerns about the commitment of their European partners to sustaining their DTIB has led to the UK favouring bilateral cooperation with France rather than through larger pan-European projects under the EDA which are more difficult to manage.⁵⁸

Conceptualizing the management of military change at the strategic level: accounting for systemic and domestic variables

Academic debate on the potential scope and effectiveness of CSDP in helping European nations manage the challenges of austerity coheres around two main theoretical positions.⁵⁹ The dominant school of thought on CSDP is constructivism. Constructivism posits that identity is central to interest formation. Norms (rules of 'legitimate' or 'appropriate' behaviour) lie at the heart of the identities and interests of policy-makers, predisposing elites to favour certain policy choices over others. National strategic cultures composed of societally and institutionally embedded norms which derive from formative historical experiences are the central determinants of states' defence policies.

While divergence in national strategic culture can hinder European defence cooperation, Constructivists argue that CSDP provides evidence of the development of a European 'strategic culture' and the 'Europeanization' of national defence and security policies.[60] From this perspective, normative convergence among national strategic cultures is taking place, around the key security challenges of the post-Cold War era and the instruments which should be deployed in response. While national strategic cultures retain a high level of rigidity, slowing down convergence, Constructivist accounts of European Defence Cooperation are broadly optimistic that the state will be able to escape the pervasive effects of anarchy by establishing supranational institutions.

The analysis presented in this chapter dovetails most closely with the insights of Neoclassical Realism, a theoretical approach that contrasts sharply with Constructivism's emphasis on the impact of ideational factors and optimism about the opportunities for cooperation. Building on the insights of Neorealist IR theory, Neoclassical Realism argues that the key variable shaping a state's defence policy is the anarchic nature of the international system which forces states to focus on increasing their power relative to other states. While Waltzian Neorealism posits that states emulate the defence reforms of the dominant power in the international system, as Resende-Santos notes, the central criterion determining military emulation is military effectiveness. Given that military power is a vital attribute of relative power, it is vital that states adhere to military 'best practice' as part of the process of 'internal balancing' against threats.[61]

The gradual emergence of a model of military 'best practice' over the post-Cold War era has been a key feature driving the convergence of the forces structures, military capabilities and doctrines of the West European great powers. As highlighted earlier in this chapter, this model has taken the form of a partial and selective emulation of the RMA and the need to develop 'balanced' forces capable of participating in expeditionary military operations which can vary quickly in intensity across the conflict spectrum. European states have identified this model of 'best practice' through a variety of mechanisms, in particular the lessons of the operational experiences and observations of the experience of alliance partners.[62] The initial post-Cold War era was characterized by a high level of uncertainty about the security threats states would face and the emerging RMA. However, the greater clarity that has since emerged about security challenges and about the implications of the RMA for defence reform has provided a stronger foundation for cross-national European defence cooperation.

Yet, correctly assessing the appropriate balance between cooperation and hedging against the risks of abandonment by Alliance partners is a difficult task. The West European Great Powers – by virtue of their similar geographical position and relative power – are exposed to similar pressures from the international security environment, incentivizing convergence around a process of bandwagoning on US power.[63] Nevertheless, as outlined above, European states must urgently reform their bandwagoning on US power to allow them to 'go it alone' by creating CSDP as a strengthened pillar of the Atlantic Alliance. While the Alliance Security Dilemma is an impediment to cooperation, as the empirical analysis has highlighted, given the pressing security challenges faced by the UK and Germany, both states are failing to identify sufficient opportunities to cooperate. As Neoclassical Realism argues, systemic variables – notably variance in the external vulnerability and the Alliance Security Dilemma – will remain enduring impediments to European efforts to pool and share military capabilities and will impede a shift from intergovernmentalism to integration in the field of defence policy.[64]

Nevertheless, a focus on systemic variables is insufficient in capturing the dynamics of UK and German policy on defence cooperation. As the empirical analysis highlights, domestic variables also play an important role in slowing down the transmission belt between security challenges and domestic policy response. The impact of ideological path dependency, executive autonomy and the political/defence–industrial nexus can also be accommodated within Neoclassical Realism. The theory argues that domestic variables play an important intervening role in slowing down convergence with the imperatives of military best practice. Neoclassical Realism appears to predict that such domestic-level variables will eventually give way to the pressures of the international security environment as these systemic forces become increasingly pressing.[65]

However, this premise stands on a misinterpretation of Neorealist theory. As Rathburn notes, Neorealist – and by extension Neoclassical Realist IR theory – allows for states to make strategic miscalculations by following the dictates of domestic political priorities or ideology.[66] The UK and Germany are increasingly behaving in a manner that threatens to be indicative of Rathburn's insights, as their policy towards CSDP appears to be led by domestic rather than by systemic imperatives. Hence the UK and Germany should heed the insights of Neoclassical Realism – for if its theoretical insights are correct then the failure of these nations to pool and share military capabilities within CSDP through enhanced intergovernmental cooperation will lead inexorably to a loss of relative power.

Conclusions

This chapter has focused on the management of military change at the national strategic level, examining, in particular, the ability of the UK and Germany to translate the imperative of closer defence cooperation through the CSDP to policy change at the national level. The chapter has found that systemic variables – the Alliance Security Dilemma and variance in external vulnerability – play an important role in limiting the scope and depth of cooperation through CSDP. Yet domestic factors are also important, notably the role played by the deeply rooted ideological frameworks of defence policy in both the UK and Germany. However, the chapter finds that the level of 'executive autonomy' enjoyed by the core executive is the central determinant of its ability to transform ideology from a constraint to a resource. Finally, cooperation under CSDP in the area of armaments is also limited by the influence of national defence industries, particularly in Germany, where the federal political system magnifies the sensitivity of politicians to the interests of defence industries in the north and south of the country.

This chapter provides an alternative perspective to the Constructivist paradigm that dominates theorizing on CSDP by arguing that Neoclassical Realism provides a strong level of analytical leverage in understanding British and German policy towards CSDP. As outlined above, Neoclassical Realism expects states to converge with the imperatives of the balance of power/balance of threat. The theory argues that if a state follows the dictates of ideology or allows other domestic political exigencies to influence defence, it will suffer a loss of relative power. The UK and Germany are presently at a critical juncture as the threat of abandonment by the US is looming ever larger, yet this has not led to a recalibration of their defence policies in favour of greater pooling and sharing of capabilities under CSDP. Given the constraints of low executive autonomy in the UK and Germany, the lack of a bold initiative at the December 2013 European Council from either nation was unsurprising. However, after the 2015 UK General Election, the UK may find in the SPD an unlikely partner in pushing ahead with adding greater substance to CSDP – an initiative that should also receive enthusiastic support from the French. The SPD – and the new CDU German Defence Minister, Ursula von der Leyen (2013 to the present) – have been active in pushing for a stronger role for the Bundeswehr in German foreign and security policy, and have emphasized the necessity for broader societal debate about the role of German foreign policy in the light of contemporary security challenges.[67] Should this window of opportunity for cooperation between the UK and Germany not be taken, Europe's demilitarization will gather pace, with serious negative implications for European security.

Notes

1 Andrew Krepinevic, 'Cavalry to Computer: The Pattern of Military Revolutions', *The National Interest*, vol. 37 (autumn 2004), 30–42. Krepinevic provides a useful definition of the RMA that recognizes the organizational and conceptual as well as technological dimensions of the concept:

> It is what occurs when the application of new technologies into a significant number of military systems combines with innovative operational concepts and organisational adaptation in a way that fundamentally alters the character and conduct of conflict. It does so by producing a dramatic increase – often an order of magnitude or greater – in the combat potential or military effectiveness of armed forces.
>
> (p. 30)

2 U.S. Department of Defense, 'Transformation Planning Guidance', 2003.
3 David Jordan et al., *Understanding Modern Warfare* (Cambridge: Cambridge University Press, 2008), 112–113.
4 Tom Dyson, *Neoclassical Realism and Defence Reform in post-Cold War Europe* (Basingstoke: Palgrave, 2010).
5 Sophie Brune et al., 'Restructuring Europe's Armed Forces in an Age of Austerity', *SWP Comments*, 28 November 2010.
6 Jon Swaine, 'British Defence Cuts are of "Critical Concern" to the Special Relationship', *Daily Telegraph*, 31 July 2013, www.telegraph.co.uk/news/uknews/defence/10212763/Britains-defence-cuts-are-of-critical-concern-to-special-relationship.html (accessed 14 October 2013).
7 'Securing Britain in an Age of Uncertainty: The SDSR', HMSO, 2010, pt. 2.15, 19.
8 Brune et al., 'Restructuring Europe's Armed Forces', 2.
9 'German Armed Forces Face Big Changes', *International Institute for Strategic Studies Strategic Comments*, vol. 16, no. 64 (2011), 1.
10 'German Armed Forces Face Big Changes', 1.
11 French defence planning assumptions currently outline the capacity to simultaneously deploy up to 15,000 troops in a major expeditionary operation, in addition to a 7,000-strong crisis-management force. 'The French White Paper on National Security 2013, 12 Key Points', Odilie Jacob, 2013, 5.
12 World Bank, 'Military Expenditure as Percentage of GDP', http://data.worldbank.org/indicator/MS.MIL.XPND.GD.ZS (accessed 14 October 2013).
13 Eric Schmitt, 'NATO Sees Flaws in Air Campaign against Qaddafi', *The New York Times*, 14 April 2012, www.nytimes.com/2012/04/15/world/africa/nato-sees-flaws-in-air-campaign-against-qaddafi.html?pagewanted=all&_r=0 (accessed 2 January 2013).
14 Sven Biscop, 'Pool It, Share It, Use It: The European Council on Defence', Egmont, Royal Institute for International Affairs, Security Policy Brief, 44, March 2013, 2.
15 Biscop, 'Pool It, Share It, Use It'.
16 Nick Witney, 'How to Stop the Demilitarisation of Europe', *European Council on Foreign Relations Policy Brief*, 40, November 2011, 7.
17 Michael O'Hanlon, 'Getting Real on Defence Cuts', *Brookings Institution*, 22 July 2012, www.brookings.edu/research/opinions/2012/07/22-defense-cuts-ohanlon (accessed 3 January 2013).
18 Sven Biscop, 'The UK and European Defence: Leaving or Leading?', *International Affairs*, vol. 88, no. 6 (2012),1297–1313, 1303.
19 Maia Davis-Cross, 'CSDP and the Transatlantic Partnership', Clingendael, Netherlands Institute of International Relations, 7 March 2013, www.clingendael.nl/publication/csdp-and-transatlantic-partnership (accessed 23 October 2013).
20 Sven Biscop and Jo Colemont, 'Military Capabilities: From Pooling and Sharing to a Permanent Structured Approach', *Security Policy Brief*, Royal Institute for International Affairs, September 2012.
21 Glenn H. Snyder, 'The Alliance Security Dilemma', *World Politics*, vol. 36, no. 4 (1984), 461–495, 461–466.
22 Galia Press-Barnathan, 'Managing the Hegemon: NATO Under Unipolarity', *Security Studies*, vol. 15, no. 2 (2006), 271–309, 271; Snyder, 'The Alliance Security Dilemma', 461.
23 Snyder, 'The Alliance Security Dilemma', 461.
24 Interview, Frau Elke Hoff, MdB, Free Democratic (Liberal) Party, Member of Bundestag Defence Committee, 11 September 2012.
25 Biscop, 'The UK and European Defence'; Clara O'Donnell, 'Britain's Coalition Agreement and EU Defence Cooperation: Undermining British Interests', *International Affairs*, vol. 87, no. 2 (2011), 419–433.

26 Sarah Brockmeier, 'Germany and the Intervention in Libya', *Survival*, vol. 55, no. 6 (2013), 63–90; Alister Miskimmon, 'German Foreign Policy and the Libya Crisis', *German Politics*, vol. 21, no. 4 (2012), 392–410, at 395.
27 Blake Clayton and Michael Levi, 'The Surprising Sources of Oil's Influence', *Survival*, vol. 54, no. 6 (2012), 107–122.
   For a country by country breakdown of German oil imports in 2010, see Eurostat, European Commission, 'Energy, Yearly Statistics, Deutschland' (2010), Table 4.1, Crude Oil and Feedstocks, 84.
28 Terry Macalister, 'Secret Documents Uncover UK's Interest in Libyan Oil', *Guardian*, 30 August 2009, www.guardian.co.uk/world/2009/aug/30/libya-oil-shell-megrahi (accessed 23 August 2013); Julian Borger and Terry Macalister, 'The Race is On For Libya's Oil with Britain and France Both Staking a Claim', *Guardian*, 1 September 2011, www.theguardian.com/world/2011/sep/01/libya-oil (accessed 23 October 2013); Roland Goetz, 'Germany and Russia: Strategic Partners?', *Geopolitical Affairs* (2007).
29 Interview, Defence and Security Policy Division, Foreign Ministry, Berlin, 9 March 2012.
30 'Energy Wars: Russia Halts Oil Deliveries to Germany', *Der Spiegel*, 8 January 2007, www.spiegel.de/international/energy-wars-russia-halts-oil-deliveries-to-germany-a-458401.html (accessed 3 March 2014).
31 Simon Tisdall, 'West's Puny Response to Ukraine Crisis Will Not Deter Vladimir Putin', *Guardian*, 2 March 2014, www.theguardian.com/commentisfree/2014/mar/02/west-response-ukraine-vladimir-putin-russia (accessed 3 March 2014).
32 Biscop, 'The UK and European Defence'.
33 William Keylor, *The 20th Century World and Beyond: An International History Since 1900* (Oxford: Oxford University Press, 2006), 306.
34 Alex Sipillius, 'Britain Will Be Weaker without the EU, Says USA', *Daily Telegraph*, 18 December 2012, www.telegraph.co.uk/news/worldnews/europe/eu/9754042/Britain-will-be-weaker-without-EU-says-USA.html (accessed 3 January 2013).
35 Tomas Valasek, *Surviving Austerity: The Case for a New Approach to EU Military Cooperation* (London: Centre for European Reform, 2011), 29–30, www.cer.org.uk/sites/default/files/publications/attachments/pdf/2011/rp_981-141.pdf.
36 World Bank, Defence Spending as Percentage of GDP, figures for 2011, http://data.worldbank.org/indicator/MS.MIL.XPND.GD.ZS (accessed 3 January 2013).
37 James Kirkup, 'Don't Mention the War: Germany Should Leave the Past Behind and be Willing to Undertake Military Action Overseas Says Defence Secretary', *Daily Telegraph*, 2 May 2012, www.telegraph.co.uk/news/worldnews/europe/germany/9241980/Dont-mention-the-war-Germany-should-leave-the-past-behind-and-be-willing-to-undertake-military-action-overseas-says-Defence-Secretary.html (accessed 3 January 2013).
38 Interview, Berlin Office, European Council on Foreign Relations, 14 December 2012; interview, SPD member of Bundestag Defence Committee, Berlin, 21 February, 2013; interview, SPD Party Headquarters, Willy Brandt Haus, Berlin, 6 November 2012.
39 Interview, SPD member of Bundestag Defence Committee, Berlin, 21 February 2013; interview, SPD Party Headquarters, Willy Brandt Haus, Berlin, 6 November 2012.
40 Anja Dalgaard-Nielsen, *Germany, Pacifism and Peace Enforcement* (Manchester: Manchester University Press, 2006); Kerry Longhurst, *Germany and the Use of Force: The Evolution of German Security Policy 1990–2003* (Manchester: Manchester University Press, 2004).
41 Alister Cole, *Franco-German Relations* (Harlow: Pearson, 2001), 11–12.
42 Tom Dyson, *The Politics of German Defence and Security: Policy Leadership and*

Military Reform in the post-Cold War Era (New York: Berghahn, 2007); Dyson, *Neoclassical Realism and Defence Reform in post-Cold War Europe*.
43 Tom Dyson, 'Condemned Forever to Becoming and Never to Being: The Weise Commission and German Military Isomorphism', *German Politics*, vol. 20, no. 4 (2011), 545–567.
44 Interview, Chancellor's Office, Berlin, 1 August 2012; interview, Defence and Security Policy Division, German Foreign Ministry, Berlin, 9 March 2012; interview, Konrad Adenauer Stiftung, 23 July 2013.
45 Interview, Chancellor's Office, Berlin, 1 August 2012; interview, Konrad Adenauer Stiftung, 23 July 2013; interview, Office of Roderich Kiesewetter, MdB, CDU/CSU, Berlin, 29 May 2013.
46 'Europa's sicherheitspolitische handlungsfaehigkeit staerken: Es ist hoechste Zeit', 30 May 2012, www.andreas-schockenhoff.de/download/120611_GSVP-Papier.pdf (accessed 3 June 2013).
47 Interview, Frau Elke Hoff, MdB, FDP Bundestagsfraktion, 11 September 2012; interview, Herr Mathias Martin, Koordinierender Referent der AG-Außenpolitik SPD Bundestagsfraktion, 22 August 2012.
48 Interview, Office of Roderich Kiesewetter, MdB, CDU/CSU, Berlin, 29 May 2013; 'Kritik an geplanter Rühe-Kommission zur Parlamentsbeteiligung an Auslandseinsätzen', *Rheinische Post*, 7 March 2014, www.presseportal.de/pm/30621/2681435/rheinische-post-kritik-an-geplanter-ruehe-kommission-zur-parlamentsbeteiligung-an-auslandseinsaetzen (accessed 7 April 2014).
49 Angelique Chrisafis, 'Sarkozy to End France's 40 Year NATO Feud', www.theguardian.com/world/2009/mar/11/france-sarkozy-nato, 11 March 2009 (accessed 11 April 2014).
50 www.france24.com/en/20090311-france-nato-commandstructure-wider-role-troops-charles-de-gaulle (accessed 24 October 2013).
51 Henry Samuel, 'Sarkozy Announces French Return to NATO After 43 Years', *Daily Telegraph*, 11 March 2009, www.telegraph.co.uk/news/worldnews/europe/france/4974756/Sarkozy-announces-French-return-to-Nato-after-43-years.html (accessed 11 April 2014).
52 Interview, Nick Witney, former EDA Chief-Executive, London, 9 May 2013; interview, Dr Hilmar Linnenkamp, former EDA Deputy Chief Executive, Berlin, 24 May 2013; interview, Division for International Armaments, Political Department, German Defence Ministry, Berlin, 29 May 2013.
53 Dyson, *The Politics of German Defence and Security*.
54 Interview, Dr Hilmar Linnenkamp, former EDA Deputy Chief Executive, Berlin, 24 May 2013; interview, Division for International Armaments, Political Department, German Defence Ministry, Berlin, 29 May 2013.
55 Interview, Division for International Armaments, Political Department, German Defence Ministry, Berlin, 29 May 2013.
56 'Defence Industry Vital for the UK', *BBC News*, 1 September 2009, http://news.bbc.co.uk/1/hi/uk/8230910.stm (accessed 23 October 2013).
57 Jocelyn Mawdsley, 'France, the UK and the European Defence Agency', Paper presented to workshop, 'Arming Europe: The European Defence Agency at 10', St Gallen, Switzerland, 25 June 2013, 10.
58 Mawdsley, 'France, the UK and the European Defence Agency', 13.
59 It is important to note that CSDP is nevertheless characterized by a highly diverse theoretical debate, with contributions from a range of theoretical perspectives.
60 Alessia Biava *et al.*, 'Characterising the EU's Strategic Culture', *Journal of Common Market Studies*, vol. 49, no. 6 (2011), 1227–1248; Bastien Giegerich, *European Security and Strategic Culture* (Baden Baden: Nomos, 2006); Christoph

Meyer, 'Convergence towards a European Strategic Culture', *European Journal of International Relations*, vol. 11, no. 4 (2005), 523–549.
61 Joao Resende-Santos, *Neorealism, the State and the Modern Mass Army* (Cambridge: Cambridge University Press, 2007).
62 Dyson, *Neoclassical Realism and Defence Reform in post-Cold War Europe*.
63 Tom Dyson, 'Balancing Threat, Not Capabilities: European Defence Cooperation as Reformed Bandwagoning', *Contemporary Security Policy*, vol. 34, no. 2 (2013), 387–391.
64 Dyson, *Neoclassical Realism and Defence Reform in post-Cold War Europe*.
65 Dyson, *Neoclassical Realism and Defence Reform in post-Cold War Europe*, 120–127.
66 Brian Rathburn, 'A Rose by Any Other Name? Neoclassical Realism as the Logical and Necessary Extension of Structural Realism', *Security Studies*, vol. 17, no. 2 (2008), 294–321.
67 'Gauck Opens Munich Security Conference with Call for More German Engagement', *Deutsche Welle*, 30 January 2014, www.dw.de/gauck-opens-munich-security-conference-with-call-for-more-german-engagement/a-17399048 (accessed 13 February 2014); interview, Herr Rainer Arnold, MdB, SPD Defence Policy Spokesperson, Bundestagsfraktion, 3 July 2013.

11 Austerity is the new normal
The case of Danish defence reform

Mikkel Vedby Rasmussen

Doing more with less has been the watchword for the Danish armed forces for 20 years. To the Danish armed forces, the age of austerity began in earnest with the end of the Cold War. From 1985 to 2004 the Danish defence budget was cut by 28.6 per cent.[1] From the 1990s the budget settled on was 25 billion Danish kroner (DKK) (US$4 billion). For that amount, the Danish taxpayer could not possibly buy a joint and combined defence in depth of the approaches as planned by Denmark and the Baltics during the Cold War. Reducing the budget thus led to a demand for reform of the armed forces to fulfil a new strategic purpose. The globalization of Western security after the end of the Cold War provided such a purpose – first, by making peace-keeping and peacemaking missions in the Balkans a new production goal for the Danish armed forces, and then, following 9/11, making deploying expeditionary forces to Afghanistan and Iraq the armed forces' raison d'être. Military reform was thus focused on producing a force structure that would deliver small unit contributions from each service. The armed forces are therefore no longer a coherent fighting force that can deliver a joint and combined effort. The Danish armed forces have become a subcontractor that can deliver individual units for individual missions within a Western alliance framework.

This military business model has prescribed a focus on unit costs, which seem set to guide Danish defence policy in the years to come, in relation to output. The defence budget for 2013 to 2017 is thus focused on cutting 15 per cent of the budget by decreasing costs associated with running the overall structure, while retaining funding for deployable units. This entrenches Denmark's position as a subcontractor. Further, though this has been a highly successful scheme for defence reform, to put it bluntly, it has turned generals into accountants who are focused on generating forces at the lowest possible price rather than on developing strategies for how to use the forces in ways that best further operational and political ends.

This chapter will describe the way in which Danish defence policy is defined by austerity and analyse the options that this leaves for the Danish armed forces.

Danish defence budget

The Danish defence budget was on average DKK25 billion (US$4 billion) between 1990 and 2012 (Figure 11.1). This high degree of budgetary stability reflects the fact that the defence budget is traditionally defined in a political agreement between the majority parties in Parliament. These agreements define procurement, force goals and other spending priorities for a five-year period. Such a budgetary practice gives Parliament a high level of direct engagement in the running of the Ministry of Defence, but also provides the defence minister as well as the military leadership with the opportunity to do long-term planning. Defence investment is thus normally distributed unevenly within the five-year budget, but with the purpose of realizing a particular 'shopping list' by the end of the five-year term. The average investment per year is DKK843 million with wide variations.

The ratio between defence expenditure and the number of troops is one way to show how capital intensive a military is. If one compares this with countries that field approximately the same number of troops as Denmark (20,000), one can see that the Danish military is comparatively 'heavy'. However, it is also clear that the Danish armed forces have not invested as heavily in military hardware as have the NATO leaders in that category. In 2010, Denmark spent US$249,222 per soldier; this is less than Norway, which spends US$319,65 per soldier, but more than the Czech Republic, which spends US$111,333 per soldier.[2] Since 1990, the expenditure per serviceman ratio has increased by 90 per cent. This is a substantial

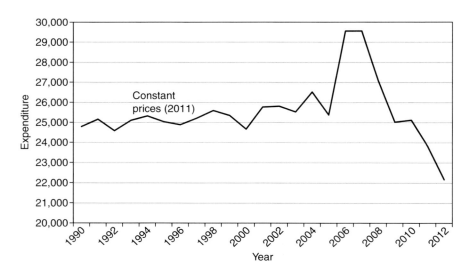

Figure 11.1 Danish defence expenditure, 1990 to 2012 (constant prices) (source: Statistics Denmark).

increase and, as shown in Figure 11.2, larger than the general 75 per cent increase among core NATO members. The United Kingdom, which had a very capital-intensive military even in 1990, has increased the ratio by only 49 per cent, whereas a number of other important NATO countries, which by the end of the Cold War had the same militaries configured for fighting in their own neighbourhood as Denmark, had increased the expenditure per serviceman ratio dramatically. Italy and the Netherlands have doubled the expenditure–serviceman ratio.

The increased capital intensiveness of the Danish armed forces is most dramatically demonstrated by the Royal Danish Navy. In 1990, 67 vessels were in commission by the Danish Navy. In 2010, 41 ships were in commission. However, the number of tons the navy displaced had increased, from 35,862 tons in 1990 to 55,550 tons in 2010.[3] These numbers reflect the restructuring of the navy from a brown water navy to a blue water navy. During the Cold War the navy focused on submarines and smaller surface combatants to defend the Baltic approaches. After the Cold War, submarines and torpedo boats were decommissioned and replaced by a new class of patrol vessel at the same time as an ambitious building programme was begun. By 2010, the programme had left the Royal Danish Navy with nine frigates (*Iver Huitfelt*, *Absalon* and *Thetis* class) and a number of smaller vessels, including Arctic patrol vessels of the *Knud Rasmussen* class. The Royal Danish Air Force increased the ratio by reducing the number of service personnel and procuring C-130J transporters. The Air Force was

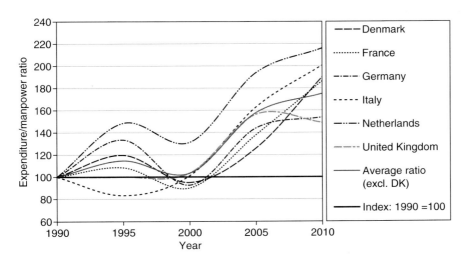

Figure 11.2 Ratio between total defence expenditure (constant 2010 US$) and armed forces strength (source: NATO, Financial and Economic Data Relating to NATO Defence, press release PR/CP(2011)027, Table 1 and Table 6).

able to reduce staff considerably because the end of the Cold War meant that air defence and the defence of air bases was no longer needed, freeing up considerable resources to restructure around expeditionary air wings. The effect was somewhat reduced, however, by the gradual reduction in the number of F-16s available to the air force in order to save money and to postpone the time when the costly procurement of a new generation of fighters was needed. The army dramatically reduced its personnel during this period. From a conscript force of 60,000 in 1990 the army was reduced to two armoured battalions. Armour was reduced congruently. In 1989 the Army fielded 491 tanks; in 2013 only 34 were still in service. These Leopard II tanks had been substantially updated however, and served, as we shall see, with distinction in Helmand.

The Danish armed forces thus became smaller and more capital intensive from 1990 to 2012. This was the result of defence reforms that transformed a territorial defence force into an expeditionary force. This transformation was focused on the development of a number of key platforms and capabilities which were configured to conduct international operations while the national force structure was cut back in order to fund these operations.

Strategy: the business model of the Danish military

The Danish armed forces' mission statement states that the mission for the armed forces is 'securing a peaceful and democratic development in the world and a secure Denmark by the ability to fight and win'.[4] This statement may seem bland, bordering on meaningless. After all, what kind of military would not require its forces to be able to fight and who would not prefer that they prevailed in conflict? Even if victory might not be practically possible, planning for anything else seems a deliration of duty. Similarly, since the Kingdom of Denmark is a founding member of the United Nations and has been a democracy since 1848, one would be more than a little surprised if a peaceful and democratic world was *not* a Danish priority and Denmark would be a peculiar democracy indeed if the Danish government did not see a connection between a peaceful and democratic world development and Danish security. Thus, in the final analysis, the Danish armed forces' mission statement defines it as a military like any other in the Western world. When the mission statement was adopted this was actually an important message for the generals to get across. Unsure of their future role in the post-Cold War world, the armed forces felt the need to underline their ability to fight and prevail in conflict as a crucial element among others in Danish foreign and security policy. This argument entrenched the armed forces' importance as a political means, but most importantly it ensured that the armed forces could not be reduced to a gendarmerie.

In other words, the mission statement defines a turf; it hardly sets a course. Thus, by default, the statement makes a general point about

Danish defence policy. It is concerned with how defence works and what defence costs rather than with the role of military force as a political instrument. 'The nature of war was never a topic for discussion in Denmark', former Chief of Defence General Jesper Helsø notes: 'in Denmark there is no tradition for debating war as the ultimate political means.'[5] As the author of the mission statement, General Helsø kept within this discourse, focusing on the mission for the armed forces themselves rather than the mission the armed forces had in relation to the politicians and Danish society in general.

Danish military strategy is pragmatic and businesslike, thus focusing on the *how* rather than on the *why*. In focusing on the ways in which 'the how' has been reformed during the period from 1990 to 2010 when the armed forces became so much more capital intensive, one clearly sees how economic constraints made reforms necessary. If one paints the picture without too many nuances, three business models for Danish defence are apparent in this 20-year time frame.

The Cold War Business Model was the point of departure for reform. This too was defined on a limited budget, even if the threat to national security was arguably much greater during that period. The bedrock of this model was Denmark's membership of NATO, which Jens Ringsmose aptly describes as 'freedom insurance'.[6] Denmark paid its premium for NATO membership in terms of a certain level of defence spending and political, strategic deference to Western security policy and overall strategy. Like any fire insurance, the purpose of the 'freedom insurance' in NATO was to receive help if disaster struck, but otherwise to keep payments as low as possible in the hope that there would never be an actual need to claim damages. In order to avoid confrontations with the Soviet Empire that could set the European house on fire, Denmark worked consistently for detent rather than deterrence.[7] NATO membership and the diplomatic focus on de-escalating policies were to allow the Danish government to prioritize welfare over warfare. Taking its point of departure as the insurance model of defence, the force structure was based on mobilization on the basis of a small, professional cadre. The force structure was organized around a large conscript army which was to defend Danish territory and NATO's line of communication. The navy and the air force were to support these efforts by defending Danish territory and by harassing Soviet naval and air assets in the Baltic approaches. The armed forces thus had a combined, joint operations plan for fighting the Third World War but, as with any insurance policy, it was kept in a safe place only to be used in an emergency. The day-to-day business of the armed forces was training for this war that never came. Patrolling the kingdom's territory in and around Greenland and the Faroe Islands in the Arctic and occasional UN-peace-keeping missions were exceptions to this, but as such they did not constitute the armed forces' core business.

The Post-Cold War Business Model was remarkably consistent with the Cold War approach to NATO as an insurance premium. The collapse of the Soviet Empire changed the tasks that needed to be performed in order to get that insurance. A 1998 White Paper on Defence concluded that 'the task of the Danish Armed Forces has changed in nature from being an element in a reactive, deterrence based guarantee of security to also being an active and confidence building instrument of security policy'.[8] The Cold War Model was defined as passive as opposed to the new model in which security and defence policy were to be active. These activities had to be done on the basis of large cuts in the defence budget. As mentioned, the defence budget was cut 28.6 per cent between 1985 and 2004.[9] Thus, the Danish government still prioritized welfare over warfare. A new, active defence policy had to be carved out of a shrinking budget. Peace-keeping missions in the Balkans became a vehicle for this new, active policy. In order be able to send troops to the Balkans, the Danish International Brigade was created. While the rest of the Cold War force structure remained largely intact, the Brigade was given a new type of mission, new equipment and recruits who signed up for extra training after their normal conscript service had ended. A student of innovation in the business sector would recognize the International Brigade as an innovative business division with the mission to reform a large company somewhat set in its ways. The International Brigade's missions challenged the insurance notion of defence, since the Brigade was actually on continuous operations in the 1990s. In the spring of 2001, then Defence Minister Jan Trøjborg concluded that peace-keeping missions were 'one of the foremost missions in our time'.[10]

The Intervention Business Model made what had been the exception the rule. This confirms the notion that the International Brigade was an innovation business division which was to experiment with changes that ultimately were to be implemented throughout the entire organization. Previously only one brigade had been 'international'. Now, the entire defence establishment would be 'international'. In conceptual terms this was a break with the notion of defence as insurance. Now, defence was rather regarded as an investment. This investment served much the same political purpose as NATO membership had done during the Cold War and its aftermath. Denmark wanted to have a close relationship with the United States and wanted to be a core member of the Atlantic Alliance. In the first decade of the 21st century this meant active engagement in the Global War on Terror. Danish defence had to invest in global capabilities to be part of the Global War on Terror. This investment served an economic purpose, since the post-Cold War Business Model had created two core missions – national and international – fighting for resources and attention as the budget was shrinking. This was an untenable situation, Defence Minister Svend Aage Jensby told Parliament in December 2002 when he had to report that the armed forces had accumulated a deficit of

DKK1 billion in 2003 and 2004. 'The armed forces budget has to be adjusted to ensure a better balance between means and ends', the Minister concluded.[11] The budget for 2004 to 2009 was to change that. Inspired by the agenda of reform set by the NATO NRF,[12] the Armed Forces were to reform the two-tier business model. The international brigade had succeeded as an experiment in a future force. The professional ethos of the armed forces was to be transformed from a national defence force to an intervention force. The army was to be able to field one to two battalions, the navy one to two surface combatants and the air force a squadron of F-16s. In total, the number of troops the armed forces were to be able to deploy was to double to 2,000.

These operational capabilities were to be paid for by a large-scale reform of the force and support structure.[13] By focusing on deployable forces, the entire set-up for mobilizing a conscript force could be scrapped. This professionalization meant that HR, procurement and other back-office functions could be centralized in joint commands in order to create economies of scale. Conscription was retained for political reasons, but cut back to a four-month introductory course from which, as had been the case with the International Brigade, soldiers could be recruited for international missions. One of two army brigades was to be staffed by professionals only.

In Danish military parlance the forces had to be 'helstøbte'[14] which was a translation of the 'essential operational capabilities' as defined by NATO's MC317/1.[15] A White Paper on Defence and Security Policy concluded that 'the characteristics of the Danish Armed Forces' involvement in international operations in the future should be the will and ability to rapidly deploy short-term and focused contributions'.[16] While international missions were to be the raison d' tre of the armed forces, the budget simply did not allow for a long-term commitment. For the force structure and the operations to fit, operations had to be conducted on a 'first-in, first-out' basis. Only that way could the price of defence remain fixed. The military planners who drove much of this process were thus focused on creating a force structure which would cost around the same – whether units were deployed or not. The goal was to field capable forces which could be deployed for national defence as international operations within the same budget to ensure a high degree of flexibility.[17] This required a buffer in the budget to pay off operational expenses. However, the Ministry of Finance was not inclined to allow the armed forces that kind of pocket-money and insisted that the funds be allocated to special drawing rights on the state budget, and thus subject to approval by the Ministry of Finance. This arrangement left the armed forces with a commitment to deploy an unprecedented number of troops, but without the financial flexibility to support their training and deployment. Furthermore, the entire budget was based on the premise that the Danish armed forces should either sustain minor missions for a long time or major missions for a short time. With deployments to Iraq (2003–2006) and Afghanistan (2006–2014) the government chose to deploy the army on major missions for a long time. Operational expenditure thus increased by 44 per cent from 2006 to 2010.[18] A 2008 White Paper on Defence concluded that 'at present there is an imbalance between the operational requirements to the armed forces and the resources allocated the armed forces'.[19] This was an echo of Minister Jensby's statement in 2002.

The deployments to Iraq and Afghanistan had made the armed forces focus on fielding an army battle group and the force and support structure had been reconfigured for that objective. The defence budget was not big enough to pay for this without prioritizing other tasks, which in practice meant down-sizing the two other services. In order to avoid this situation, Defence Minister Søren Gade was able to secure DKK1 billion more for the defence budget for 2010 to 2014. Unfortunately for Mr Gade, the financial crisis meant that the government had introduced a number of cuts and the Ministry of Finance used this as an opportunity to reduce the defence budget by 15 per cent, thus annulling the increase in the defence budget which the Ministry of Finance had opposed all along. The armed forces had no choice but to prioritize, but in the meantime the commitments in Afghanistan were drawing down which reduced operational costs considerably, thus creating budgetary breathing space. However, the fundamental choice remained. If it wanted to maintain a broad range of 'helstøbte' capabilities, the Danish armed forces had to choose between minor and long missions or major and short missions. In the 2013 to 2017 budget agreement the 15 per cent cut in the budget was implemented by reducing the essential operational capabilities. Capabilities were simply not as 'helstøbte' had previously been planned for. Furthermore, back-office functions were cut yet again, which put a certain perspective on the insistence in the budget agreement that there would be an 'increased focus on the ability to conduct joint operations'.[20] Actually, the agreement entrenched the focus on individual capabilities by defining the armed forces' core business in terms of a list of deliverables. The agreement thus lists a number of capabilities: a battalion-sized battle group (300–800 troops), two naval units (or one unit for longer operations), and three air force capabilities (C-130J, helicopters or F-16). These were to be supplemented by other contributions, such as a task force for humanitarian operations, staff contributions, etc.[21]

Since the end of the Cold War the Danish armed forces have been reformed in order for the task to fit the budget. The budget has been cut as a reflection of the way in which consecutive governments have defined other items in the national budget as more important. The need for increased welfare spending is one such need. Figure 11.3 offers a comparison between social expenditure and defence expenditure and is illustrative of the deprioritization of defence. The impetus for change has thus been budgetary austerity which meant that every new investment had

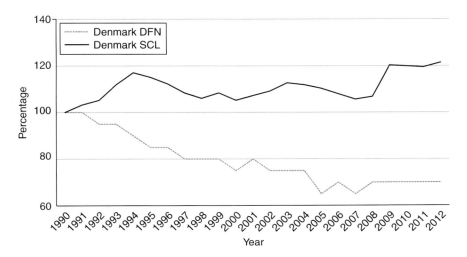

Figure 11.3 Index 100 of defence expenditure in per cent of BNP vs. social expenditure in per cent of BNP (source: Defence expenditures as percentage of GDP: www.sipri.org/research/armaments/milex/milex_database and social expenditures as percentage of GDP, 'Social Expenditure: Aggregated data', OECD Social Expenditure Statistics (database)).

to be financed by cutting existing capabilities or structure. The Danish armed forces have introduced a number of business models that have gradually replaced the 'insurance logic' of the Cold War with an 'investment logic'. Following this logic the Danish armed forces have focused on developing a set of specific deployable capabilities within each service, sacrificing joint operational capabilities and a large force structure in order to pay for this. Reforming defence in order to develop capabilities under austere conditions has thus become a signature feature of Danish defence policy. The next section will describe the model Denmark has adopted to do this.

Benchmarking

Danish generals are fond of describing the development of the business models for the Danish armed forces in terms of assembly and movement. Most presentations on yet another defence restructuring will be accompanied by a metaphor describing this reform as assembling a plane while flying it or changing the motor of a car while driving it. The fact that this is physically impossible says a lot about how daunting the generals find defence reform. It also underlines the urgency that conditions of austerity create for the Danish armed forces. There is no time to carefully develop a new vehicle, test it and then put it on the road. The budget will only allow for structural change and operations to take place at the same time. This leaves huge processual risks in implementing the five-year budget plans. In order to overcome these risks the Danish armed forces have learned to validate the processes themselves as well as set targets by benchmarking to other defence establishments as well as private companies. This benchmarking has reduced processual risk, but should also be regarded as a result of the pragmatic, value-for-money-centred Danish discourse on defence. An argument based solely on defence requirements simply will not cut it in this discourse, and the defence leadership has to point to standards in business or from other countries in order to get the validity of their arguments recognized.

This benchmarking approach is used on specific projects. The Ministry of Defence has established an in-house lean office to guide individual sections within the Ministry and the armed forces on how to achieve better business practices. This focus on business practices is reflected in the widespread use of private consultancies to benchmark the armed forces' performance with similar private or public organizations. Deliotte has thus been tasked with validating the process for procuring so-called Replacement Fighters, while Mckinsey has analysed defence economics in general and specific offices (e.g. HR). Furthermore, the Centre for Military Studies at the University of Copenhagen has, for example, provided data and comparative analyses for a MoD working group on conscription and the researchers have produced an analysis of the conditions for defence planning in a time of austerity, which provided the conceptual background for the 2012 budget debate.[22] The entire approach to defence reform in Denmark is a rejection of the notion that the 'military business' is unique and is thus an argument for subjecting the armed forces to the same rules of productivity and effectiveness as other organizations, public or private. In the 1990s this approach was clearly normative in the sense that the Ministry of Defence wanted the armed forces to be more businesslike and therefore benchmarked it with other organizations. In time, however, benchmarking has become constitutive in a sense; in comparing itself to private businesses the military has become more like a private business.

Benchmarking is not only used in terms of strategic and business practices but also as a way to set a standard for military performance. The armed forces mission and vision statement thus declares that:

> Prioritized alliance- and coalition partners should find that the Danish armed forces is a sought-after partner. In other words, we must be able to measure up to our partners. At a minimum, this means the United States and the United Kingdom. If we can measure up to them then we have very good chances for collaborating with other partners. If we are good and attractive partners on the military level, then the political level will have the greatest possible freedom of manoeuvre. Then Denmark has the possibility of participating in operations.[23]

The Danish armed forces were not to be measured by a set of national requirements but by international standards, standards set by the foremost military powers. This was no mean ambition for a country with a US$8 billion defence budget. The ambition does reflect the huge influence NATO has on the Danish armed forces. The very notion of setting standards for the armed forces was, as mentioned above, born from NATO's notion of essential requirements that were tested in the NRF. In the early 2000s, the NRF served to demonstrate to the Danish armed forces what it would actually take to develop truly expeditionary capabilities as opposed to the in-place force structure the armed forces had had during the Cold War. In the 2003 budget, the armed forces embraced this challenge and wanted to develop capabilities that could measure up to the United States and the United Kingdom. By setting the bar that high, the Danish generals avoided a discussion about quantity in favour of a discussion about quality. The armed forces were cutting the number of platforms and service personnel, but since the quality of the remaining capabilities was growing, this was supposed to offset concerns about the drop in quantity. By focusing on the highest level of quality, the armed forces' leadership avoided a discussion about different levels of quality within the armed forces. They did not want to discuss whether, for example, the navy should have platforms less advanced than those procured for the army. A debate about priorities would have been natural in the austere budgetary environment, but benchmarking to the highest possible standard closed that debate. It was assumed that forces able to operate at the highest and hardest end of the mission spectrum would be so well trained that they would be able to conduct missions at the lower and softer end of the spectrum as well.[24]

This 'lesser included'logic was prevalent in most Western armed forces at the time. The wars in Iraq and Afghanistan seem to have demonstrated that while well-trained units are easier to train for other functions simply because they have learned to learn, the training they receive for high-end missions does not in itself prepare them for low-end missions. In fact, lesser missions are not included; it is rather the case that high-end missions are precluded if soldiers are trained and equipped only for low-end missions. The end of operations in Afghanistan thus left the armed forces with the need to address notions of quality in other ways than setting a universally high bar. Different types of missions, different types of forces and different levels of readiness were simply necessary if one wanted the budget to fit. The way in which the British army restructured, by modularizing and using regular and reserve forces on different levels of readiness to maintain capability even as the budget was cut by 20 per cent, was a clear inspiration to Denmark. It followed quite naturally from the armed forces' focus on benchmarking that when the United States and, especially, the United Kingdom changed their doctrine and force structure, Denmark would follow.

The defence reforms of its allies served as an inspiration to Denmark, but the Danish armed forces followed the British lead for operational reasons as well. Benchmarking was a way to achieve better performance, but it was also a way to ensure that performance was possible in the first place. The Danish armed forces have realized that the ambition for fielding forces in international missions is only achievable if Danish forces are deployed as part of a larger allied task force. Denmark defines itself as a 'subcontractor' of defence.

The continued reference to business practices in the Danish defence is instructive here. A subcontractor's relationship with the larger contracting firm is well documented. McDonald's is a well-known example of how to actively use suppliers as well as retailers to develop quality. McDonald's defines in minute detail how to conduct its own operations and that translates to the demands for certain kinds of meat, bread, potatoes, etc.[25] Because McDonald's is such a large customer the firm is able to shape the conduct of meat producers, bakeries, potato growers, etc. The Danish military may be compared to an Ohio potato farm in relation to McDonald's cooperation. When McDonald's needs a new type of chip, the farmer grows a new type of potato. Yet, in order to get the best possible chips, McDonald's needs to be in dialogue with the farm about what type of potato to use and how to harvest and transport it, etc. The more commitment McDonald's can generate among suppliers like the Ohio potato farmer, the better the quality of the product. The Danish armed forces have a similar relationship with the British and the American military. The Danish armed forces get the greatest possible benefit by organizing their forces in ways that fit into allied structures; in military jargon Denmark has to be able to 'plug and play'. This presents two challenges for Danish defence planning. First, it has to ensure that Danish forces know how to play. Benchmarking is used to ensure this. Second, it has to ensure that the allies actually have plugs which Denmark can plug into. Having relevant and well-trained forces is not enough to ensure this. Denmark has to show political flexibility and flair to be part of the force generation process among allies. The result is a kind of political benchmarking where Danish foreign policy goals are being attuned to allies. Defence reform has thus been driving the political agenda setting.

Austerity has made benchmarking a central management tool within the Danish armed forces. It has provided a template for reforming business practices and this template has, in time, transformed the way in which the armed forces do business. The Danish armed forces increasingly think of their practices in terms familiar to other large private or public enterprises – so much so that the generation of the military forces themselves has been benchmarked. This military benchmarking is a result of Denmark's role as subcontractor, but it has also reinforced the tendencies to develop military capabilities as stand-alone contributions to international missions rather than as coherent elements in a joint Danish force structure. In the next section, I will address the mechanics of this subcontractor force structure in further detail.

Subcontractor

The extent to which austerity sets the agenda for Danish defence policy is illustrated by the fact that the best public accounts of the economics of the armed forces are provided as background material for budget cuts. The Danish armed forces do not publish time series on key economic and performance indicators in the way the Americans and the British do. The Danish armed forces are much more focused on the here and now, so accounts are presented in relation to specific challenges. In 2012, the Ministry of Defence published an economic analysis carried out with the assistance of McKinsey, which for the first time gave figures for the costs associated with either maintaining or deploying military capabilities.[26] The fact that these numbers were presented as a means to debate priorities in defence clearly demonstrated that the mission and vision statement's notion of pointing to a general high quality in order to avoid a discussion about quantity would no longer work. McKinsey's work followed the template of O'Hanlon, which in 2009 had produced similar numbers for the American armed forces. As O'Hanlon points out, estimating the cost of a capability is inherently difficult because capabilities support each other and it depends on the support structure, which is hard to calculate.[27] In other words, such numbers are no guide to defence cuts. However, they are a good illustration of the huge marginal costs, as well as the wide variation in marginal costs, associated with the armed forces and their use. In the following section, I will use these numbers to illustrate the mechanics of the subcontractor force structure.

The overall difference between a high level of activity and a maintenance level is 41 per cent or DKK2 billion. Merely maintaining the armed forces' capability thus comes with a substantial cost, but actually using the forces increases this cost even more (see Figure 11.4). These marginal costs vary hugely, however; and this huge variation is the pivotal economic factor for a defence establishment where austerity is a normal condition. Maintenance cost is almost the same (around DKK1.5 billion) for each service. Maintaining the two army brigades and second naval squadron costs around DKK0.5 billion. The real difference is in operational costs. A high level of activity makes the cost of the Brigade quadruple while the second squadron only increases its cost by 40 per cent. This variation in capabilities is reflected in the overall costs of the services. Army costs are more variable than navy and air force costs. Army cost increases by 327 per cent, whereas navy cost increases by 52 per cent and air force cost by 108 per cent. This explains the financial stress under which two – by Danish standards – large ground operations in Iraq and Afghanistan put the Ministry of Defence. The problem is not only the

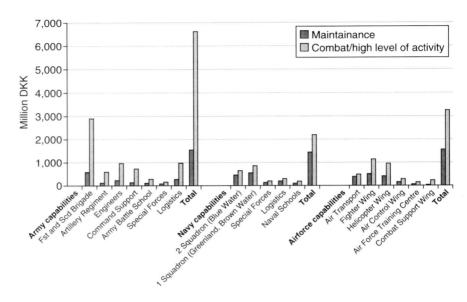

Figure 11.4 Cost of Danish capabilities (in millions of DKK) (source: Udvalget for analyser af forsvaret, *Effektiviseringer og bedre økonomistyring. Afrapporting af budgetanalyser af forsvaret* (Copenhagen: Forsvarsministeriet, 2012)).

amount itself, but the stress that the difference between the maintenance and the high-activity cost level puts on the organization.

Since deploying an army battle group has proved such a cost factor, it makes sense to go deeper into the way such a deployment works on the Ministry of Defence's balance sheet. In order to pursue its ambition to deploy units benchmarked with the United States and Britain in terms of quality and to be able to conduct independent operations, the Danish Army has deployed a battle group (DANCON, Danish Contingent) of two companies of armoured infantry supported by a contingent of Leopard II tanks. Combat engineers, fire support, ordnance disposal and a civil–military relations team have supported these forces. Occasionally, special forces have been deployed in support and medical teams, etc. have also supplemented the British effort from time to time. This small force has provided a remarkable punch. Teeth-to-tail ratio is a measure of military effectiveness in the sense that how much of a force is dedicated to actual combat and how much is taking care of 'back-office functions', like administration and logistics. In 2011, the teeth-to-tail ratio for the Danish battle group in Helmand was 2:1 with 65 per cent of the Danish forces in teeth functions and 35 per cent in the tail.[28]

This is a remarkably high ratio of combat troops. A US combined brigade during the same period managed 43 per cent teeth and 57 per cent tail. Thus, in terms of benchmarking, the Danes had clearly outdone the great powers they were supposed to learn from. From a Danish perspective this was a demonstration of how a continuous focus cost had enabled the Danish armed forces to increase quality for less money. This was also true in terms of the costs of the operation. In 2009, Minister of Defence Lillelund-Bech told Parliament that the annual cost of the Afghanistan operation was DKK2 billion,[29] which means that the 750 soldiers in Afghanistan that year had an individual operating cost of DKK2.6 million or US$452,000. This figure is cheap compared to the US estimate of US$1 million per American soldier for the US deployment to Afghanistan.[30]

From one perspective this is evidence of how effective the Danish Army has become, yet from another perspective it is an example of doing war on the cheap. Deploying a battalion means that the Danish Army does not have to provide for the staff, support and logistics on a brigade or division level. In effect, the Danes provide teeth and let others pay for the tail. The British taxpayer had already paid for the tail of Task Force Helmand, so accommodating Danish and other allied units might actually make good economic sense, since it means that the sunk costs of establishing the operation count for more. But this is only the case if the allied units provide fire power which makes up for what they do not provide in terms support, enablers and tail. The Danish contribution of Leopard II tanks to an operation where the British forces did not bring any tanks should be seen in this regard. 'At the moment in Afghanistan, we quite happily rely upon the Danish to provide their own Leopard II tanks', the presenter of a video on the future of the British Army noted as an introduction to a discussion on how much the British Army could rely on allied capabilities and how many capabilities (especially tanks) the army needed itself.[31]

For a Danish defence establishment defined by austerity it hardly matters whether the low teeth-to-tail ratio is evidence of effectiveness or piggybacking the capabilities of larger allies. In Copenhagen the salient point is that the Danish armed forces are able to conduct international missions even under conditions of austerity. The subcontractor model for doing so, however, means that the Danish armed forces focus on battalions, squadrons and vessels. The way capabilities are listed in the defence budget and priced in the economic analysis underscores how austerity has forced the Danish armed forces to focus on small, individual capabilities. The result is an armed force which is a collection of individual army, naval or air units rather than a coherent, joint fighting force. This unit focus actually increases the distance between administration and support functions and combat capabilities because administration and support have become more and more centralized and joint in order to create more effective civilian benchmarked services.

Costing military models

The focus on small, individual capabilities gives the Danish armed forces a modular set-up. Focused on civilian benchmarking as they are, the Danish generals would probably not mind being compared to Lego. The Danish approach to force generation is a Lego model where individual capabilities can be combined with each other and with the capabilities of allies. One might identify four such Lego models for future Danish force generation:[32]

1. A long-term stabilization force.
2. An international assistance force.
3. A humanitarian deployment force.
4. A defensive force.

The tasks of the Danish armed forces as a *long-term stabilization force* would be to deploy forces capable of stabilizing areas of conflict or failed states that have become bases for terrorist activities or have threatened to destabilize neighbouring countries or regions in the absence of intervention. This model reflects an attempt to strengthen Denmark's foreign policy profile with the help of ongoing military activism and would mean a continuation of operations like those about to end in Afghanistan. The tasks of the Danish armed forces as *an international assistance force* would be to contribute to the enforcement of sanctions and embargoes, to create peaceful conditions and to participate in anti-terrorist operations within the framework of an international coalition for the purpose of maintaining international law and the resolutions of the UN Security Council. This would involve the use of naval assets and planes which should be able to be deployed at short notice. The army would focus its international contribution on smaller units and special operations forces suitable for carrying out effective actions at short notice. This requires logistics and transport to and within operational areas as well as army or naval detachments that could create a framework for actions by special operations forces.

Instead of high-end operations the armed forces could specialize in the tasks of *a humanitarian deployment force*, which would be to contribute military capabilities that could ensure the successful performance of UN missions, including civil missions. The Danish contribution could comprise the elements of a mission that are decisive for its performance, such as logistics, intelligence gathering, training or special operations forces. These contributions could be provided in collaboration with other Nordic countries with a high UN profile. A humanitarian deployment force is probably the model which would be the most suitable to pave the way for comprehensive Nordic defence cooperation.

The military models described above focus on how to set up various ideal types of force, optimized to perform various international tasks. An alternative could be to focus on *territorial defence*. For the past ten years, NATO has taken its point of departure in the idea that the capacity for collective defence began with deployable forces, but Denmark could also decide to take its point of departure in the defence of its own territory and in the appropriate capabilities in this connection. Although this would constitute a marked break with Danish defence policy since the end of the Cold War, it is nevertheless a possibility. The nature of such a case is that a defensive force would be of less interest to Denmark's allies, and the contribution of the armed forces to foreign policy would be significantly reduced.

The tasks of a defensive force would be to defend Danish territory and integrity against direct threats. The Danish armed forces are thus to be organized with a view to defend national territory and are only occasionally to be deployed in the Baltic region (e.g. air policing in the Baltic countries).

The McKinsey Report referred to in the previous section offers us the opportunity to put a cost on these models. Obviously, such numbers should be used with extreme caution because they do not account for the cost borne by the structure and tend to obfuscate joint and combined efforts. These numbers work best when compared to one another, since the comparison internalizes many of the measurement issues. Used carefully, however, this approach offers an opportunity to compare the costs of the alternative models.

Figure 11.5 demonstrates that the army's commitment is the largest cost driver. This is even truer in terms of the difference between the costs of

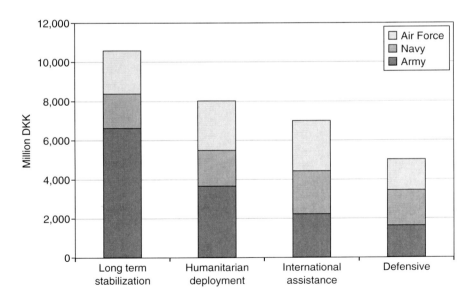

Figure 11.5 Costing military models (in millions of DKK) (source: Udvalget for analyser af forsvaret, *Effektiviseringer og bedre økonomistyring. Afrapporting af budgetanalyser af forsvaret* (Copenhagen: Forsvarsministeriet, 2012)).

maintaining army capabilities on a day-to-day level and using them for high-tempo operations (e.g. international deployments). Army costs increase by 327 per cent while navy operations costs only increase by 52 per cent, whereas air force costs increase by 108 per cent. This demonstrates that navy and air force expenditure is, primarily, the down payment for a capability, whereas the main costs of the army are operational. This is probably true of most Western armed forces, but in a small state defence budget huge differences in costs can be crippling, since the budget is not of a size where shocks are easily absorbed. The Danish defence budget was thus put under considerable strain during the deployment to Afghanistan, where the large army expenditure made it difficult for the entire organization to operate. This financial strain was aggregated by the fact that the entire system was under pressure because of the logistical and manpower needs of Afghanistan.

The austerity design

Strategy, Richard Rumlet notes, is 'scarcity's child'.[33] Because the Danish armed forces have been on an austerity budget since the end of the Cold War, Danish generals and politicians have been forced to continuously develop defence strategy. This strategy development has been focused on the business side of the armed forces rather than on grand strategy. National security considerations have served to define the 'market' for the Danish military rather than constituting the framework for a structured discussion of national interests and national security priorities. For example, discussions about Russia have focused on what kind of force structure Denmark would need to develop: first, in response to the collapse of the Soviet Union, and second, in response to a resurgent Russia under Putin. These discussions have largely ignored other security concerns – for example, energy security – in relation to Russia. The Russian intervention in Ukraine in 2014 fundamentally challenged this technocratic and pragmatic approach to force structure and planning. With Russian power looming more threateningly, the usefulness of the armed forces would not merely be defined by the market of allied operations. Now, national security may play a greater part. But, since Denmark had focused on a business strategy at the expense of a political-military strategy, even discussing a new threat assessment proved difficult. The armed forces awaited guidance from NATO that would redefine the 'security market' and set the aims for which Denmark should benchmark its forces.

Rumlet would argue that 'many effective strategies are more designs than decisions'.[34] An organization defines a way of doing its thing and thus it defines what that thing is and, in turn, acquires an identity from doing it. Thus, the Danish armed forces are subcontractors, and proud subcontractors, not the least because subcontracting fits the organizational culture of the Danish armed forces well. The legacy from the conscript army of the Cold War is still present in the ethos of the officer corps which is educated with special focus on training and motivating personnel in small units. Thus, the Danish armed forces comprise platoons, tank crews, planes and ships. This creates a strong low-level cohesion which is perhaps the foremost reason why the organization has been able to increase quality at the same time as it has gone through two decades of reform. The focus of small units was naturally translated into a focus on how to deploy them when international operations became the name of the game. Replacing mobilization with mobility was a huge challenge, but the unit ethos made it possible to make this transition one deployment at a time. This created huge frustrations because, in the absence of one overarching vision, units readying themselves for deployment often found that they did not have the right equipment to train with or encountered various bureaucratic obstacles from the rest of the organization. The support functions were transforming one unit at a time as well, but often this transformation was not in sync with the units they were supposed to support. The resulting differences in perspective created huge frustrations between those faced with the operational challenges and those challenged with implementing reforms on the support structure. Cutting services to reduce costs on one end of the organization was not necessarily helpful for those needing those services in order to facilitate deployment.

The focus on individual units created an effective business model that could deliver operative forces for coalition warfare at an affordable price. This business model did come with a price, however. What made good business sense became a long-term strategic challenge. By focusing on sub-contracting individual units, the armed forces have found their niche in a very particular market defined by the military needs of the great powers. Starting with the NRF, the Danish armed forces have done their best to live up to the expectations of NATO, the United States and the United Kingdom in particular. For the past 10 to 15 years, these expectations have been defined by stabilization operations in places like Iraq and Afghanistan. Now, priorities are changing in the United States and the United Kingdom, and NATO seems increasingly divided on the notion of what the Alliance should do after Afghanistan. Even if events in Ukraine have put Article Five Operations and conventional defence higher on the agenda, this does not remove the fundamental question of how to organize deployable forces able to defend the entire NATO area, which has grown so large that for Western allies like Denmark, any Article Five Operation would entail deployment far from home base.

This leaves the Danish armed forces in a position well known to students of business strategy. Clayton Christensen notes that 'when big companies fail, it's often not because they do something wrong but because they do everything right'.[35] Christensen reaches this conclusion based on work he and Joseph Bower did on IBM. In their view, IBM came close to collapse in the 1990s because IBM's management followed the management textbook and delivered what its customers wanted. The customers wanted big machines for big data and were not particularly interested in microcomputers, and IBM thus neglected that business area with the consequence that in a space of 10 years microcomputers had taken most of IBM's business away.[36] The Danish armed forces run the same risk by focusing so much on individual capabilities which will be deployed in a coalition of the willing that they have disregarded the demand for new types of military 'products'. What if the United States and the United Kingdom do not need Danish Army units for the next 15 years? What is the Danish Army's raison d'être then? What is true of mission types is also true of the entire subcontractor model adopted by the Danish armed forces. Denmark depends on enablers, support and logistics from larger partners – what happens when these are not there? That moment is actually approaching fast.

Structurally, the Danish subcontracting model has worked because of a post-Cold War surplus of transport and other enablers. By the 2040s the US Air Force (USAF) will fall over the 'the airlift cliff' when C-17As and C5Ms are to be retired. In 20 years, they will all be retired and the USAF will have to recapitalize its transport fleet. However, given the budgetary situation, allies should not expect much surplus capacity to be bought. In other words, no more free rides with Uncle Sam.[37] Does this mean that countries like Denmark will have to invest in their own tail rather than merely provide teeth? This will profoundly change the calculus behind defence austerity and force the Danish armed forces to prioritize in ways it has not done before. A high level of quality on the 'product' from all services in all types of missions will simply not be economically possible under these circumstances. This serves to show how vulnerable a small defence force like Denmark's is to procurement decisions. On a severely limited budget the margin for error will always be small, but when the number of capabilities one can depend on changes because of the state of affairs in other military establishments then the ability to make the right decisions becomes severely challenged. These challenges are enforced by the unit culture of Danish defence.

With the focus on small individual units for operations and the centralization of support functions, a few joint milieux make planning the future of the entire force extremely difficult. This is much more of a challenge because the operational future may be defined in very different terms from those of the army deployments of the past ten to 15 years. A renewed focus on conventional deterrence in Europe will be one case, but perhaps the most pertinent example of this is the Arctic. The Kingdom of Denmark includes Greenland, which is in a process of change because of climate change in the Arctic and the new possibilities for economic development and geopolitical rivalry this change creates.[38] The Arctic places a new type of demand on the armed forces. These are operations where Denmark is the sole contractor rather than a subcontractor and where the effort therefore is joint rather than based on individual contributions. In short, the Arctic places a new premium on coherent planning, logistics and operational redundancy which was not necessary before. This means that the armed forces have to focus more on tail functions and find the money to pay the considerable price associated with such functions. Taken together, this means a new focus on defining strategy in order to set procurement priorities. A subcontractor does not have to think harder than it wants to about the big picture – the Ohio potatoes farmer does not need to consider McDonald's corporate strategy. When Denmark is conducting operations in relation to its own territory, as the sole contractor in cooperation with its neighbours, then no one else will set the strategic priorities.

Notes

1 Jens Ringsmose, *'Frihedens assurancepræmie': Danmark, NATO og forsvarsbudgetterne* (Odense: Syddansk universitetsforlag, 2008).
2 NATO, Financial and Economic Data Relating to NATO Defence, press release PR/CP(2011)027, table 1 and table 6.
3 'Søværnet skifter ham', *Forsvaret* (2007), 5, årgang, no. 2, 6–8.
4 Ministry of Defence, 'Årsrapport', *Forsvaret* (2012), 4, www.ft.dk/samling/20121/almdel/fou/bilag/156/1248166.pdf (accessed 15 May 2013).
5 Interview with the author, 8 April 2010.
6 Ringsmose, *'Frihedens assurancepræmie'*.
7 Poul Villaume, *Allieret med forbehold. Danmark, NATO og den kolde krig – et studie i dansk sikkerhedspolitik, 1949–1961* (Copenhagen: Eirene,1995).
8 The Danish Defence Commission of 1997, *Defence for the Future, English Summary* (Copenhagen: Statens Information, 1998), 3.
9 Ringsmose, *'Frihedens assurancepræmie'*.
10 Jan Trøjborg, 'Hvem er Fjenden?', *Berlingske Tidende*, 9 March 2001.
11 'Milliarddunderskud i forsvaret', *Ritzau*,12 December 2002.
12 The Security Policy Conditions for Danish Defence (2003), Executive Summary of the Report August 2003, Royal Danish Ministry of Foreign Affairs (at www.um.dk) (accessed 30 September 2003).
13 Notat vedrørende forsvarets operative kapaciteter (Memo Concerning the Operative Capabilities of the Armed forces), AG05 Secretariat, 18 June 2003, 11.
14 Danish Defence Agreement, 2005–2009.
15 John R. Deni, 'The NATO Rapid Deployment Corps: Alliance Doctrine and Force Structure', *Contemporary Security Policy*, 25 (2004), 508–511.
16 The Security Policy Conditions for Danish Defence, 2003.
17 Notat vedrørende forsvarets operative kapaciteter, 3.
18 Forsvarskommissionen af 2008, Beretning (Copenhagen: Forsvarsministeriet, 2009).
19 Hækkerup i FKOM08.
20 Ministry of Defence, Defence Agreement 2013–2017, November 2012, 2.
21 Ministry of Defence, Defence Agreement 2013–2017, November 2012, 3.
22 Mikkel Vedby Rasmussen *et al.*, *An Analysis of Conditions for Danish Defence Policy – Strategic Choices 2012* (Copenhagen: Centre for Military Studies, 2012).

23 Chief of Defence and Chief of Defence Staff, Defence Mission and Vision, www2.forsvaret.dk/omos/Publikationer/Documents/Forsvarets%20Mission%20og%20Værdier.pdf (accessed 8 May 2013).
24 Notat vedrørende forsvarets operative kapaciteter, 7.
25 Nigel Slack and Michael Lewis, *Operations Strategy*, 3rd edn (Harlow: Prentice Hall, 2011), 345–346.
26 Udvalget for analyser af forsvaret, Effektiviseringer og bedre økonomistyring. Afrapporting af budgetanalyser af forsvaret (København: Forsvarsministeriet, 2012).
27 Michael O'Hanlon, *The Science of War: Defense Budgeting, Military Technology, Logistics, and Combat Outcomes* (Princeton, NJ: Princeton University Press, 2009).
28 The numbers are based on the Danish Army Staff's calculations, which breaks down as follows: DANCON ISAF11 (January 2011–July 2011) 693 troops in total, 456 in teeth functions, including two companies of infantry, tanks, engineers, fire support, ordnance disposal and a civil–military relations team, and 237 in tail functions, including staff, logistics and military policy. DANCON ISAF12 (July 2011 to February 2012) was 699 troops in total, with 458 in teeth functions and 241 in the tail.
29 Folketinger S1899, Samling 2009–2010: Om udgifter til missionen i Afghanistan.
30 Todd Harrison, 'Estimating Funding for Afghanistan, 1 December, 2009', the Center for Strategic and Budgetary Assessments.
31 RUSI Land Warfare Conference 2011, Rusi.org https://www.rusi.org/events/past/ref:E4D3EE746532B8/info:event/ (accessed 21 April 2014).
32 Rasmussen *et al.*, *An Analysis of Conditions for Danish Defence Policy*.
33 Richard Rumlet, *Good Strategy/Bad Strategy. The Difference and Why it Matters* (London: Profile Books, 2011).
34 Rumlet, *Good Strategy/Bad Strategy*, 129.
35 'Clayton Christensen Wants to Transform Capitalism', Wired Business, Wired.com. www.wired.com/business/2013/02/mf-clayton-christensen-wants-to-transform-capitalism/all/ (accessed 15 May 2013).
36 Joseph L. Bower and Clayton M. Christensen, 'Disruptive Technologies: Catching the Wave', *Havard Business Review* (January–February1995), 44.
37 'The Airlift Cliff', *Airforce Magazine* (April 2013), 21.
38 Mikkel Vedby Rasmussen *et al.*, *Keep Cool* (Copenhagen: Centre for Military Studies, 2013).

Part III
Military change in the United States and NATO

12 Military change in NATO

The CJTF concept – a case study of military innovation in a multinational environment

Paal Sigurd Hilde

Changing the organizational, doctrinal and material set-up of national military forces is challenging. In a multinational setting, obstacles to change are multiplied. In most forms of international security and defence cooperation, at least those involving democratic states, decision-making is based on the principles of sovereignty and consensus. When combined with diverging threat perceptions and interests, possible outcomes are invariably limited and change becomes difficult. Yet, like national militaries, multinational military establishments must adapt to changing requirements in order to remain relevant.

In the post-Cold War era, the allies have successfully transformed the North Atlantic Treaty Organization (NATO), in many ways fundamentally. NATO heads of state and government have adopted important reform agendas at all 13 (full) summits since the end of the Cold War. The September 2014 summit in Cardiff, Wales, was no exception. NATO has seen 20 years of engagement in and emphasis on crisis response operations outside NATO's territory. The Wales Summit adopted a military reform plan – the Readiness Action Plan (RAP) – primarily motivated by the need to enhance NATO's ability to react to military challenges at home. Arguably, the most important part of the plan is the establishment of a Very High Readiness Joint Task Force (VJTF). This decision was, however, mainly one of principle: that the Alliance will establish a force about 4,000 strong. At the time of writing (December 2014) details remained to be settled, including the VJTF's relation to the NATO Response Force (NRF). The eventual success or failure of the plan will depend on the degree to which the allies are able to overcome both political and military obstacles to reach consensus on a workable, concrete and detailed set-up of the VJTF.

The main aim of this chapter is to contribute to our understanding of military change in NATO and to provide a case study of military innovation in a multinational environment. It will attempt to do so by examining the rise and fall of the Combined Joint Task Force (CJTF) concept. The CJTF concept was adopted at the NATO Brussels Summit on 10–11 January 1994. The summit declaration heralded the concept as a centrepiece of NATO's post-Cold War military transformation. As an approach to organizing NATO's military headquarters and Alliance forces, the concept was to address three core challenges that the Alliance faced in the post-Cold War security environment. The first was to enable NATO-led, out-of-area[1] crisis response operations, the second to strengthen the European pillar in NATO by enabling European-led operations using NATO assets, and the third to enable the participation of NATO's new partners in NATO-led operations. On paper, the CJTF concept represented a brilliant catch-all solution to key aspects of NATO's adaptation. One may gauge the significance attributed to CJTF from NATO ministerial communiqués.[2] From the January 1994 Brussels Summit until the end of the decade, CJTF was on the agenda at all meetings of both defence and foreign ministers.

Seventeen years after its birth, on 21 February 2011, the North Atlantic Council quietly and unceremoniously buried the CJTF concept with the adoption of a new Military Committee (MC) policy for the use of allied forces in operations, MC 586. The grand idea of the 1990s – and one of the parts of NATO's post-Cold War military adaptation that had received the most attention in academic and other writing – was put to rest with a bureaucratic pen-stroke.

This chapter traces the origins, development and ultimate failure of the CJTF concept. It examines what factors shaped its development, and what prevented it from fulfilling the hopes invested in it in the 1990s. After a brief historical overview, a political and a military perspective is taken to analyse the evolution of the CJTF concept. This approach allows for conclusions on two levels: the systemic and the specific. The political perspective examines the development of CJTF in terms of the intra-alliance, political debate. The military perspective examines the utility and suitability of the CJTF concept in terms of the development of the military requirements of the Alliance. Taken together, the two perspectives allow an assessment of the extent to which the system – the political complexity of the multinational setting – or specific shortcomings in the CJFT concept were important in preventing its successful implementation. The distinction between these two conclusions is important, as it informs how one should approach military change in NATO. If the main obstacle was systemic, military change is hard to achieve and ambitions should be tailored mainly to create consensus. If the main obstacle was concrete, ambitions could remain high and the problem is more to attune the concrete aspects of reforms to NATO's military heterogeneity and decision-making process.

The CJTF concept

In the run-up to the January 1994 Brussels Summit, the United States in traditional fashion set the agenda for developments in NATO. On 20–21 October 1993, at an informal meeting of NATO defence ministers in Travemünde, Germany, US Secretary of Defence Les Aspin presented two proposals aimed at adapting NATO to meet the new opportunities and challenges the Alliance faced in the post-Cold War era.[3] One proposal, termed Partnership for Peace, addressed NATO's ambition to widen the Alliance's relations with the former Warsaw Pact countries from dialogue to practical cooperation, including in peacekeeping operations. The second proposal, the Combined Joint Task Force concept, aimed to address several challenges related to the development of NATO's integrated command-and-control system, and the organization of the forces NATO was to lead.

At the January 1994 Brussels Summit, Alliance heads of state and government adopted both Partnership for Peace and CJTF as central pillars of what was later termed NATO's external and internal adaptation.[4] On CJTF, the allies directed the North Atlantic Council (NAC) in Permanent Session:

> to examine how the Alliance's political and military structures and procedures might be developed and adapted to conduct more efficiently and flexibly the Alliance's missions, including peacekeeping, as well as to improve cooperation with the [Western European Union] and to reflect the emerging European Security and Defence Identity. As part of this process, we endorse the concept of Combined Joint Task Forces as a means to facilitate contingency operations, including operations with participating nations outside the Alliance.[5]

As the declaration made clear, CJTF was to address three key aspects of NATO's military adaptation: to adapt NATO's command structure to new requirements, to enable sharing of NATO military capabilities with the Western European Union (which in December 1991 became the EU's 'military arm'[6]), and to open for participation in operations by partner militaries.[7] Of the three, the last was the least complex and controversial. The first aim, to create deployable headquarters within NATO's integrated military command structure and give the Alliance an organic ability to conduct deployed land operations, was seemingly challenging but manageable. It quickly became clear that the second aim was the most controversial and difficult: to find a mechanism for allowing the use of NATO assets, notably the deployable headquarters but also assets such as early warning and strategic lift, in European-led operations.

A NATO CJTF was defined 'as a multinational, multiservice task force consisting of NATO and possibly non-NATO forces'.[8] US thinking on joint task forces was likely the conceptual model for NATO's approach.[9] Combined and joint task forces were, however, alien to neither NATO, nor allies other than the United States.[10] What was 'unique – unprecedented in military doctrine –' was NATO's attempt to 'incorporate the task force concept, which is traditionally used for ad hoc coalitions, as a *modus operandi* of a standing alliance'.[11] NATO did not attempt to establish complete, standing task forces. Rather, in line with the standard practise in NATO, the Alliance establishes and maintains headquarters. Upon a decision in the NAC to engage the Alliance militarily, and the subsequent transfer by member states of authority over national military forces to NATO commanders, forces are assigned to and normally put under the operational control of a NATO headquarters. This procedure was (inevitably) applicable also to CJTF headquarters. Detailed work in NATO on the CJTF concept therefore focused mainly on command-and-control aspects and headquarters capabilities.

While initial work on the military aspects proceeded quickly following the Brussels Summit, consensus proved hard to reach on several political issues. A breakthrough only came by the end of summer 1996, when the initiative partly passed from the political to the military side in NATO. Its work on preparing the implementation of the concept, including trials in exercises to support its further development, commenced in autumn 1996.[12] In January 1997, the NAC approved the Military Committee's first version of the capstone CJTF document, MC 389. The MC initiated a three-phased implementation plan, with final implementation scheduled to start in 1999. Phase two, the assessment phase, started in September 1998.[13]

Political issues again proved an obstacle, however. Only at the Washington Summit in April 1999 was further consensus achieved and progress in implementing CJTF possible. Work on the military side proceeded at pace. In December 1999, phase three, the final implementation phase, commenced, and in June 2000 the Military Committee issued a revised version of MC 389 (MC 389/1).[14] Further political complications were resolved by December 2002, paving the way for the framework for employing CJTFs (and other NATO assets) in EU-led operations to be finalized in March 2003. By May 2004, the last pieces had been put in place and the MC adopted a final version of MC 389 (MC 389/2).[15] Ten years after its official adoption by NATO, the CJTF concept was set to become operational in the NATO command structure. Nine years later, however, the concept disappeared without ever being fully implemented.

The political perspective

Many if not most military issues involve political considerations. This is the case both in a national and, even more so, in a multinational setting. Given the ambition to address several fundamental issues connected to NATO's role and organization, reaching consensus on the CJTF was unavoidably difficult. With the United States and France as the main protagonists, the issue of how to enable European-led – WEU-led – operations using NATO assets was by far the most challenging. The dispute was based on differing perceptions of what roles NATO and the EU should have in the post-Cold War era on the one hand, and on the other hand, what role the United States should have within NATO and in European security in general.

There were both pull-and-push factors behind the emergence of the aim of enabling WEU-led operations with NATO assets. The pull came from European countries, notably from France. In the 1980s, with increasingly concrete plans for closer cooperation in a European union, the ambition emerged for a greater European role also in the field of security and defence. Meeting in Rome in October 1984 to commemorate the thirtieth anniversary of the modified Brussels Treaty, WEU foreign and defence ministers agreed to reinvigorate cooperation.[16] As the 1987 WEU Hague Platform later made clear: 'We are convinced that the construction of an integrated Europe will remain incomplete as long as it does not include security and defence.'[17] The WEU's first concerted military actions came with *Operation Cleansweep*, the multinational mine-clearance operation in the Persian Gulf commencing in 1988, followed by WEU coordination during *Operation Desert Storm* in 1990/1991.[18]

With the end of the Cold War, the aim of strengthening European security and defence cooperation became a central part of the wider effort to construct a post-Cold War European security architecture. European ambitions to enhance the new European Union's role in foreign, security and defence issues were anchored in the Treaty on European Union, the Maastricht Treaty, signed on 7 February 1992, and the level of ambition for European-led operations set in the WEU's 'Petersberg tasks' of 19 June 1992.[19] Initially, these European ambitions materialized in the form of the aim to build a European Security and Defence Identity (ESDI). The unfolding civil war in Yugoslavia spurred European efforts, but also became a major test for them. The conflict clearly showed the WEU's need to draw on NATO resources, which to a large degree were US resources, in order to engage in larger military operations.

The push factor was the approach of the United States, particularly during the Clinton administration from 1993.[20] During the Cold War, the United States had consistently urged its European partners to shoulder a greater share of the collective defence burden. In addition, in the early 1990s, US thinking emphasized burden-sharing. The United States provided most of the forces for operations *Desert Shield* and *Desert Storm* in 1990–1991. As the crisis in Yugoslavia unfolded, it also saw growing pressure from European allies for greater US military involvement. Stung by its 1992 to 1994 intervention in Somalia however, the US was reluctant to commit ground forces. The Somalia experience 'heightened awareness that, in the post-Cold War environment, there might be cases when it would be in the interests of the United States for Europe to have the capability to employ military forces without the direct involvement of American troops'.[21] The United States thus proposed that the CJTF 'allow nations to stand aside from future non-Article 5 operations without exercising a veto over operations that are favoured by most other allies'.[22]

The United States thus encouraged European efforts to build a European ability to conduct military operations based on both a desire to increase transatlantic burden-sharing and reluctance to put US soldiers in harm's way. There were, however, two caveats to US support. The first was that the European assumption of a greater burden should not lead to wasteful duplication and the emergence of a competitor to NATO. Already the 1994 Brussels Summit declaration emphasized that CJTF should be implemented 'in a manner that provides separable but not separate military capabilities that could be employed by NATO or the WEU'.[23]

The second US caveat was that the United States wanted to retain overall control over the use of NATO assets – which in several areas consisted of mainly US assets made available to NATO – even in European/WEU-led operations. The preferred vehicle for maintaining such control was to retain a role for the Supreme Allied Commander Europe (SACEUR), always a four-star US officer double hatted as commander of US European Command. This was unacceptable to France, however, because it was not part of NATO's military structure and because it placed greater emphasis on direct political control of military operations than the United States.[24] For France, the US insistence on basing the CJTF and thus the command-and-control arrangements for non-Article 5 operations firmly in the NATO command structure, under SACEUR, seemed aimed at marginalizing French influence. From the US perspective, the French insistence on specifically designing command-and-control arrangements for individual operations, based on the composition of the specific force, would undermine NATO.[25] The terms under which NATO assets were to be made available to the WEU, thus became the object of a protracted dispute.

The failed European response to the civil war in Bosnia-Hercegovina, the general lack of French success in promoting the creation of an independent European defence capability, and the victory of Jacques Chirac in the 1995 French presidential election gradually shifted French policy towards NATO.[26] Under President Chirac, French aims were not abandoned but rather refocused: the aim became 'the Europeanisation of NATO and not the setting up of a separate military structure within the WEU'.[27] On 5 December 1995, France announced that it would rejoin the NATO Military Committee and meetings of NATO defence ministers.[28] Combined with a willingness among the allies to accommodate France, this allowed for a political rapprochement and a breakthrough in negotiations over CJTF.

At their meeting in Berlin on 3 June 1996, NATO foreign ministers agreed to build ESDI within NATO and to an overall 'politico-military framework' for the CJTF. Most significantly, the framework allowed for European-led CJTFs in the NATO command structure. These were to be enabled through 'dual hatting' European officers in the command structure and allowing the participation of 'all allies' (i.e. including France and

Spain, which were not part of NATO's military structures) in the CJTF nuclei (see below).[29] The agreement also entailed the establishment of a permanent Policy Coordination Group (PCG) to accommodate French calls for tighter political control over non-Article 5 operations.[30] The PCG was to 'meet the need, especially in NATO's new missions, for closer coordination of political and military viewpoints'.[31] Hailed as the 'first significant change in the way the alliance does business since 1966', the historic significance of the agreement is seen in that it gave name to the Berlin Plus arrangement in 2003.[32]

The Berlin breakthrough again shifted the emphasis of NATO's internal work on CJTF back to military channels. Political issues still served as breaks on overall progress, however. Two issues in particular were important. The first was the failure of France to rejoin the NATO command structure. The Berlin agreement set France on the path of military reintegration in NATO (with the exception of NATO's nuclear cooperation). Against this background, since 'everything indicates that both France and Spain will fully participate in the new military structure', the special arrangements created by the July 1996 defence ministers' meeting were expected to become 'unnecessary'.[33]

The conditions France set for rejoining the command structure included a French commander of NATO's southern command, AFSOUTH. As the United States had refused to surrender this command to the United Kingdom in the 1950s, it also rejected the French claim.[34] French President Chirac argued for the French cause in a letter to US President Bill Clinton. President Clinton and the United States refused to budge on AFSOUTH, both due to its increasing interests in the Middle East and, as in the 1950s, to its insistence upon retaining the Sixth Fleet under a US NATO commander.[35] As a result, while Spain on 18 December 1996 announced its integration into the NATO command structure, France 'announced prior to the [1997 Madrid] summit that NATO had not yet changed sufficiently to permit it to enter into full military cooperation with a new NATO'.[36] This brought renewed emphasis on the special arrangements decided in the summer of 1996.

The second issue that served as an impediment on progress with CJTF implementation was a battle played out mainly outside NATO. While France and the United States were the main opponents in the struggle within NATO, the United Kingdom led resistance to French ideas in the European Union. The division between the French-led Europeanist and the British-led Atlanticist visions stood at the core of the dispute. For the Atlanticists, NATO was the primary institution and should hold a 'right of first refusal' when military operations were considered.[37] European institutions could conduct operations using NATO assets if NATO (i.e. notably the United States) was unwilling to commit. For Europeanists, European-led operations were a goal unto themselves. Such operations should be enabled through the independent use of NATO assets when required.

Throughout the 1990s, European leaders debated how to construct the Common Foreign and Security Policy called for by the Maastricht Treaty. This included, notably, the extent to which the EU should institutionalize security and defence issues, and the nature of its relationship with the WEU. As Rafael Estrella concluded at the time: 'It is difficult to imagine the effective implementation of CJTF HQ using NATO resources by European Allies without a clarification of the relationship between the WEU and the EU.'[38] The 1997 Treaty of Amsterdam marked a milestone – 'the actual start of the EU's transformation into a security and defence political actor'.[39] A real breakthrough came only, however, with the British–French agreement of St-Malo on 4 December 1998.[40] This agreement opened the door for the creation of a European Security and Defence Policy within the EU, and for further agreement at the European Council meetings in Cologne in June and in Helsinki in December 1999, the latter yielding the 'Helsinki headline goal', or the level of ambition for EU-led military operations.

The US's reaction to St-Malo was reserved, but positive. Three days later, on 7 December 1998, US Secretary of State Madeleine Albright stated that the United States welcomed 'the call from Tony Blair ... for Europeans to consider ways they can take more responsibility for their own security and defence'. Albright framed the conditions for US support in three Ds: 'decoupling, duplication and discrimination'.[41] European decision-making should not be decoupled from NATO, building European capability should not lead to duplication of defence efforts, and there should be no discrimination of non-EU NATO allies. US acceptance opened for agreement at NATO's April 1999 Washington Summit on taking NATO–WEU and NATO–EU relations further both in overall terms, and specifically related to the CJTF concept.

In essence, the key parts of the CJTF concept were finalized in 1999. This was most evident in the start of phase 3, the full implementation phase, in December. NATO in 1999 claimed that CJTF was to be fully implemented, 'including the acquisition of necessary headquarter ... equipment', in 'late 2004'.[42]

While the military implementation of CJTF forged ahead, it was to take more than three years before the adapted Berlin framework for allowing European-led operations using CJTF (and other NATO assets) – the Berlin Plus agreement – was finalized. This took place on 17 March 2003 in the form of an exchange of letters between the EU's High Representative for the EU's Common Foreign and Security Policy Javier Solana, and NATO Secretary General Lord Robertson.[43] Only a few days later, the EU established *EUFOR Concordia* in the Republic of Macedonia by taking over NATO's *Operation Allied Harmony*. In 2004, *EUFOR Althea* in Bosnia-Hercegovina was similarly established, based on NATO's SFOR operation.[44]

The most important reason for the yet further delay was an issue more or less unrelated to the main debates of the 1990s: the role of non-EU NATO allies, and that of Turkey in particular. At the WEU Council of Ministers' meeting in Rome in 1992, the WEU admitted Iceland, Norway and Turkey as associate members.⁴⁵ However, with the agreement after St-Malo to build the ESDP and transfer the responsibilities and bodies of the WEU into the EU, a process that started on 13 November 2000, non-EU NATO allies were left on the outside.⁴⁶ Turkey reacted to 'the loss of the privileged rights of participation to the decision-making and operational activities of the WEU' and 'the danger that [the EU could become engaged] in Turkey's immediate vicinity without Turkey's involvement or consent' by blocking the NATO–EU agreement.⁴⁷ Only on 16 December 2002 was consensus reached on the involvement of non-EU European NATO members in EU-led operations in the joint 'EU–NATO Declaration on ESDP'.⁴⁸ This opened the door for the finalization of the Berlin Plus arrangements in March 2003.

In May 2004, the EU admitted the Greek part of Cyprus as a new member. As a consequence, Turkey blocked the modification of the NATO–EU security agreement to include Cyprus, effectively halting further use of the Berlin Plus framework and strictly limiting NATO–EU relations.⁴⁹ The Turkish move removed the option of using CJTF for European-led operations, but it did not hinder its use by NATO in its own operations. To understand why this did not happen either, we must look at the military perspective.

The military perspective

With the end of the Cold War, NATO allies reduced defence spending and the size and readiness of their military forces. A reorganization and reform of NATO's military structures also quickly came on to the agenda. In the late 1980s, NATO's integrated command structure was a massive, mainly static structure consisting of about 24,500 personnel in 78 headquarters of varying size and significance. NATO command structure (NCS) reform has since taken place in four major rounds: 1991 to 1992, 1994 to 1997, 2002 to 2003 and 2010 to 2011.⁵⁰ In addition, there were smaller but significant adaptations in between these rounds.

The 1991 to 1992 round of reforms heralded the advent of a 'streamlined' command structure, 'adapted to the new environment' as set out in the 1991 Alliance's Strategic Concept.⁵¹ It quickly became evident, however, that the reforms were inadequate in dealing with emerging requirements. The tragedy in the former Yugoslavia drew NATO into an increasingly active military role. NATO's command-and-control structure could lead maritime and air operations over and off the western Balkans, starting in 1992, from its static locations. It was evident, however, that the command structure was ill equipped to take command of a larger land operation outside the Treaty area.

The January 1994 Brussels Summit therefore launched a more substantial revision of the NCS. At the core of the new reforms stood the CJTF concept. Further work on reform of the size and composition of the NATO command structure took place in traditional fashion under the leadership of the Military Committee (in the form of the Long Term Study), and under the auspices of the Defence Planning Committee. For work on the CJTF, special provisions were necessary to accommodate France and Spain, which were not part of NATO's military structures. A dual-track process of command structure reform was thus created 'to ensure that French officials would have a full say in the development of CJTF arrangements'.[52]

As noted above, the initial military work on the CJTF proceeded quickly. Already in March 1994, an initial 'draft operational concept for CJTF' command and control was developed 'under the executive agency of' SACEUR. In September 1994, several 'follow-up studies to define many aspects of the headquarters concept in greater detail' were presented to the MC.[53]

While many aspects would evolve over the years, the basic set-up of CJTF headquarters remained constant. In order to avoid building separate structures for collective defence and non-Article 5 operations, and to limit costs, CJTF headquarters were to be formed on the basis of existing NATO commands. The number and location of the headquarters organized and equipped for CJTF operations was a source of some debate. The ambition was to field both land- and sea-based CJTFs. Consensus was quickly reached on basing the single sea-based CJTF in one of Allied Command Atlantic's (ACLANT) functional subordinate commands, Strikefleet Atlantic. After the 2002 to 2003 reorganization of ACLANT into Allied Command Transformation, the CJTF sea-based headquarters was assigned to Joint Forces HQ Lisbon.

The number of land-based CJTFs was more hotly debated. The United States initially proposed to have three, associated with the three regional subordinate commands under Allied Command Europe (ACE).[54] The reorganization of the command structure that was agreed upon with the approval of the Long Term Study in 1997 resolved the issue, however. ACE was left with only two subordinate commands: Allied Forces Northern Europe in Brunssum, the Netherlands, and Allied Forces Southern Europe in Naples, Italy. There were therefore only two headquarters capable of accommodating CJTF structures. The reorganization of ACE into Allied Command Operations, and the regional commands in Joint Forces Commands, decided in 2002 to 2003, did not change this structure.

Within the designated, existing NATO headquarters, a permanent CJTF nucleus was to be established.[55] The nucleus consisted of designated personnel from all essential branches of the headquarters double hatted as members of the core CJTF HQ staff. An even smaller key nucleus staff was made responsible for maintaining CJTF functions on a daily basis. The nucleus, consisting of about 500 personnel, was not designed as a fully fledged headquarters. When activated, it relied on augmentation 'with modules (resources provided by other NATO or national sources) and individuals (personnel)'. Modular (i.e. complete, fully staffed headquarter and support components) and individual augmentation was initially to come 'from within the Alliance's military structure before seeking additional augmentation from nations'.[56]

Fully developed and set up, the land-based CJTF HQ came to number an estimated 2,000 personnel, and was to be in command of the subordinate component commands (land, sea, air, special forces) necessary to conduct a corps-size land operation. While the CJTF concept emphasized flexibility, including the flexible format and size of CJTF headquarters, it seems to have become synonymous with corps-size land operations.[57] This is in many ways not surprising. NATO's main operational experiences from the Western Balkans in the 1990s, with IFOR/SFOR in Bosnia-Hercegovina and KFOR in Kosovo, were of corps-sized operations (peaking at 60,000 strong in IFOR/SFOR and 50,000 in KFOR). Moreover, the ability to conduct a corps-size operation – consisting of 'up to 15 brigades or 50,000–60,000 persons' – was one of the 1999 Helsinki headline goals.[58]

Given the ambitions set out in the 1994 Brussel Summit decision, CJTF was to have three employment possibilities: 'a pure NATO CJTF, a NATO-plus CJTF that would include some non-NATO states, and a European-led/WEU CJTF'.[59] Moreover, while CJTF was primarily designed for non-Article 5 operations, employment for collective defence purposes, in a pure NATO configuration, was not excluded. The CJTF was thus constructed as a seemingly very flexible, multi-format and cost-effective command-and-control system.

The relatively detailed proposals presented by NATO's military authorities in the autumn of 1994 subsequently became the object of political scrutiny. Several issues needed political agreement. The most difficult were how to accommodate French and Spanish participation in the planning for and conduct of non-Article 5 CJTF operations, and how CJTFs (and other NATO assets) should be made available for WEU-led operations.

As described above, NATO foreign ministers reached a breakthrough agreement on the overall politico-military framework for CJTF at their meeting in Berlin on 3 June 1996. Meeting ten days later, on 13 June 1996, NATO defence ministers (including the French minister for the first time in 30 years) took the decision of their foreign minister colleagues one step further.[60] This notably included the decision to establish two organizational entities and other special arrangements to accommodate the special needs of France and Spain, and those of NATO-enabled, WEU-led operations. At NATO Headquarters in Brussels, a Capabilities Coordination Cell was to support the Military Committee in liaising with WEU bodies, and in developing military guidance for the planning of non-article 5 contingencies, including specifically CJTF employments and WEU-led operations. It was established as an independent entity within the International Military Staff and would include French and Spanish officers. At SHAPE in

232 *P.S. Hilde*

Mons, a Combined Joint Planning Staff, also including French and Spanish officers, was co-located with Allied Command Europe. Its role was to perform centralized CJTF HQ planning and coordination, including the implementation of the CJTF concept in lower level commands, to liaise with the WEU Planning Cell, and to conduct planning for CJTF operations and exercises.

After the Berlin and Brussel meetings in summer 1996, work could again proceed on a range of aspects of both WEU–NATO cooperation in general and on the implementation of the CJTF framework.[61] The Policy Coordination Group played a key role in this work along with the Military Committee. The aspects included WEU–NATO work on identifying NATO assets and capabilities that could be made available for WEU-led operations. The modalities of the release, monitoring and return or recall, notably in case an Article 5 operation so required, of such assets and capabilities was another. Other strands of work were mechanisms for consultation and information sharing, the adaption and harmonization of NATO and WEU planning processes and exercises, and the modalities for WEU involvement in NATO defence planning processes.

The most challenging aspect was, however, the need to identify European command arrangements for WEU-led CJTF operations. In early 1996, 'the French and British governments proposed what became known as the "Deputies proposal." They suggested that the Deputy SACEUR, traditionally a senior European officer, and other European officers in the NATO command structure, wear WEU command hats as well as their NATO and national command hats.'[62] While the proposal saw opposition from the US Joint Chiefs of Staff and SACEUR General George Joulwan, the allies quickly adopted it.[63] On 26 March 1997, the Military Committee issued MC 403, which set the terms of reference for Deputy SACEUR (DSACEUR) both for his NATO role, but also, more importantly, for his ESDI role. While commanders of specific operations, including commanders of WEU-led CJTF operations would be selected on a case-by-case basis, DSACEUR became the key, day-to-day European commander in NATO.

By the 1999 Washington Summit, the allies could:

> note with satisfaction that the key elements of the Berlin decisions are being put in place. These include flexible options for the selection of a European NATO Commander and NATO Headquarters for WEU-led operations, as well as specific terms of reference for DSACEUR and an adapted CJTF concept.[64]

NATO Secretary General Javier Solana confidently concluded that:

> The adjustments made to the command structure now allow for European-led NATO operations and the Combined Joint Task Force

Military change in NATO 233

initiative, soon to be fully implemented, will allow European allies to use NATO assets without necessarily involving the North American allies directly.[65]

From December 1999, the CJTF implementation phase saw the issuance of a series of documents directing NATO's command structure to adjust to the CJTF concept. In June 2000, as noted above, the Military Committee issued a revised version of MC 389 (MC 389/1).[66] It was followed in March 2002 by a draft version of a CJTF operational concept developed by the two NATO strategic commanders, SACEUR and Supreme Allied Commander Atlantic (SACLANT, traditionally a US admiral). In December 2002, the MC presented the minimum military requirements for the initial CJTF operational capability (termed 'foundation capability'). On 3 April 2003, the MC adopted adapted terms of reference for DSACEUR (MC 403/1), making DSACEUR 'NATO's Strategic Coordinator for the EU with respect to European issues'.[67] In May and September 2003, the NAC approved funding for the so-called capability packages the MC had recommended for making CJTF headquarters deployable. Finally, as noted, in May 2004, the MC issued the final version of the CJTF capstone document, MC 389/2.

As seen above, the employment of the CJTF in European-led operations was blocked by Turkey after Cyprus became a member of the EU. NATO had, however, a fully developed, seemingly flexible and cost-effective instrument in the CJFT concept for its own crisis response operations. Already in April 2006, however, the Military Committee decided to postpone CJTF initial (foundation) capability to 2015 and full (tailored) capability to 2018 – that is, indefinitely.[68]

A key reason for the MC's 2006 decision is found in the decision at the 2002 Prague Summit to establish the NATO Response Force (NRF). Although scalable and nominally very flexible, the CJTF concept was not designed to support the NRF. Relying on augmentation to become fully functional, CJTF headquarters were both too big and at too low readiness to serve the NRF.[69] To address the needs of the NRF, NATO created an adapted CJTF model, termed the Deployable Joint Task Force (DJTF). A DJTF was a small but functionally complete standing HQ that would see the deployment of the NRF commander with the force, and the use *reach back* – the extensive use of secure, deployable communication and information systems (DCIS) – to draw on the resources of the parent headquarters.[70] This reliance on DCIS equipment brought the NRF-DJTF concept into competition with the plans for the full CJTFs for available equipment and scarce investment resources. It was a competition which NRF and DJTF won. This economic aspect seems to have been the key reason for the 2006 decision to postpone the implementation of CJTF.

Why the CJTF concept failed and lessons learned

The above analysis of the political and military aspects suggests that there were both structural/political and specific/military reasons for the failure of the CJTF concept. Clearly, Turkey's 2004 decision to block the hard-fought consensus on the employment of CJTF headquarters (and other NATO assets) in WEU/EU-led operations is important. While WEU and EU employment of CJTFs represented only one aspect of the concept, most of the energy NATO and allies had spent on agreeing on the concept had been on this aspect. This included, as described above, hammering out consensus on a wide range of issues ranging from broad questions like the relationship between the EU and WEU, to specific challenges like how to enable the participation of French (and Spanish) officers in CJTF headquarters. By the early 2000s, it is conceivable that NATO and allied decision-makers had come to see the CJFT as primarily aimed at enabling EU-led operations, while the significance of other aspects had waned. The decision to postpone CJTF implementation indefinitely, taken two years after the Turkish decision to block Berlin Plus, suggests that this may have been a factor.

Overall, however, the political complexity inherent in reaching agreement in NATO on far-reaching reform only slowed down work on the CJTF concept. It took ten years, but the Alliance did in the end reach consensus on both the concept and its use in European-led operations. In this context, it is worth considering the political complexity of the CJTF concept. It is hard to conceive of a politically more challenging aim than to make the military assets of an alliance of 16 members, increasing to 19 in 1999, available to a different international organization – of which not all allies are members. In addition, the CJTF had to accommodate both strong French-led ambitions for a European security and defence role, and the French non-membership in NATO's integrated command structure. Few if any other reforms in NATO have been of a comparable complexity. This shows that the systemic factor – the multinational nature of NATO – does not render military change impossible. Given NATO's persistence and adaptation in the Cold War period, this is hardly a surprising conclusion. The CJTF experience does show, however, that the multinational nature of NATO introduces the need to accommodate special interests through institutional or other forms of adaptation. It also shows that the speed at which military change can take place is limited. Even in cases where there is broad political consensus, as, for instance, in the establishment of the NRF, political considerations will slow down and impede progress.[71]

When asked why the CJTF concept failed as a concept for making the NATO command structure more flexible and deployable, NATO insiders invariably pointed to the CJFT being too big and too expensive as the most important reason.[72] As noted above, the Alliance's experience from operations in the 1990s put emphasis on corps size, joint headquarters and forces. In the early 2000s, however, NATO gradually moved away from the ambition of rapidly fielding such large joint headquarters. As noted above, one reason was the need to cater for the command-and-control needs of the NATO Response Force. A further, possible reason for the shift may be the longer term experience from operations. In both SFOR and KFOR, and later in the ISAF operation in Afghanistan, operation-specific headquarters – composite headquarters in NATO terminology – were established.[73] While these operations required a land headquarters to deploy into the area of operation initially – in both IFOR/SFOR and KFOR the British-led Allied Rapid Reaction Corps – the larger, joint mission headquarters was built up over time. Moreover, as noted above, technological advances allowing virtual integration of headquarter resources, through *reach back*, lessened the relevance of rapidly deploying a large joint headquarter structure like CJTF.

In 2006, allies confirmed the changed operational emphasis in setting a new military level of ambition for NATO. Instead of the existing ambition of three major joint operations (MJO), basically corps-size operations, *Ministerial Guidance 2006* – the political guidance for NATO defence planning – set NATO's level of ambition to two MJO and six smaller joint operations.[74] Also in 2006, the *Comprehensive Political Guidance* – an attempt by NATO to give strategic guidance without drafting a new strategic concept – emphasized the global nature of threats. NATO thus shifted its focus away from major operations in and around Europe, to smaller operations globally. This made fielding two CJTFs only part of the answer. In 2008, due to the need to address the ambition to conduct six smaller operations, NATO came up with a new concept called the Deployable Joint Staff Elements (DJSE).[75] While the DJTF approach was essentially constructed on the same principles as the CJTF, and therefore compatible, the DJSE was not. Its adoption was thus a clear sign that the CJTF, though nominally still alive, survived in a zombie-like existence.

The overall conclusion which emerges is that the CJTF failed primarily because by the time the Allies reached consensus on making NATO assets available to European-led operations, a key part of the rationale for the concept, it had become outdated. Not only had the military requirements changed, or at least started changing, but new ideas and concepts had emerged that pushed the CJTF down the list of priorities.

What lessons may we draw from the CJTF experience on how to approach military change in a multinational setting? One lesson is that while they may be politically attractive, catch-all solutions are likely to involve the need to reach consensus on a range of both political and military aspects prior to implementation. This is likely, as in the case of the CJTF, to lead to substantial delays in implementation and to make the proposed reform more vulnerable to political obstacles. Reforms thus run the risk of becoming outdated before being implemented. If broad reforms are necessary, allies should seek, to the extent possible, to isolate sensitive issues from the overall implementation of the reform. In the case of CJTF, such an approach would have entailed allies seeking consensus on and implementing the CJTF in its pure NATO and NATO-plus configuration separately from tackling the complicated issue of enabling WEU/EU operations. While such separation may in itself be politically challenging, it would likely enable a more speedy implementation of military change.

A second lesson is that given the almost inevitably slower pace of reform in a multinational setting, evolution rather than revolution may often be the better choice. Evolution was not an option at the time the CJTF concept was born, but the introduction of DJTF as a variant of CJTF for the NATO Response Force represents a seemingly successful, if relatively short-lived evolution. In NATO, as in national and international politics in general, political leaders often try to shape the agenda through launching grand new reforms under catchy slogans. NATO Secretary General Anders Fogh Rasmussen has favoured this approach. While politically attractive, and sometimes effective in generating political will, new ideas and concepts also generate the potential for political obstacles. Adapting or enhancing existing, consensus-based arrangements may therefore be more conducive to change, and should be carefully considered. The decision at the 2014 Wales Summit to create a Very High Readiness Joint Task Force, or Spearhead Force, may become an example of a good compromise. A catchy political initiative that emerged in response to the crisis in Ukraine, the VJTF is presented as a new force – supposed to reach 'initial operational capability in the fall of 2015 and full operational capability in early 2016'.[76] Once created, however, the VJTF seems set to become part of the NATO Response Force. While the history of the NRF has been far from unproblematic, including the new force under the NRF umbrella may ease its birth, as it should limit the need for new agreements on rules and procedures, and perhaps even the need to revise old ones. Whatever the end result for the VJTF, it is a reminder that NATO both has and needs to continue to adapt to changing political and military requirements. In doing so, NATO's leaders should not only have bold ambitions, but also seek out and heed the lessons of past reforms. Institutional memory and lessons learned are not highly prioritized in NATO. They should be.

Notes

1 Out-of-area at the time meaning outside the area defined in Article 6 of the North Atlantic Treaty.
2 See also Anthony Cragg, 'The Combined Joint Task Force Concept: A Key Component of the Alliance's Adaptation', *NATO Review*, vol. 44, no. 4 (July 1996). Online edition at www.nato.int.
3 On meeting see: NATO, 'Press Statement Meeting of NATO Defence Ministers Travemunde 20th–21st October 1993', 20 October 1993. All public NATO documents referred to in this chapter are available at www.nato.int/cps/en/natolive/official_texts.htm.

4 NATO, 'Declaration of the Heads of State and Government Participating in the Meeting of the North Atlantic Council ("The Brussels Summit Declaration")', 11 January 1994. On PfP see also NATO, 'Partnership for Peace: Framework Document', 10–11 January 1994.
5 NATO, 'The Brussels Summit Declaration', paragraph 9.
6 Western European Union: *Declaration on the Role of the Western European Union and its Relations with the European Union and with the Atlantic Alliance*, 10 December 1991. WEU documents are available at www.weu.int.
7 See also e.g. Sloan, 'Combined Joint Task Forces (CJTF) and New Missions for NATO', *CRS Report for Congress 94–249 S*, 17 March 1994, 2–3.
8 Charles Barry, 'NATO's Bold New Concept – CJTF', *Joint Forces Quarterly*, summer 1994, 47.
9 In US doctrine, a Joint Task Force was and is a group of forces from two or more of the armed services (land, sea and air forces), put together to accomplish missions with specific, limited objectives. A Combined Task Force is (implicitly) defined as a force composed of military components from more than two countries. US Joint Chiefs of Staff, *Joint Publication 3.0. Joint Operations*, 1995 edition, II-13, available at www.bits.de/NRANEU/others/jp-doctrine/jp3_0(95).pdf.
10 Sloan points to the Allied Command Europe Mobile Force (AMF) established in 1960 as an example of a CJTF in 'Combined Joint Task Forces (CJTF) and New Missions for NATO', 2.
11 Sloan, 'Combined Joint Task Forces (CJTF) and New Missions for NATO', 46.
12 Lt. Gen. Mario da Silva, 'Implementing the Combined Joint Task Force Concept', *NATO Review*, vol. 46, no. 4 (winter 1998). Online edition.
13 Da Silva, 'Implementing the Combined Joint Task Force Concept'. Another source gives the following breakdown of the phases: Phase I – establish initial capability 1996 to March 1999; Phase II – assessments June 1998 to October 1999; Phase III – full implementation March 1999 to January 2000. See Peter L. Jones, *NATO's Combined Joint Task Force Concept – Viable Tiger or Paper Dragon* (Fort Leavenworth, KA: US Army War College, 1999), 46–47.
14 NATO, 'Final Communiqué Meeting of the North Atlantic Council in Defence Ministers Session held in Brussels', 2 December 1999.
15 Parts of this chapter are based on background interviews and consultation of NATO documents. The interviewees were with a small number of highly experienced, present and former International Staff and International Military Staff members, as well as national representatives.
16 See WEU, *Rome Declaration*, 24 October 1984. The WEU members were Belgium, France, Luxembourg, the Netherlands, the United Kingdom, the Federal Republic of Germany and Italy. The first five were the original signatories of the Brussels Treaty. West Germany and Italy joined with the signing of the 1954 Paris Agreement that amended the Brussels Treaty.
17 WEU, *Platform on European Security Interests*, The Hague, 27 October 1987.
18 See Arie Bloed and Ramses A.Wessel (eds), *The Changing Functions of the Western European Union (WEU): Introduction and Basic Documents* (Dordrecht: Kluwer Academic, 1994), xxvii–xxviii.
19 The *Treaty on European Union* may be accessed at: http://old.eur-lex.europa.eu/LexUriServ/LexUriServ.do?uri=OJ:C:1992:191:FULL:EN:PDF. For Petersberg tasks see WEU, *Petersberg Declaration*, Bonn, 19 June 1992, part II.
20 Stanley R. Sloan, 'NATO's Future: Beyond Collective Defense', *McNair Paper 46* (Washington, DC: National Defence University, 1995), 11. Available at: www.dtic.mil.
21 S. Nelson Drew, 'NATO From Berlin to Bosnia: Trans-Atlantic Security in Transition', *McNair Paper 35* (Washington, DC: National Defense University, 1995), 21–22.

22 Sloan, 'NATO's Future: Beyond Collective Defense', 4.
23 NATO, 'The Brussels Summit Declaration', paragraph 9.
24 On this see Lawrence S. Kaplan, *NATO United, NATO Divided: The Evolution of an Alliance* (Westport, CT: Praeger, 2004), 123; Rob de Wijk, *NATO on the Brink of the New Millennium: The Battle for Consensus* (London: Brassey's, 1997), 99, 126–127.
25 Drew, 'NATO From Berlin to Bosnia', 18.
26 De Wijk, *NATO on the Brink of the New Millennium*, 124; Jeremy Ghez and F. Stephen Larrabee, 'France and NATO', *Survival*, vol. 51, no. 2 (2009), 77–90, at 78.
27 De Wijk, *NATO on the Brink of the New Millennium*, 125.
28 Stanley R. Sloan, 'NATO Adapts for New Missions: The Berlin Accord and Combined Joint Task Forces (CJTF)', *CRS Report for Congress 96–561F*, 19 June 1996, 4.
29 NATO, 'Final Communiqué, Ministerial Meeting of the North Atlantic Council', 3 June 1996, point 7.
30 Sloan, 'NATO Adapts for New Missions', 3.
31 NATO, 'Final Communiqué, Ministerial Meeting of the North Atlantic Council', point 6. The Provisional PCG established in 1994 had worked out the proposals adopted in Berlin.
32 US Ambassador to NATO Robert Hunter quoted in Sloan, 'NATO Adapts for New Missions', 1.
33 Rafael Estrella, 'CJTF and the Reform of NATO', NATO Parliamentary Assembly Draft General Report AN 230 DSC (96) 8 rev. 1, 24 October 1996. Available online at: www.nato-pa.int/archivedpub/comrep/1996/an230dsc.asp.
34 On the 1950s, see Gregory W. Pedlow, *The Evolution of NATO's Command Structure, 1951–2009*, 4. Available online at: www.aco.nato.int.
35 For a US perspective on the dispute, see INSS, *Allied Command Structures in the New NATO* (Washington, DC: National Defense University, 1997). Available online at: www.ndu.edu.
36 Stanley R. Sloan, 'NATO: July 1997 Madrid Summit Outcome', *CRS Report for Congress 97–443 F*, 14 July 1997, 3.
37 Jennifer Medcalf, 'Cooperation Between the EU and NATO', in Joachim Krause, Andreas Wenger and Lisa Watanabe (eds), *Unraveling the European Security and Defense Policy Conundrum* (Bern: Peter Lang, 2003), 103–105.
38 Estrella,'CJTF and the Reform of NATO'.
39 Hanna Ojanen, 'EU–NATO Relations after the Cold War', in Basil Germond, Jussi Hanhimäki and Georges-Henri Soutou (eds) *The Routledge Handbook of Transatlantic Security* (London: Routledge, 2010), 182.
40 See 'Joint Declaration on European Defence', reprinted in Maartje Rutten (compiled) *From St-Malo to Nice. European Defence: Core Documents*, Chaillot Papers no. 47 (May 2001), 10–11. Available online at: www.iss.europa.eu.
41 Madeleine K. Albright, 'The Right Balance Will Secure NATO's Future', *Financial Times*, 7 December 1998, Rutten, 'Joint Declaration on European Defence', 8–9.
42 NATO, 'The Combined Joint Task Force Concept', *Press Factsheet*, available online at: www.nato.int/docu/comm/1999//9904-wsh/pres-eng/16cjtf.pdf.
43 Ojanen,'EU–NATO Relations after the Cold War', 185.
44 Nick Hawton, 'EU Troops Prepare for Bosnia Swap', *BBC News*, 23 October 2004.
45 WEU, 'Council of Ministers Communiqué', 20 November 1992. Greece became a full member and Denmark and Ireland observers. Sweden and Finland became WEU observers in 1995 when they joined the EU.
46 WEU, 'Marseilles Declaration', 13 November 2000.

47 Sinem Akgul Acikmese and Dimitrios Triantaphyllou, 'The NATO–EU–Turkey Trilogy: The Impact of the Cyprus Conundrum', *Southeast European and Black Sea Studies*, vol. 12, no. 4 (December 2012), 562.
48 'EU-NATO: The Framework for Permanent Relations and Berlin Plus', available online at: consilium.europa.eu/uedocs/cmsUpload/03–11–11%20Berlin%20Plus%20press%20note%20BL.pdf.
49 Acikmese and Triantaphyllou,'The NATO–EU–Turkey Trilogy', 564.
50 See Paal Sigurd Hilde, 'Lean, Mean, Fighting Machine? Institutional Change in NATO and the NATO Command Structure', in Andrew A. Michta and Paal Sigurd Hilde (eds) *The Future of NATO: Regional Defense and Global Security* (Michigan, OH: University of Michigan Press, 2014), 135–144.
51 *Defence Planning Committee Final Communiqué*, 12–13 December 1991, paragraph 8.
52 Sloan, 'Combined Joint Task Forces (CJTF) and New Missions for NATO', 3.
53 Charles Barry notes 13 studies, 'most of' which were ready by September. Charles Barry, 'NATO's Combined Joint Task Forces in Theory and Practice', *Survival*, vol. 38, no. 1 (spring 1996), 84.
54 Sloan,'Combined Joint Task Forces (CJTF) and New Missions for NATO', 3.
55 Da Silva, 'Implementing the Combined Joint Task Force Concept'.
56 Da Silva, 'Implementing the Combined Joint Task Force Concept'.
57 NATO also considered establishing smaller scale CJTF HQs capable of commanding smaller forces. It is unclear what came of these plans. Work in NATO on CJTF seems to have focused almost exclusively on the top-level CJTF formations. Development of the sea-based CJTF was seemingly more straightforward and less controversial.
58 EU, 'About CSDP – Military Headline Goals', available online at: eeas.europa.eu/csdp/about-csdp/military_headline_goals/index_en.htm.
59 Barry, 'NATO's Bold New Concept – CJTF', 50.
60 The meeting was called a meeting of the NAC (traditionally the foreign ministers' forum) 'in defence minister session'. NATO, 'North Atlantic Council in Defence Minister Session Final Communiqué', 13 June 1996, point 1.
61 For sources for this and the following paragraph, see n. 15.
62 Sloan, 'NATO Adapts for New Missions', 4.
63 Sloan, 'NATO Adapts for New Missions', 4.
64 NATO, 'Washington Summit Communique: An Alliance for the 21st Century', 24 April 1999, point 8.
65 Javier Solana, 'The Washington Summit: NATO Steps Boldly into the 21st Century', *NATO Review*, vol. 47, no. 1 (spring 1999). Online edition.
66 For sources see n. 15.
67 See e.g. Dieter Farwick, 'Interview with Deputy SACEUR Admiral Rainer Feist', *World Security Network*, 25 January 2004, available online at: www.globalsecuritynews.com/Europe-NATO/dieter-farwick-1/Exclusive-WSN-Interview-with-Deputy-SACEUR-Admiral-Rainer-Feist.
68 Craisor-Constantin Ionita, 'DJSE – A New Concept Regarding a Deployable Command and Control System at Operational Level', *Romanian Military Thinking*, no. 2 (2010), 31.
69 For an early criticism of the CJTF concept see Thomas Cooke, 'NATO CJTF Doctrine: The Naked Emperor', *Parameters* (winter 1998), 124–136.
70 See e.g. Michael L. McGinnis, 'A Deployable Joint Headquarter for the NATO Response Force', *Joint Forces Quarterly*, issue 38 (2005), 60–67.
71 On NRF see e.g. Jens Ringsmose, 'Caught Between Strategic Visions: The NATO Response Force', in Michta and Hilde, *The Future of NATO*, 155–175.
72 See n. 15.

73 The understanding was not new. As Charles Barry pointed out in 1994: 'over time, long-running operations … must necessarily be converted to more permanent arrangements' (Barry, 'NATO's Bold New Concept – CJTF', 88).
74 See Hilde, 'Lean, Mean, Fighting Machine?', 138–139.
75 See Ionita, 'DJSE – A New Concept Regarding a Deployable Command and Control System at Operational Level'. The need to find a role for the two land component commands in the command structure, in Madrid and Heidelberg, was also an important driver for the DJSE.
76 Interview with Lt. Gen. Frederick Hodges in Michelle Tan, 'New Quick-reaction NATO Force to Stand Up Next Year', *Military Times*, 23 October 2014, available online at: www.militarytimes.com/article/20141023/NEWS08/310230050/New-quick-reaction-NATO-force-stand-up-next-year.

13 A new wary titan

US defence policy in an era of military change, Asian growth and European austerity

Austin Long

The United States remains the most powerful nation-state in the world. Yet the years following the terrorist attacks of 11 September 2001 have been extraordinarily difficult for it. Smashingly successful invasions of Afghanistan and Iraq gave way to protracted conflict and a global financial conflagration led to massive increases in US debt. At the same time, militaries in Asia, a region of substantial and increasing US interest, have grown even as traditional European allies have experienced enormous economic challenges, limiting their ability and willingness to generate military power.

This chapter will explore the implications of military change, Asian growth and European austerity for United States defence policy. It does so in four parts. First, it outlines the current state of US defence policy in the context of US domestic politics and economics to provide context. Second, it examines military change, which has been a major factor in the United States for more than a decade. Third, it discusses the effect of Asian growth on US defence policy. Fourth, it describes the impact of European austerity on the transatlantic alliance and the implications for the United States.

US defence policy, politics and economics

The events of the past decade have cast a pall over US defence policy. Significant cuts in the defence budget, as part of an overall effort to reduce US federal government spending, seem inevitable. Reports on defence policy and budgeting with titles like *Healing the Wounded Giant* and *Chaos and Uncertainty* are proliferating in Washington.[1]

However, there are bright spots in the gloom. First, during more than a decade of war, the United States has been extraordinarily successful in implementing military change. As described in the following section, this change has been more evolutionary than revolutionary, yet its consequences are profound as it has created capabilities that are global in reach and, for the foreseeable future, unique to the United States and a handful of its key allies.

Second, overall US military and economic primacy remains fundamentally intact. This is true both regionally (i.e. in Asia and Europe) and globally. While the utility of this primacy is hotly debated by scholars, all but the most sceptical acknowledge that primacy is likely to continue at least for the medium term should the United States continue to pursue it. US retrenchment, if it happens, will be as a result of choice rather than requirement.[2]

It is important to note that 'primacy' or 'hegemony' is not synonymous with 'omnipotence'. This may seem obvious but some analysts believe that because states like China have increased their capabilities, both economically and militarily, and have therefore lessened the relative dominance of the US, it is no longer correct to talk of US primacy. This misses the point of primacy, which remains a fact even if the relative advantage of the United States is less than in 1991. In terms of China, for example, Thomas Christensen's formulation of 'posing problems without catching up' remains accurate.[3] Thus the United States does not need to restore primacy in Asia but rather to maintain it.

Thus, rather than being a 'weary policeman' or 'weary titan', the United States is better thought of as a new 'wary titan', to borrow John Hobson's description of Great Britain in the late 19th and early 20th century.[4] The United States, while still able to maintain primacy, may simply be unwilling to do so, just as Great Britain was able but unwilling to maintain primacy (or something like it) in the earlier period. This is an important contrast to Great Britain in the mid-20th century, which was barely able to remain a great power as it retreated from 'east of Suez' due to economic constraints.

A brief review of the economics of US defence demonstrates that the burden of primacy is relatively light. For fiscal year (FY) 2013, the total expenditure on national defence activities, including ongoing military operations in Afghanistan and elsewhere, was over US$650 billion.[5] This is a staggering sum, yet closer analysis reveals that it is not a crushing burden. This was just under 18 per cent of total federal government spending and about 11.4 per cent of total US government spending (which includes federal, state and local). More importantly, this represents only 4.1 per cent of US gross domestic product (GDP).[6]

A comparison with US history as well as that of other great powers demonstrates that this is an eminently sustainable burden. France, for example, sustained a 4 per cent of net national product average for defence spending for the period 1870 to 1913.[7] However, the French relied heavily on conscription to provide manpower, which saved them substantial sums as conscripts were paid relatively little. Thus the 'real' cost of French defence was undoubtedly higher.[8] Both the United States and Great Britain spent more than 4 per cent of GDP on defence every year for the entire period of the Cold War, with the peak years around 10 per cent.[9]

Yet the fact that these defence burdens were sustainable may not make them wise. An argument, advanced by Paul Kennedy among others, that excessive defence expenditure undermines economic growth and ultimately the ability to maintain primacy means these expenditures may have been foolish in the long run.[10] Many argue that this was true of the Soviet Union, which spent as much as 40 per cent of GDP on defence and related industry.

While this argument may be true in extreme cases like the Soviet Union, the defence burdens just cited do not appear to have reached the threshold to undermine economic viability. France, despite losing territory and population to Prussia following the Franco-Prussian War, managed an average economic growth rate of just over 1 per cent for the period 1870 to 1910.[11] During the Cold War both US and British real GDP growth was somewhat higher, with an average rate of just over 2.5 per cent for the period 1950 to 1990.[12]

If defence spending per se is not crippling, perhaps total government spending is. In FY13, for example, US total federal spending is estimated to have been about 21 per cent of GDP, while total US government spending was probably around 37 perr cent of GDP. However, federal taxation only brought in about 16.7 per cent of GDP, producing a federal deficit of just over 4 per cent of GDP (as noted above, this is about the level of defence spending).[13]

The concern is that such deficit spending is unsustainable, as US total debt has sky-rocketed, so US government spending, including defence spending, must be curtailed. Yet this deficit, far from being insurmountable, could be closed by raising taxes so that total US government taxation matched spending; in other words, if total US taxation was approximately 37 per cent of GDP. This would be slightly over the Organization for Economic Cooperation and Development (OECD) average taxation rate as a percentage of GDP.[14] While significant, this level of taxation is unlikely to strangle economic growth by itself. Germany, for example, has approximately this total tax rate for the two decades following reunification and it averaged just under 1.5 per cent annual real GDP growth.[15] This growth was despite the significant difficulties and costs imposed by reunification as well as the global financial crisis.

Inevitably this leads to the conclusion that the central issue for the United States today, as was the case with Great Britain in the decades before the Great War, is less inability to pay for primacy and more unwillingness to pay. The origin of this gap between expenditures and revenues is the strong anti-statist and to a lesser extent anti-militarist ethos that has been central to the United States from its beginning. As Aaron Friedberg has described, this ethos is the core limitation on US resource extraction for defence. While this ethos eroded during the Cold War due to the perceived level of threat, the United States still refused to come close to matching the Soviet Union in resource extraction for military ends.[16]

This anti-statist ethos failed to fully return following the end of the Cold War due in part to other changes in American society. From the 1930s to the 1960s, two differing visions of liberty contended in the US political sphere. Brendan Green, following the nomenclature of political philosopher Isaiah Berlin, refers to these visions as 'positive' and 'negative' liberty. The former views an expansive role for the state in providing opportunities for individuals to exercise choice (e.g. by providing a social safety net) as critical to liberty, while the latter views the freedom from state constraint (e.g. taxes and regulations) as the defining feature of liberty.[17]

By the 1960s, positive liberty had come to dominate the US political sphere, as the New Deal of the Roosevelt Administration was solidified and the Great Society of the Johnson administration was born. Subsequent attempts to resuscitate negative liberty in the 1980s were unsuccessful, as the notionally anti-statist Reagan and Bush administrations presided over significant increases in government spending (in part driven by a desire to expand defence spending). The George W. Bush administration, which though nominally conservative expanded the role of the state in a variety of ways, seemed to indicate that positive liberty would continue to dominate US politics, as both social spending and defence spending increased.

Yet, following the US elections in 2008, the concept of negative liberty rose from the dead. Embodied in the diverse anti-statist Tea Party movement and particularly in a set of Congressmen elected in 2010, this revived concept of negative liberty is critical to understanding the current state of the US government and, by extension, the US defence budget. The idea of raising taxes is fundamentally anathema to these groups, a position that has been adopted by most of the Republican Party. Yet, unlike the previous two generations of Republicans, these new proponents of negative liberty are unwilling to countenance anything less than the significant contraction of the size of the federal government, including defence spending (though this is not their favourite target).

The extent to which this rebirth of negative liberty as a political force in the United States will endure is at present unclear. If it continues as a significant force, then the United States could conceivably return to something akin to the mid- to late 1990s when defence budgets declined below 4 per cent for much of the decade. However, the overall decline of the demographics underlying the movement (principally older and white) may mean that the rebirth is short-lived.

Regardless of the long-term viability of renewed negative liberty, the US defence budget will be tighter in the near term than it has been in the previous decade. As discussed in the following section, the US military has developed a substantial number of capabilities and assets during these 'fat' years, but will face significant trade-offs between maintaining size, particularly in the grounds forces, and improving the quality of capital assets (aircraft, ships, etc.), some of which are still of Cold War vintage. Most notably the US Air Force continues to operate B-52 heavy bombers that are decades old and plans to continue operating them for decades to come.

The exact form these trade-offs will take is as yet unclear but the most likely course will involve substantial decreases in personnel size. A major contributing factor to this is that personnel costs have come to consume an ever larger share of the defence budget as pay and benefits for both active personnel and retired veterans have grown significantly faster than almost any other expense. According to Secretary of Defense Chuck Hagel, these costs now account for approximately half the US defence budget.[18]

This growth is again in part due to politics, as Congress has consistently voted to support generous military pay and benefits, sometimes more generously than the Department of Defense or the military services would like. As Cindy Williams notes:

> By 2007, annual increases in pay and allowances had more than made up for the relatively lower raises of earlier years, and the Pentagon began asking Congress for more modest annual raises. Under pressure from military and veterans' associations and fearful of stinting the military in a time of war, however, Congress persistently added more than the department requested.... Since 2006, the Defense Department has asked repeatedly for permission to raise health insurance fees on military retirees. Changing the cost-sharing arrangement would of course help defray the government's costs, but more important, it would also make the government's insurance plan less attractive to retirees who have other good health coverage options. Yet Congress, under the same pressure it has faced about pay, has turned down all but the tiniest rise in fees.[19]

This Congressional reluctance to rein in personnel costs could combine with the impact of a renewed streak of negative liberty to produce a double squeeze on US defence budgets. Even a modest decline in total budgets will have a significant impact if personnel costs continue to climb. While US$650 billion is a vast sum, if more than half goes to paying for people, many of whom may not have worn a uniform in decades, then the ability to pay for operations, procurement, and research and development will nonetheless become substantially constrained.

Finally, US domestic politics have had a significant impact not just on defence budgets but also on defence policy, particularly on the large-scale use of force. The attempt to 'lead from behind' in the 2011 intervention in Libya and the reluctance of both Congress and the executive branch to use force in Syria in 2013 demonstrate the increasing consensus among many Americans that large-scale intervention is seldom worth the cost. Even the proponents of extensive US global presence (termed 'deep engagement') nonetheless argue that, in large part due to the experience of the war in Iraq, major US use of force is increasingly unlikely for the near term.[20]

This reinforces the image of the United States as the new 'wary titan'. While still capable of paying for primacy and of undertaking military inventions on a scale impossible for almost any other country, the United States is increasingly reluctant to do so. There may certainly be critical interests that may still merit major use of force (an Iranian attempt to actually build a nuclear weapon could be one) these will be few and far between. This then provides the context for understanding the implications for the United States of military change, Asian growth and European austerity.

Military change: evolution with revolutionary characteristics

The debate about military revolutions, described in the editors' Introduction to this volume, has receded into the background in the United States as a result of at least three major factors. The first is that many of the technologies that were believed to be revolutionary, particularly precision strike and pervasive intelligence, surveillance and reconnaissance (ISR) capabilities, have become integrated into existing force structure in ways that enhance capability without fundamentally challenging organizational essence. The second is that the challenges of the years following 11 September 2001 were substantially different from the challenges that 'network centric warfare' and other concepts associated with the so-called 'revolution in military affairs' of the 1990s were intended to combat. These challenges, principally those posed by counter-insurgency and counter-terrorism, forced changes and adaptations by the US military that have had far-reaching implications, but again have not fundamentally changed the organizational essence.[21] Third, in part due to increasing reliance on computer networks, is the emerging concern about so-called 'cyber' war (i.e. computer network attack and intrusion). The debate about the implications of cyber has now come to occupy the intellectual energy that in the 1990s was devoted to military revolutions.[22]

The result of the three factors described above has been a major evolution in US capabilities without a revolutionary reordering of how the US military generates and employs force. In the frame of reference that the editors use in this volume the evolution has been more than just modernization, as it has produced capabilities substantially different from those that existed before. Yet it stops short of transformation as existing capabilities have not been displaced.

I term this intermediate step an evolution with revolutionary characteristics. The principal expansion of US capabilities has been in the ability to apply discrete and discriminate force on a global scale. Discrete in this context means the effective use of force apart from a large-scale military campaign. Discriminate means directed against a very limited target set, indeed in many cases against single individuals with limited collateral effects. The remainder of this section outlines the elements of this capability, which has implications for both Asian and European militaries.

First, as noted above, the vision of 'network centric warfare" has largely been realized in the US military, particularly in air and naval forces. For example, the US Air Force continues to develop a set of fielded technologies termed Network Centric Collaborative Targeting (NCCT). NCCT allows the Air Force:

> to horizontally and vertically integrate ISR sensor systems both within and across intelligence disciplines (for example SIGINT to SIGINT or GMTI to SIGINT). NCCT software applications employ machine-to-machine interfaces and Internet Protocol (IP) communications to coordinate sensor cross-cues and collection activities. NCCT correlation and fusion services ingest collection data to produce a single, composite track (geolocation and ID) for high-value targets.[23]

Programmes such as NCCT are supported by complementary efforts in the US intelligence community. The National Security Agency (NSA) developed a similar SIGINT network targeting tool in response to the challenges of Iraq and Afghanistan. Known as Real Time Regional Gateway, this system collates SIGINT collected on a given signal from all platforms (ground, airborne or satellite) and rapidly fuses those data to provide geolocation. While the exact capabilities remain classified, retired Air Force Colonel Pete Rustan who helped develop the system offered a general description:

> The Real Time Regional Gateway concept started in Iraq as a concept that I helped to develop with our partners. Imagine that you are in Iraq. You have insurgents. They are on the telephone, making phone calls. That signal would be intercepted by ground [antennas], by the aircraft network and by the space network. If you're smart enough to combine all that data in real time, you can determine where Dick is out there. He's in block 23 down there, and he just said he's going to place a bomb. That's the real-time regional gateway: the ability to integrate the signals [for] geolocation.[24]

These network targeting systems have been supported by additional technological and organizational infrastructure to orchestrate the targeting process. The critical links in the targeting process may be defined in a variety of ways but one set of definitions that has become widely used in the US military and intelligence community describes five critical links. These are find, fix, finish, exploit and analyse; collectively referred to as the F3EA cycle.[25]

'Find' requires identifying where the target is in a broad geographic sense as well as providing a rough idea of what the target does. 'Fix' applies additional intelligence resources to refine the target to a more specific location. Eventually the target is 'fixed' when the intelligence has been refined sufficiently to allow a 'pattern of life' to be identified for the target.

Once the target is 'fixed' it can be 'finished'. Finished means precisely located and then either killed with a precision strike or captured by military or other forces. After a target is finished, the intelligence which results from that finished target is 'exploited'. Exploitation varies depending on how the target is finished. If the target is finished by a strike then exploitation relies heavily on such things as observing any funeral that is held for the target. If the target is killed by a ground force, exploitation also relies on the collection of material (e.g. computers or papers) from the target. If the target was captured, then exploitation includes interrogation of the target.

The final link in the chain is to 'analyse' the intelligence gained from the previous 'exploitation' in the context of other intelligence. This builds up the existing intelligence picture on the broader organization or organizations to which the target belonged. This improved intelligence picture allows the cycle to begin again with the 'finding' of a new target.

The infrastructure required for this is enormous and elaborate. Signals intelligence (SIGINT) is a major component. Without SIGINT, which is central to finding, fixing, and in some cases exploiting targets, there would simply be no F3EA as currently practised. While not a new capability, the current SIGINT infrastructure that underpins F3EA for the United States (and key allies) is unprecedented in scale and integration, making it a key element of this evolution with revolutionary characteristics.

The centre-piece of this global SIGINT capability is the UKUSA agreement, which dates to the beginning of the Cold War, and created an intelligence-sharing alliance between the United States and four Commonwealth countries: the United Kingdom, Australia, Canada and New Zealand. This nearly 70-year-old alliance has allowed for the creation of a SIGINT capability that will be extraordinarily difficult for any other country, much less a non-state actor, to replicate.[26]

Each of the alliance's five members, which are often referred to collectively as 'Five Eyes' (reflecting the 'eyes-only' nature of the intelligence), has its own SIGINT organization with unique attributes. The largest and far-away the best funded is the US NSA. Its vast size and budget allows it to draw from SIGINT resources ranging from a constellation of enormously expensive satellites to a diverse array of land- and air-based platforms. NSA also maintains extensive cryptographic and computer expertise in its workforce, along with thousands of linguists.[27]

The Commonwealth countries, though lacking the funding of NSA, are nonetheless critical to the global capability of the alliance. The UK's Government Communication Headquarters (GCHQ) is the second largest of the Five Eyes intelligence organizations. Although much smaller than

NSA, it has considerable expertise going back to the great British SIGINT successes of the Second World War. In addition, the British provide a variety of strategic outposts for SIGINT collection. These outposts, relics of the British Empire, are crucial to Five Eyes collection in several parts of the world.[28] Australia's Defence Signals Directorate (DSD), New Zealand's Government Communications Security Bureau (GCSB) and Canada's Communications Security Establishment (CSE) are all smaller than either NSA or GCHQ. Yet they also offer critical geographic basing, with Australia and New Zealand providing coverage of critical areas in Asia (especially for satellite links) while Canada provides geographic access in the far northern hemisphere.[29]

Without the Five Eyes alliance the US SIGINT capability would be greatly reduced, though not eliminated. This underscores the importance not just of advanced SIGINT technology but also of alliance politics. In the United Kingdom, for example, the intelligence community is more restricted in the use of lethal force than in the United States. This has led to legal challenges about Five Eyes sharing of intelligence, as UK intelligence may be used for lethal targeting by the United States.[30]

In addition to the UKUSA, there are a number of bilateral treaties with so-called 'third parties'. The third parties, which allegedly included Norway (at least at one time), provide geographic access for SIGINT as well as regional expertise.[31] The importance of allies for this capability is a key point for discussion in a subsequent section of the implications of European austerity.

In addition to SIGINT, the United States operates a large fleet of ISR and precision strike drones to support F3EA.[32] Volumes have been written on drones so I will not elaborate extensively on this development, except to note that the US military has integrated the drone without fundamentally changing the nature of its forces.[33] For example, the vast majority of US Air Force acquisition is still devoted to the purchase of manned systems.

In all, these advances in network technology and organization have produced a tremendous ability to rapidly track and target everything from enemy surface to air missiles to suicide bombers. Yet there have been a number of sceptics who believe that these changes would fundamentally alter the nature of war or even the essence of the US military. One early critique was by Lieutenant Colonel (now Lieutenant General McMaster), who argued in 2003 that while improvements in ISR and networking were important, they would not fundamentally remove uncertainty from war.[34]

The second major factor in military change in the United States is that McMaster and other critics were at least partially vindicated by the events following 11 September 2001. Neither the initial invasions of Afghanistan and Iraq nor the subsequent counter-insurgency campaigns were exactly what advocates for network centric warfare imagined. The initial invasion of Afghanistan showed both the promise of precision strike but also its limits.³⁵ Likewise, the initial invasion of Iraq demonstrated that even extensive US ISR and mobility also had its limits, particularly against irregular adversaries such as the so-called Saddam Fedayeen.³⁶

Yet these challenges paled in comparison to the subsequent challenge of counter-insurgency. Counter-insurgency proved to be just as much of a challenge to the highly networked US military of the 21st century as it was to the less networked US military of the 20th century.³⁷ In neither case has the US effort to embrace change to confront counter-insurgency been entirely successful and the US military has emerged from Iraq and Afghanistan without having experienced anything like a total transformation.

A partial exception alluded to above is that the US military has adapted much of its networked targeting capability to finding insurgents and terrorists. This capability is most notably resident in the US special operations community, parts of which have undergone extensive organizational change in order to enable the effective prosecution of the F3EA cycle against these types of targets. These changes involved integrating into inter-agency targeting teams that brought together intelligence from many sources (including human intelligence, which is a notable weakness of network centric warfare as originally envisioned) and then provided that intelligence to special operations units in as close to real time as possible.³⁸

At the same time as these organizational changes were taking place, the overall size and capability of US special operations forces grew substantially. This continued and accelerated a process that began more than three decades ago. Following the failed rescue of Iranian hostages in 1980, a commission headed by retired Admiral James Holloway argued for the creation of a multi-service (joint) special operations task force in order to provide the United States with a standing, integrated capability to conduct special operations, specifically with a focus on counter-terrorism.

The commission's recommendation was accepted and the US Joint Special Operations Command (JSOC) was established. In 1987, still unsatisfied with the state of US special operations, Congress enacted legislation which created a four-star senior command over all special operations forces. US Special Operations Command (USSOCOM or SOCOM) created a high-level advocate for all special operations, not just those under the authority of JSOC. JSOC in turn became a subordinate command under SOCOM. In addition, Congress would also create a Major Force Program (MFP) for special operations unique equipment, in effect giving SOCOM its own budget as if it were an independent military service, leading some to term SOCOM 'the fifth service'.³⁹ After 11 September, special operations forces were the centre-piece of US use of force. SOCOM was made the lead organization for synchronizing what was termed the Global War on Terror and by 2009 had expanded to over 55,000 total personnel.

The combination of networked targeting with a vast special operations force supported by an even vaster intelligence community has produced a truly revolutionary capability to project discrete and discriminate force globally. It is easy to forget in an era of weekly if not daily operations against individuals across the globe, from Libya to Somalia to Afghanistan and Pakistan, that even 15 years ago targeting a single individual, namely Osama bin Laden, was extraordinarily difficult.

This revolutionary capability has limits. Despite more than a decade of targeting, al Qaeda still exists. In neither Iraq nor Afghanistan has successful targeting on a massive scale led to victory.[40]

Yet the option to use force in a way that does not require massive mobilization of the US military has changed the political calculus about force fundamentally. While, as discussed previously, the United States seems deeply unwilling to commit itself to major military operations, there is only mild debate about the global use of discrete and discriminate force. For example, in the first week of October 2013, only weeks after balking at the use of force in Syria, the United States employed special operations personnel against two targets, one in Libya and one in Somalia. While the latter was unsuccessful, neither provoked very much debate in the United States.[41] These operations, along with the general use of ISR and precision strike, have become so commonplace that they no longer merit much debate. As noted above, this has drained a substantial amount of the intellectual energy from the discussion of military revolution in the United States.

However, the final factor in the decline in debate about military change has been the emergence of extensive debate about cyberwarfare. While a full discussion of the debate is beyond the scope of this chapter, the focus of the debate is on whether cyberwarfare is truly a major new type of warfare or 'merely' a major development that expands on the traditional capabilities of SIGINT, covert action and electronic warfare.[42] This debate is ongoing in all of the major US academic journals of security studies as well as inside the Department of Defense.[43]

As noted above, this debate is driven in part due to the success in adopting network centric systems. The US military realizes that these systems, while enormously powerful, are also a potential Achilles heel against attackers that can penetrate the networks. This fear, as noted in the following section, is largely though not exclusively about Chinese capabilities, even though the Department of Defense is officially circumspect about Chinese capabilities and intentions.

However, the debate is also driven in part by US capabilities in this regard. For example, the alleged US Air Force program known as Suter enables the US to intercept data transmissions between components of an enemy air defence system and even to introduce false data into the system. This capability, supposedly demonstrated in the Israeli raid on a Syrian nuclear reactor in 2007, turns networked systems against the user, breeding concern (some might say paranoia) in parts of the US military that the same could be done to US systems.[44]

Military change thus continues to be a major topic in the United States, though it has shifted from the examination of network centric warfare to cyberwar. This shift has taken place in part due to the general adoption of networked systems, which has been more than modernization but less than transformation. However, these changes have some revolutionary characteristics that do affect the implications of Asian growth and European austerity for the United States, as described in the following two sections.

Asian growth: implications for the United States

The United States, despite its growing wariness of major defence expenditures and use of force, nonetheless remains deeply committed to Asia. The much noted 'pivot to Asia' may or may not represent a serious reordering of US defence priorities but it does underscore the extent to which the United States recognizes the importance of Asia and the US role in the Pacific. Indeed the growth in many Asian economies has fuelled the development of military forces in many countries. Those developments have significant implications for US defence policy, not only regionally but also globally.

Indeed, the reality is that the United States has had more continuity in its approach to Asia than is commonly appreciated. The central feature of the past two decades of Asia policy has been the management of the rise of China and the concomitant reassurance of regional allies, especially Japan. This marked a major shift from the previous two decades when the United States and China were aligned against the Soviet Union.

As with the Cold War, when the overall policy towards the Soviet Union oscillated between confrontation and collaboration, the US policy towards China has oscillated, albeit never becoming anything as close to confrontational as the US–Soviet relationship. Indeed, unlike the Soviet Union, China has become a vital trading partner of the United States and many of its allies (perhaps most notably Australia). Periods of confrontation, such as those following the 1995 to 1996 Taiwan Straits crisis or the 2001 EP-3 incident, have typically been followed by more relaxed periods.

Yet the central goal has remained one of managing rather than containing China, seeking to integrate it into the global order while ensuring that growing Chinese power does not threaten US key interests in the region.[45] Thus efforts to reassure and bolster allies have gone hand in hand with growing economic integration. In addition, the United States has sought to cultivate new allies.

The central difficulty in this balancing act is that it requires reassuring China and US allies simultaneously; yet actions to reassure one can cause trepidation in the other. This is compounded by the region's history and resulting mistrust, particularly between China and Japan. The so-called pivot should then be seen as part of this overarching policy which is intended to reassure allies that have, rightly or wrongly, felt neglected during the height of the so-called War on Terror.

With this overall policy framework in mind, this section briefly discusses the implications for Asian growth and military change by focusing on three categories of states. The first are potential peer or near-peer competitors, which at this point is only China. The second are established allies and partners, which in the context of this volume include Japan and Singapore. The third are potential or nascent partners, which in this context include India and Indonesia. These are not the only aspects of Asia that have implications for the United States – the Koreas, for example, are another major factor – but these other aspects are beyond the scope of the present volume.

The potential for China to become a peer or near-peer competitor to the United States has been an ongoing debate for more than a decade. Some argue that China is or soon will be such a competitor.[46] Others argue that there are extraordinary obstacles to China becoming a true peer while acknowledging that it can still pose regional challenges to the United States.[47]

This view is in striking contrast to the view which Dennis Blasko argues (Chapter 2, this volume) that the Chinese have of their military capability. The Chinese view, embodied in the 'two incompatibilities', seems the inverse of that of the United States, believing that best case they can pose regional challenges to the United States. Worst case, the Chinese view seems to be that their limited technology combined with their lack of combat experience and NCO development could mean that they would be overwhelmed by the types of networked capabilities that the United States has cultivated over the past decade.

This divergence in perceptions has implications for US overall policy of managing China, especially in terms of defence policy and regional stability. If both sides believe that their capabilities against the other are uncertain this will likely lead to efforts to avoid military confrontation. For the United States, this presents an opportunity to refocus on Asia without unduly risking escalation in crisis, assuming this refocus does not lead to a change in either side's caution.

At a more specific military technology level however, the prospects for crisis stability are grimmer. Much of the debate about Chinese military change in the US military centres on Chinese capabilities for what the United States has termed 'anti-access/area denial' (A2/AD). These capabilities, which include advanced surface to air missiles, quiet diesel submarines and a variety of anti-ship missiles, seek to prevent the US military from approaching the Chinese littoral and to deny any forces that do penetrate the ability to operate effectively.

In order to potentially combat such technologies, the US Navy and Air Force are promoting the development of a new doctrine known as AirSea Battle. While AirSea Battle is at least partly an attempt to justify a significant share of the defence budget for those services, it also marks a real effort, both organizational and technological, to integrate and network air and naval forces to an even greater degree. It relies heavily on many of the technologies described earlier that emerge from network centric warfare concepts of the 1990s as refined during the 2000s in Iraq, Afghanistan and elsewhere.

Neither A2/AD nor AirSea Battle represents transformational change. Instead they represent a modernization of the same sort of confrontation between Soviet naval and land-based assets and the US fleet during the late Cold War. The US Navy sought to project power from the North Atlantic into Soviet naval 'bastions' as a means of influencing both the conventional and nuclear balance. This approach was referred to as the 'Maritime Strategy', and AirSea Battle is a modernization of this concept combined with the addition of a major role for air force assets. Chinese A2/AD capabilities are likewise a modernization of many similar Soviet efforts.[48]

However, the interaction of A2/AD and AirSea Battle has the potential to be destabilizing in a crisis. While both sides have studiously avoided declaring these capabilities as directed at one another, at least officially, the reality is that in a crisis both sides would be anticipating the potential impact of allowing the other side to employ these capabilities pre-emptively. This could potentially make both sides, cautious in peacetime due to uncertainty, more aggressive in a crisis as uncertainty about the other's capabilities exacerbated fears that unless one acted promptly the other side might act with decisive results – a scenario referred to during the Cold War nuclear contest as 'use or lose'.

Nowhere is this potential more clear than in the Chinese development of long-range anti-ship ballistic missiles (ASBM). These systems potentially threaten US ships, particularly the massive and expensive aircraft carriers of the Pacific Fleet, at ranges of more than 1,000 kilometres with destruction in very short order as the weapons could cover this distance under 20 minutes. While the technology is unproven, there are real fears that a large salvo of such missiles may overwhelm US missile defences and sink billions of dollars of US ships in a matter of minutes.

However, targeting these weapons will require successful Chinese adoption of at least primitive forms of networked warfare. Sensors, whether over the horizon radar or SIGINT systems, will have to locate and track ships hundreds of miles from the Chinese coast and transmit these data to missile launchers in near real time. The United States invested heavily in these capabilities beginning in the 1970s and it still took considerable time and effort to bring them to fruition, both organizationally and technically.

The requirement to develop so-called 'over-the-horizon' targeting capability is not only challenging but also induces vulnerability. If these systems for targeting US ships or the links that connect them to the launchers can be destroyed or disabled, whether kinetically or electronically, then the Chinese investment is for naught. Conversely, if the missiles are launched before these systems can be disabled then the US Navy may suffer its most serious losses since the Second World War. Both sides in a deep crisis therefore may face significant pressure to strike first, fearing that if they do not they will suffer greatly from the other side's pre-emption.

Further, layered on top of these conventional capabilities are the possibilities raised by cyber operations. While cyber may not be a decisive capability or a truly new form of warfare systems like the Suter program described earlier it may enhance the perceived and/or actual utility of a first strike. For example, cyber operations may disrupt Chinese command and control of A2/AD systems covertly and instantaneously, and thus greatly improve follow-on conventional attacks to permanently destroy the systems. Yet the same calculus may affect the Chinese side as well, compounding the 'use-or-lose' incentives.

The implications of Chinese military growth and change are thus mixed. Both sides may have parallel fears about the other's capability that may induce both caution and aggression depending on the circumstances. Cultivating the former can help reduce the possibility of the latter, yet paradoxically, as each side comes to understand the other better, it may be more rather than less difficult to prevent the emergence of crises as both sides become less uncertain.

The second major component of US Asia policy is reassuring allies. In this section I focus on Japan and Singapore, though these are not the only allies in the region. The growth and change of the military in Japan, a long-standing US ally, and Singapore, which while not a formal ally is a significant strategic partner, have very different implications for the United States. Both are viewed heavily but not exclusively through the lens of China, further refracted through US domestic politics.

As Isao Miyaoka notes in this volume, Japan's defence posture has changed substantially. There is now a significant emphasis on Japanese ability to project power at least sufficiently to defend outlying islands as well as to meet new threats, ranging from cyber to potential instability in other regions. This represents a significant if gradual shift from the Cold War, where the main concern of the Japanese self-defence forces was on preparing for the Soviet threat.

On the face of it, this is very good news for the wary titan. Japan's economy, though long in the doldrums, remains one of the world's largest and most consistent investment if even 1 per cent of this economy can produce substantial capability. A willingness to shoulder more of the burden in providing for Japanese security defined more broadly than that of defence against invasion could amplify the effect of the US pivot to Asia while assuaging the concerns among those advocating negative liberty in US domestic politics that the United States is the lone policeman of the Pacific.

However, the fact that Japan and China have unresolved territorial disputes is a potential cause for concern about a more expansive definition of Japanese security. As both China and Japan develop more robust capability to project power some distance from their respective littorals, the chance for conflict over disputed territory may increase. The United States, which remains firmly committed to the alliance with Japan, could thus be pulled into a conflict with China in the future as a result of both Japanese and Chinese military growth and development.

The November 2013 Chinese declaration of an Air Defense Identification Zone (ADIZ) that is expansive enough to encompass disputed territory has already provided the first concrete example of the mechanism by which such a catalytic conflict could begin. The United States, as part of a previously declared position and existing military manoeuvres, promptly violated the Chinese ADIZ. While it is important not to overstate the significance of this event, which at present has limited potential to cause a crisis, it is an example of the potential for friction over disputed territory that leads to conflict between improving Japanese and Chinese militaries to quickly implicate the United States.

Singapore, as Bernard Fook Weng Loo notes (Chapter 5, this volume), has followed a similar trajectory, evolving from a 'poisonous shrimp' focused on defence against invasion to an agile 'dolphin' seeking to embrace new technologies while remaining cognizant of the concern about the potential for internal threats and terrorism. As with Japan, this change holds some promise for the United States. The development in 2005 and after of a strategic framework between the United States and Singapore, which includes the ability under certain circumstances for US Navy ships to use Singaporean facilities, means that a more capable Singapore may be a useful partner in many contingencies.

However, the contingency where the United States is likely to need the most help, one involving conflict with China, is also one where Singapore may be most reluctant to become involved. There are already discussions in the United States about the potential to blockade Chinese energy imports, particularly at critical choke points such as the Strait of Malacca, as an alternative to AirSea Battle. Singapore's military and military facilities could contribute significantly to such an effort, yet becoming caught between two titans is likely to be a terrible place for a small state like Singapore, and thus not a position in which Singapore, not formally a US ally, would likely place itself.

The United States is therefore likely to ask regional allies, especially Japan, to maintain capabilities while restraining their ambitions. The US is cognizant of the local interests of its allies and, as in the ADIZ incident noted above, is willing to support them to some extent. Regional operational planning reflects this, as in February 2014 when US–Japan exercises focused on amphibious warfare (as would be required to seize or recapture islands).[49]

Yet the United States remains wary of the potential to be dragged into a major conflict over minor issues. Given the domestic politics of Japan the restraint of ambition may be difficult. US policy must therefore walk a fine line between reassuring allies against Chinese aggression and emboldening allies to provoke China.

Finally, in addition to established allies and partners, military change in the potential or nascent partners of India and Indonesia have perhaps the greatest promise and greatest risk for US defence policy. India, estranged during the Cold War, has over the past decade been moving closer to the United States. As the two largest democracies in the world, some in both countries argue that such movement is natural and to be expected.

Moreover, the two share several strategic interests. Most notable is a mutual concern about a rising China. India, like Japan, has outstanding territorial disputes with China that, while not actively contested for some time, remain somewhat ominously in the background. Another shared interest is in countering terrorism, which, as Vivek Chadha notes (Chapter 7, this volume), is part of a broader Indian concern about internal security. Cooperation with an Indian military that is successful in growth and modernization thus has the potential to greatly benefit the United States defence policy in the region.

However, an increasing partnership with India holds substantial risk for the United States. The issue of Pakistan is at the forefront of this risk as the United States has attempted to perform the difficult balancing act of maintaining strong defence ties to these rivals (if not enemies) at the same time. The US's relationship with Pakistan is much more contentious than that with India, yet it remains a critical one for combating terrorism and, at least for now, supplying the war in Afghanistan. Any movement to further increase partnership with India would be viewed with extreme suspicion by the Pakistanis, potentially jeopardizing the relationship.

At the same time, an Indian military that is more capable of conducting operations against Pakistan may also make nuclear use by Pakistan more likely. While, as S. Kalyanaraman notes (Chapter 6, this volume), the Indian political leadership has little appetite for conflict that risks such nuclear use, the Indian military has increasingly sought ways to fight limited war in the shadow of the Pakistani arsenal. A change in Indian political leadership towards more hawkish postures combined with greater Indian military capability following successful growth and change could therefore confront the United States with a potential for conflict between two partners that holds significant risk of nuclear use.

Even if Indian military action against Pakistan did not lead to nuclear use, the destruction of significant portions of the Pakistani military might destabilize the country, which is fundamentally held together by the military. This would create a scenario perhaps even more nightmarish, since Pakistan's nuclear arsenal, well guarded at present, may become vulnerable to militants. Thus, as with much of Asian military growth and change, India holds both promise and peril for the United States.

Indonesia's military, in contrast to all the previous militaries, is one that has not been a significant factor in US defence policy in recent decades.

258 A. Long

In truth, the United States has been willing in some sense to subcontract to the Australians the task of engaging with Indonesia. As Indonesia's military grows and alters, this is beginning to change. Many in the United States believe Indonesia will be the next major military player to emerge in Asia. Yet the implications of this are unclear, particularly given the ongoing political transition following the election of President Joko Widodo, the first president of Indonesia without strong ties to the military or political elite.

European austerity

If the implications of Asian growth are new opportunities and new risks for the United States, along with significant uncertainty, the implications of European austerity seem clearer. Our oldest and closest allies seem unlikely to develop and maintain the same or even similar capabilities that the United States has over the past decade.

At the same time, the United States would like to see European allies doing more rather than less. While this has been a near constant refrain throughout the life of NATO, it has taken on somewhat greater than usual significance given the domestic politics of the United States described earlier and the desire to pivot to Asia. In recent operations in Libya and Mali, the United States has sought to put Europe in the forefront, even if in the case of Libya the United States was ultimately required to do much of the heavy lifting. This is closer to a division of labour (as in the case of Mali) than NATO/Europe as a mere adjunct to the United States but acknowledges that the United States will have to provide some critical enabling capabilities such as airlift and aerial refuelling.

All of the authors writing on Europe, covering countries ranging from the relatively large United Kingdom and Germany to the relatively small (but resourceful) Denmark, concur that it is unlikely that the United States will see much of this desire to have Europe at the forefront. Only if European countries are willing to cede sovereign rights, as both Tom Dyson and Sven Bernhard Gareis note (Chapters 10 and 8, this volume respectively), is there a chance that Europe may be able to generate sufficient capability to assume significant leadership in major military operations? The implication then is that major military action originating in capitals on either side of the Atlantic is unlikely in the near term. Libya may have been the last gasp of such major military actions for some time.

Yet Europe is playing a significant role in the use of discrete and discriminate force by providing highly capable special operations and intelligence personnel to multinational missions.[50] While little discussed, the emergence of a NATO Special Operations Headquarters (NSHQ) and the shared experience of more than a decade of operations in Afghanistan have greatly bolstered the capability of NATO special operations forces (SOF) to operate together. While there are still serious challenges, this represents a major potential way for European militaries to remain highly relevant even under conditions of austerity. This penultimate section briefly outlines the promise and problems of NATO SOF and intelligence coordination under austerity.

NSHQ is a fairly recent organizational change at the Supreme Headquarters Allied Powers Europe (SHAPE). Established in 2009, NSHQ is intended to provide common training standards and shared understanding of the nature and purpose of SOF throughout the NATO SOF community. The headquarters has made significant progress in this respect, no small feat considering that some of the key Eastern European SOF leaders began their careers at a time when their mission was to attack the vulnerable rear of NATO forces.

However, even before the formal establishment of NSHQ, NATO SOF had significant experience in training and operating together. In the post-Cold War environment, the first major set of operations conducted by some NATO SOF was the hunt for those individuals indicted for war crimes in the Balkans. These operations laid the foundation for interoperability and a certain shared understanding of how such SOF operations should be conducted.

This cooperation took a major leap forward with the war in Afghanistan, as special operations forces were a significant portion of the troops' contributions of almost all NATO members to the International Security Assistance Force (ISAF). Operating under a command known appropriately as ISAF SOF, these units operated unilaterally against terrorists and insurgents at first but over time became responsible for the development of Afghan special operations units organized by the Ministry of Interior to conduct crisis response and high-risk arrest operations. However, the United States maintained most of its special operations forces in Afghanistan under a separate chain of command along with a few other members of NATO. Thus the potential for shared operations and intelligence sharing between most NATO countries and the United States was relatively limited for most of the war.

This changed in 2012 when the NATO Special Operations Component Command-Afghanistan was established jointly with the US Special Operations Joint Task Force Afghanistan. These two commands were led by a single dual-hatted US special operations officer holding the rank of major general and for the first time all special operations in Afghanistan were brought under one roof (at least on paper). While this structure is very much work in progress, it offers a very clear model for future NATO missions that are heavily or exclusively composed of SOF units. Indeed, NSHQ intends to generate and maintain the ability to establish one such command on a rapid basis to respond to future contingencies.

The advantages of this arrangement are significant for both the United States and European allies. The development and adoption of two secure communications (the NATO-only BICES and the ISAF-wide CENTRIX) have begun to create an environment where SOF from all over NATO can communicate regularly. NATO SOF, which often lack the vast resources of US SOF, are able to use these communications systems to receive intelligence and also to coordinate use of Coalition SOF assets such as ISR platforms and helicopters. Thus NATO SOF receives benefits in the form of physical capital.

The United States, in contrast, benefits most in terms of human capital. While the US special operations community dwarfs any of the NATO allies, it is not infinite. The effort to expand US SOF after 11 September has also, according to some, led to dilution of the overall quality of US SOF. By partnering with NATO SOF, the pool of high-quality personnel is significantly expanded.

The same is true of intelligence collection. As noted earlier, the United States provides the bulk of SIGINT resources shared among its allies but benefits substantially from the UKUSA and other bilateral treaties. However, the playing field is much more level in terms of human intelligence. While the US Central Intelligence Agency (CIA) is large, this provides much less advantage in human intelligence and the CIA relies heavily on liaison with other countries' intelligence services, including NATO members.

Moreover, there is potential for greater coordination between NATO SOF and intelligence in terms of regional specialization. Cultivating regional expertise, including language skills, cultural understanding and personal relationships, is one of the most challenging and difficult to maintain capabilities of SOF. If each NATO member assumed responsibility for specializing in a few countries from one region, then the overall capability of NATO SOF would be vastly enhanced, as the lead country could coordinate operations, intelligence and relationships while other countries would provide any required 'muscle'.

Finally, the relative lack of political sensitivity towards SOF operations in the United States appears to be true to some extent in European countries. Indeed, many NATO countries are significantly more secretive about SOF operations than the United States so there is even less public knowledge and therefore public concern about SOF operations. This makes them potentially as politically attractive for European NATO members as for the United States.

At the same time, there are a number of obstacles to the continued effectiveness of NATO SOF under conditions of austerity. First, SOF units, while not as expensive as acquiring a new fleet of fighter aircraft, are not cheap. Maintaining SOF skills requires not only continual training to a high level but also the ability to recruit and retain the highest quality personnel. Under austerity, European NATO members will have to consciously prioritize SOF in order to maintain this capability. The same is true of intelligence collection, whether SIGINT or human intelligence.

Second, intelligence sharing remains a problem within NATO. This is most notable in the difference between intelligence that is classified as

NATO Secret (meaning any NATO member that holds the equivalent of a US Secret clearance and is read into NATO intelligence by their home country) and that which is classified for only Five Eyes access. The latter category includes a substantial amount of SIGINT, which as noted is critical to the sorts of F3EA targeting operations SOF have increasingly come to carry out. The result is a two-tiered system where everyone not in Five Eyes can only share intelligence that is not particularly sensitive, while Five Eyes have access to much more sensitive intelligence. This also requires Five Eyes countries to maintain additional communications systems, as Five Eyes-only intelligence cannot be transmitted on either BICES or CENTRIX.

One possible solution is to overhaul the UKUSA and NATO agreements, both of which date to the early Cold War. The goal would be to create an intermediate classification between NATO Secret and full access to Five Eyes intelligence for NATO SOF operations. Gaining and holding such a clearance would not be automatic and would be based on adherence to an agreed standard for counter-intelligence and operational security. This would make for complicated alliance politics, as some countries would be very obvious to include (e.g. Norway), while others, particularly some of the former Warsaw Pact, would be much more contentious. Moreover, following the revelations of Edward Snowden, it may simply be too politically difficult to renegotiate these treaties, though if done in the context of improving transparency on US intelligence to NATO members it might be possible.

Thus a major implication of European austerity for US defence policy is that there may be an opportunity to make a virtue out of a vice. Austerity may provide a forcing function for NATO to prioritize the development of integrated SOF that in turn will be able to leverage US capital assets. This synergy could ultimately magnify the most revolutionary of the capabilities the US has developed over the past decade despite the wariness of both US and European domestic audiences.

The importance of these capabilities is underscored by the events in Ukraine in late 2013 and 2014. Russian intervention has relied heavily on clandestine capabilities such as intelligence and special operations.[51] NATO SOF can and should be a major component of the response to this and any future Russian intervention, which is more likely to follow this model than the Cold War-era drive to the Fulda Gap (though Russian action in late 2014 is increasingly overt).

Conclusion

The United States, despite its continued primacy, faces challenges on multiple fronts. Even as the wars in Iraq and Afghanistan ended, Asia and Europe have only grown in significance as China continues to rise, European allies feel the pinch of austerity and Russia reasserts itself. Yet fundamentally the

United States has increased its military capabilities for both symmetric and asymmetric threats, thanks to the expenditures and experience of the past decade. The central limitation on the United States is domestic, both politically and economically. It is unclear if the current primacy will endure or if the new 'wary titan' will cast off its burden.

Notes

1 Michael O'Hanlon, *Healing the Wounded Giant: Maintaining Military Preeminence while Cutting the Defense Budget* (Washington, DC: Brookings Institution, 2013); Todd Harrison, *Chaos and Uncertainty: The FY 14 Defense Budget and Beyond* (Washington, DC: Center for Strategic and Budgetary Assessment, 2013).
2 For an overview of this debate see Stephen G. Brooks, G. John Ikenberry and William C. Wohlforth, 'Don't Come Home, America: The Case against Retrenchment', *International Security*, vol. 37 no. 3 (winter 2012/2013), 7–51; Campbell Craig *et al.*, 'Correspondence: Debating American Engagement: The Future of U.S. Grand Strategy', *International Security*, vol. 38 no. 2 (autumn 2013), 181–199.
3 Thomas Christensen, 'Posing Problems without Catching Up: China's Rise and Challenges for U.S. Security Policy', *International Security*, vol. 25, no. 4 (spring 2001), 5–40.
4 See Dana Allin and Erik Jones, *Weary Policeman: American Power in an Age of Austerity* (London: International Institute for Strategic Studies, 2012); John Hobson, 'The Military–Extraction Gap and the Wary Titan: The Fiscal-Sociology of British Defence Policy, 1870–1913', *Journal of European Economic History*, vol. 22 no. 3 (winter 1993), 461–506.
5 Office of the Undersecretary of Defense (Comptroller), *National Defense Budget Estimates for FY14* (May 2013), 10. This publication is almost universally referred to as the 'Green Book' due to its cover and will be cited as Green Book FY14 in subsequent notes.
6 Green Book FY14, 274.
7 Hobson, 'The Military–Extraction Gap and the Wary Titan', 479. While there are differences in how net national product and gross national product are calculated, they do not profoundly alter the comparison.
8 Hobson (ibid., 491) notes that French conscripts were paid about 1 per cent of the average civilian wage.
9 See Green Book FY14, 272–274 for the United States data; and www.ukpublicspending.co.uk/spending_chart_1950_1990UKp_13c1li011mcn_30t for Great Britain data.
10 Paul Kennedy, *The Rise and Fall of the Great Powers: Economic Change and Military Conflict From 1500 to 2000* (New York: Random House, 1987).
11 Estimated from data in N.F.R. Crafts, 'Gross National Product in Europe 1870–1910: Some New Estimates', *Explorations in Economic History*, vol. 20 (1983), 389.
12 Estimated from US data in Gene Smiley, *The American Economy in the 20th Century* (Cincinnati, OH: Southwest Publishing, 1993), and UK data in Roger Backhouse, *Applied U.K. Macroeconomics* (New York: Blackwell, 1991).
13 See data in FY14 Green Book, and http://money.cnn.com/2013/10/30/news/economy/deficit-2013-treasury/.
14 The average OECD total tax rate for the period 1999 to 2009 is approximately 34.8 per cent of GDP. See 'Total Tax Revenue' in *OECD Factbook 2013: Economic, Environmental and Social Statistics*.

15 See 'Total Tax Revenue' and GDP data at http://research.stlouisfed.org/fred2/graph/?id=DEURGDPR.
16 Aaron Friedberg, *In the Shadow of the Garrison State: America's Anti-Statism and Its Cold War Grand Strategy* (Princeton, NJ: Princeton University Press, 2000).
17 Brendan Green, 'Two Concepts of Liberty: U.S. Cold War Grand Strategies and the Liberal Tradition', *International Security*, vol. 37 no. 2 (autumn 2012), 9–43.
18 Jim Garamone, 'DOD Must Control Rising Personnel Costs, Hagel Tells NCOs', *American Forces Press Service*, 6 November 2013.
19 Cindy Williams, 'Accepting Austerity: The Right Way to Cut Defense', *Foreign Affairs* (November/December 2013).
20 See Brooks *et al.*, 'Don't Come Home, America', 33; see also John Mueller, 'The Iraq Syndrome', *Foreign Affairs* (November/December 2005).
21 For example, see discussion in James Russell, *Innovation, Transformation, and War: Counterinsurgency Operations in Anbar and Ninewa Provinces, Iraq, 2005–2007* (Stanford, CA: Stanford University Press, 2010); and Christopher Lamb and Evan Munsing, *Secret Weapon: High-Value Target Teams as an Organizational Innovation* (Washington, DC: National Defense University, 2011) for discussion on innovation in Iraq in the counter-insurgency and counter-terrorism domains respectively.
22 For three of the most lucid critiques of the furore over cyber that also provide a good overview see Jon Lindsay, 'Stuxnet and the Limits of Cyber Warfare', *Security Studies*, vol. 22, no. 3 (2013), 365–404; Eric Gartzke, 'The Myth of Cyberwar: Bringing War in Cyberspace Back Down to Earth', *International Security*, vol. 38, no. 2 (autumn 2013), 41–73; and Thomas Rid, *Cyber War Will Not Take Place* (New York: Oxford University Press, 2013).
23 US Air Force FY13 Budget Justification for PE 0305221F, *Network Centric Collaborative Targeting*, p. 1.
24 Interview with Pete Rustan, *C4ISR Journal*, 8 October 2010. See also *C4ISR Journal*'s description of the system in the 'Big 25 Awards' section of the October 2010 issue.
25 See Michael T. Flynn, Rich Juergens and Thomas L. Cantrell, 'Employing ISR: SOF Best Practices', *Joint Forces Quarterly*, no. 50 (2008), 56–61; and Jeffrey A. Builta and Eric N. Heller, 'Reflections on 10 Years of Counterterrorism Analysis', unclassified extracts from *Studies in Intelligence*, vol. 55 no. 3 (September 2011).
26 The original agreement, 'British–U.S. Communications Intelligence Agreement and Outline', 5 March 1946, was declassified in 2010 and is available online at www.nsa.gov/public_info/declass/ukusa.shtml.
27 See Matthew Aid, *The Secret Sentry: The Untold History of the National Security Agency* (New York: Bloomsbury, 2009).
28 See Richard Aldrich, *GCHQ: The Uncensored Story of Britain's Most Secret Intelligence Agency* (New York: HarperCollins, 2011).
29 See Desmond Ball, *Signals Intelligence in the Post-Cold War Era: Developments in the Asia Pacific Region* (Singapore: Institute for Southeast Asian Studies, 1993); James Cox, 'Canada and the Five Eyes Intelligence Community' (Calgary, Canada: Canadian Defence & Foreign Affairs Institute, 2012).
30 Duncan Gardham, 'Does MI6 Have a Licence to Kill?', *Telegraph* (UK), 3 December 2012.
31 See Matthew Aid and Cees Wiebes, 'Conclusions', in Matthew Aid and Cees Wiebes, *Secrets of Signals Intelligence During the Cold War and Beyond* (London: Frank Cass, 2001).
32 The term 'drone' is somewhat loaded, with the US military typically preferring the term 'remote piloted vehicle' (RPV). For simplicity and clarity I use drone.
33 See Thomas Ehrhard, *Air Force UAVs: The Secret History* (Washington, DC: Air Force Association, 2010); Adam Stulberg, 'Managing the Unmanned Revolution

in the U.S. Air Force', *Orbis* (spring 2007); Daniel Byman, 'Why Drones Work: The Case for Washington's Weapon of Choice', and Audrey Kurth Cronin, 'Why Drones Fail: When Tactics Drive Strategy', both in *Foreign Affairs* (July/August 2013).

34 H.R. McMaster, 'Crack in the Foundation: Defense Transformation and the Underlying Assumption of Dominant Knowledge in Future War', US Army War College, November 2003.
35 Stephen Biddle, 'Allies, Airpower, and Modern Warfare: The Afghan Model in Afghanistan and Iraq', *International Security*, vol. 30, no. 3 (winter 2005/2006), 161–176.
36 Stephen Biddle, 'Speed Kills? Reevaluating the Role of Speed, Precision, and Situation Awareness in the Fall of Saddam', *Journal of Strategic Studies*, vol. 30, no. 1 (February 2007), 3–46.
37 There is an extensive and growing literature on counter-insurgency in Iraq and Afghanistan. See, inter alia, Austin Long, *Doctrine of Eternal Recurrence – The U.S. Military and Counterinsurgency Doctrine, 1960–1970 and 2003–2006* (Santa Monica, CA: RAND, 2008); David Ucko, *The New Counterinsurgency Era: Transforming the U.S. Military for Modern Wars* (Washington, DC: Georgetown University Press, 2009).
38 See description in Lamb and Munsing, *Secret Weapon*, as well as Mark Urban, *Task Force Black: The Explosive True Story of the SAS and the Secret War in Iraq* (London: Little, Brown, 2010).
39 For an overview see Colin Jackson and Austin Long, 'The Fifth Service: The Rise of Special Operations Command', in Harvey Sapolsky, Benjamin Friedman and Brendan Green (eds), *U.S. Military Innovation Since the Cold War: Creation Without Destruction* (New York: Routledge, 2009).
40 See Austin Long, 'Whack-a-Mole or Coup de Grace? Institutionalization and Leadership Targeting in Iraq and Afghanistan', *Security Studies*, vol. 23, no. 3 (2014), 471–512.
41 David Kirkpatrick, Nicholas Kulish and Eric Schmitt, 'U.S. Raids in Libya and Somalia Strike Terror Targets', *New York Times*, 5 October 2013.
42 See Lindsay, 'Stuxnet and the Limits of Cyber Warfare'; Rid, *Cyber War Will Not Take Place*, and Gartzke, 'The Myth of Cyberwar'.
43 *International Security*, *Security Studies* and the *Journal of Strategic Studies* have all published at least one article on cyber in the period 2012 to 2013.
44 For a description of the program as well as its alleged role in the strike, see Richard Gaspare, 'The Israeli E-tack on Syria', Parts I and II, *Air Force Technology*, March 2008.
45 For a recent articulation and refinement of this approach see Ashley Tellis, *Balancing Without Containment: An American Strategy for Managing China* (Washington, DC: Carnegie Endowment for International Peace, 2014).
46 Ibid.
47 For an overview and critique of the debate on China's rise see Michael Beckley, 'China's Century? Why America's Edge Will Endure', *International Security*, vol. 36, no. 3 (winter 2011/2012), 41–78.
48 See Barry Posen, *Inadvertent Escalation* (Ithaca, NY: Cornell University Press, 1991); Owen Cote, *The Third Battle: Innovation in the U.S. Navy's Silent Cold War Struggle with Soviet Submarines* (Naval War College Newport Paper no. 16), and Norman Friedman, *Seapower and Space: From the Dawn of the Missile Age to Net-Centric Warfare* (Annapolis, MD: Naval Institute Press, 2000). On AirSea Battle see Aaron Friedberg, *Beyond Air–Sea Battle: The Debate Over US Military Strategy in Asia* (London: International Institute for Strategic Studies, 2014).
49 Helene Cooper, 'In Japan's Drill With the U.S., a Message for Beijing', *New York Times*, 22 February 2014.

50 This section draws heavily on Austin Long, 'NATO Special Operations: Promise and Problem', *Orbis* (forthcoming), and author experience in Afghanistan in 2013.
51 Andrew Higgins, Michael Gordon and Andrew Kramer, 'Photos Link Masked Men in East Ukraine to Russia', *New York Times*, 20 April 2014.

14 Conclusion

Security, strategy and military change in the 21st century

Jo Inge Bekkevold, Ian Bowers and Michael Raska

The diverse, cross-regional perspectives in this volume reaffirm a traditional adage – military change is a complex process shaped by a number of internal and external factors, enablers and constraints. Thus finding a single theoretical perspective is problematic in capturing the dynamics of military change – namely the interaction of a multitude of sources, components, dimensions and risks. We draw upon three contending schools of thought which explain the variances in military change: (1) neorealist (structural realist) perspectives; (2) organizational theory perspectives; and (3) cultural factors. We argue that, individually, each school is incomplete in explaining the trajectory and pathway of military change which each country pursues. By combining these theoretical perspectives however, we can better explain and project key variables that shape the pace, character and magnitude of military change.

The neorealist structural perspective

The neorealist or structural realist perspective is bound to enduring strategic competition in the context of anarchic international relations, and predicts that the external security environment plays the primary role in shaping a state's strategic choices. In particular, it is the insecurity, or search for security, that provides key incentives for states to innovate in relation to changes in their external environment. In this view, states with higher levels of insecurity will have stronger incentives to innovate and change their military affairs. At the same time, states with expanding international interests and ambitions will also have stronger incentives to pursue military innovation in order to defend their interests, shape the international environment, and gain strategic advantage by power projection.

In the process, the competitive nature of international relations will drive states to emulate the military capabilities and successful innovations, organizational forms, practices and technologies of their rivals or superior powers in the system. As Kenneth Waltz notes, 'the possibility that conflict will be conducted by force leads to competition in the arts and instruments of force. Competition produces a tendency toward the sameness of competitors.'[1] Neorealists would therefore explain military change trajectories in this volume in the context of the continuity and change in the distribution of capabilities between states in the international system that shape their power and evolving perceptions of insecurity, and provides strategic rationale to pursue military change.

As the case studies in this volume demonstrate, nations are adjusting their operational posture and policies in line with the balance of power at both the global and regional levels, and perceived future and current threats. In Asia, this is most evident in the gradual strategic and operational changes being made by Japan. Miyaoka illustrates in Chapter 3 how Japan is adjusting its strategies, doctrines and posture to meet the challenge of China. The United States is rebalancing its military, economic and diplomatic resources to the Asia-Pacific region. In developing a new doctrine – the AirSea Battle concept – the US is preparing to counter challenges to its long-established dominance of the maritime theatre in East Asia. Indonesia, the sleeping giant in South East Asia, is also pursuing a path of military modernization, as Jakarta emphasizes the need to keep up with Southeast Asia's force modernization. However, as Schreer argues in Chapter 4, international prestige and not strategic necessity is a key determinant in the nature of the TNI's military change.

Both Gareis (Chapter 8) and Dyson (Chapter 10) writing on European defence conclude that the US 'pivot' to Asia is a new reality that Europe has to grapple with, and that this is a development that will continue despite Russia's more assertive foreign policy. Zysk's analysis in Chapter 9 describes how Russia's own modernization programme is based on perceptions of great power politics and long-standing fears of NATO encirclement and European marginalization. Grand strategy, strategy and doctrine are all geared towards ensuring a Russian sphere of influence in Europe.

Conversely, European nations have demonstrated a degree of strategic ennui in the absence of a definable threat and strategic goal, and financial austerity adds further pressure on already tight European defence budgets. According to Gareis, pooling and sharing and closer European defence cooperation is a cost-effective solution safeguarding European security and relevance. However, as Gareis, Dyson, and Hilde (Chapter 12) all elaborate upon, European military integration has proven to be a rather slow process. One obvious reason has of course been the absence of an immediate threat since the end of the Cold War. In addition, as Dyson points out, a security dilemma between European states to some extent still remains, essentially restricting close European pooling as nations maintain their own individual capabilities.

Organizational theory perspectives

Organizational theories focus on internal political and military dynamics to explain why and when military change occurs and, more importantly, its success or failure. Thus civil–military relations, intra-service competition, economics and inter-service competition all play a role in military change.[2] A state's political system, governance, legal and regulatory frameworks, civil–military relations, industrial and technological resources, converging interests and conditions for the pursuit of innovation can trigger, accelerate or hinder the military change process.[3]

The civil–military perspective emphasizes the level of societal cohesion and the place of the military within society as a key determinant of military innovation and its effectiveness. As Rosen argues, 'an innovative military that extracts resources but is isolated from society may not be able to sustain that innovation in periods of prolonged conflict.'[4] In this perspective, military change is less likely to occur in divisive societies, as the state tends to focus on internal problems and to channel fewer resources to the military. As the findings in this volume demonstrate, this factor does not seem to be a significant issue in nations with long-established and cohesive political systems. However, as Schreer points out in Chapter 4, the military, despite accepting the validity of democratic rule in Indonesia, is still finding its place within this new civil–military dynamic. This uncertainty is an important factor in the arduous process of military change in Indonesia.

Linked to the importance of civil–military relations is the intra-service institutional perspective which argues that military organizations are inherently conservative, entrenched and resist innovation, as they are more concerned with the internal distribution of status, gains and power in terms of budget, manpower and domain than with organizational goals. The result is a short-term orientation on problem solving, rather than long-term planning, and military change can be stimulated only through civilian intervention, major operational failures, or persistent resource constraints forcing the organization to innovate.[5] This issue becomes apparent in militaries unused to civilian interference or initiating significant military change. Zysk finds that opening the Russian military to outside civilian influence and breaking the organizational rigidity in the armed forces in the early and mid-2000s was a major step in the ongoing process of military modernization. Likewise, as Blasko (Chapter 2) and Miyaoka (Chapter 3) note, both China and Japan have recently created overarching, largely civilian bodies (National Security Commission and National Security Council) to facilitate military change.

Inter-service contenders argue that military change can be facilitated internally, without external civilian interference. This is because military organizations are driven by professional ethos to provide security for the state, which stimulates competition among branches of military services to pursue innovation.[6] Vivek Chadha emphasizes in his study of the Indian army's counter-insurgency operations (Chapter 7) the importance of bottom-up experiences and operational adaptation to achieving success and organizational change. In this case officers and men responded to the need to be operationally successful through the incremental implementation of military change.

While these perspectives are not in evidence in every case study, they do highlight the importance of the inward view. To bookend the importance of society and the military's place within it, every case study emphasizes how alternative economic priorities constrain defence spending. How and why militaries may change and innovate is due to both external and internal factors, but the dominance of spending power creates a powerful boundary and incentive for militaries. As Miyaoka writes, Japan does not have the luxury to increase defence spending, but such a budgetary constraint actually promotes military change by creating the need to rationalize and streamline defence forces.

Cultural factors

Last but not least, different cultural factors may explain the variance in military change in this volume through differences in strategic cultures – i.e. distinctive, consistent and persistent views on how states and their military organizations think about warfare.[7] Dyson argues that the UK's Atlanticist ideology and Germany's culture of restraint in the use of force help to explain military change in these two countries and how these they are meeting the challenge of US disengagement from Europe. Among the wide range of challenges to developing multinational military structures in Europe, Gareis identifies different leadership cultures, military traditions and identities among nations but notes that the real challenge in creating a European army lies in the development of a 'European' military culture. Miyaoka finds that anti-militarism, a worldview that gradually became embedded in Japanese strategic thinking after the Second World War, is now eroding, contributing to a changing Japanese threat perception. According to Miyaoka, the erosion of anti-militarism is not only driven by China's rise, but is also influenced by Japan's growing willingness and ambition to contribute to international peace and stability missions, and to play a more pro-active role in international politics and global security. Hence, we see a development where changes in the balance of power and threat perceptions are putting pressure on the strategic culture in Japan, Germany and in Europe as a whole, with consequences for military change.

Pathways of military change

Each chapter in this volume provides an insightful analysis of how select militaries adapt to new strategic realities, operational and technological challenges, as well as internal political and socio-economic changes. In providing a cross-regional examination of military change, this book provides some indications as to the direction in which military change is heading.

We are entering a new era in which future conflicts will be increasingly defined by the convergence of traditional conventional threats with asymmetric, low-intensity and non-linear security challenges. A key question is how and to what extent militaries are reacting to this new era. Illustrated by Long in Chapter 13, the US has created capabilities that are global in reach for both symmetric and asymmetric threats. The contributions by Kalyanaraman and Chadha confirm that India, due to its security environment, has also been forced to maintain capabilities to deal with this full spectrum of potential threats. Dyson argues in Chapter 10 for the importance of the European great powers to develop balanced forces that can undertake high-intensity warfare as well as land-based stabilization operations.

The increased cost of modern armed forces and weapons platforms, combined with the next wave of information and technological changes reshaping the use of force, means that only the United States and great powers like China, Russia and India have the resources to independently develop capabilities along both dimensions. Loo acknowledges that even for an economic powerhouse like Singapore, managing the pace of technological change proves to be very costly. He describes how Singapore views collaborative procurement and weapons research programmes as one way to manage this challenge. As Rasmussen highlights in Chapter 11, Denmark is a prime example of a small country providing a value-added military through creating plug-and-play capabilities along one of the dimensions without a high level of defence expenditure. This is possible because Denmark is part of an integrated package of military and security strategies within the framework of NATO, trusting larger powers to support and buttress their own armed forces for national defence and when on deployment. Long (Chapter 13) makes the argument that Europe already plays and can potentially play an even more significant role in the use of discrete and discriminate force, intelligence and special operations.

The increased complexity of military operations today requires interoperability, joint operations, and command and control. Again, the US is at the forefront. While undertaking a decreased role in Europe, the US will continue to backstop the European powers. Essentially, interoperability is also the trend in Europe, using technology derived from US concepts. The US is also looking to gain advantage from such a willingness to cooperate, with surveillance, intelligence and special operations key pathways for closer military integration. As explained by Loo (Chapter 5), Singapore is an interesting case study with regard to this trend, traditionally emphasizing its unique geo-strategic vulnerabilities posing a potential risk to the survival of the state. The city state is attempting to maintain an independent and operationally potent defence capability while increasing operational and technological interdependence with the US. Here, technology, strategy and geo-politics are combining to drive Singapore's military change towards the US. A bloc is clearly forming where militaries, primarily in Europe but increasingly in Asia, are moving towards operations more closely aligned with the US.

China has just released a new joint operations doctrine, and improving joint operation command is a top priority in Beijing. However, as pointed out by Blasko, the PLA has not conducted a true joint campaign since 1955 and operational learning is a key component for the success of such endeavours. Zysk highlights that the needs for a unified command and control and joint operations were among the main conclusions Russia drew from the 1999 to 2000 war in Chechnya and the 2008 war in Georgia. How Russia will create the necessary structures and technologies is unknown. However, in this respect, the US, its allies and partners are increasingly integrated and collaborative, and seemingly far in advance of states managing such issues independently.

The strategic environment and the future

The authors in this volume agree that military change today, in a time of rapid changes and continuously high technological innovation, is characterized by evolution, not revolution. Extreme cases, like Pakistan going nuclear, changing the balance of power and threat overnight in a revolutionary way, and enforcing military change on both sides of the India–Pakistani border, are exceptions to the rule.

Despite the fact that the United States has undertaken major military changes over the past decade, Long argues that this process has been more evolutionary than revolutionary. Focusing on RMA, he gives three explanations for why it has receded into the background in the US. First, network centric warfare associated with RMA is now fully integrated into existing force structures without having fundamentally changed or challenged existing organizations. Second, COIN forced military change in a different way, but again without fundamentally changing the structure and organization of the armed forces. Third, even though the emerging concern about cyberwarfare dominates the debate today, there is no clear opinion on whether cyber truly is a new type of warfare or not.

Globally, the maritime theatre of East Asia is emerging as the new focal point of international security. However, neither China's A2/AD strategy nor the US AirSea Battle doctrine represents revolutionary change, as they constitute modern versions of well-known concepts from the Cold War. Indeed, according to Hilde, the lesson learned from his analysis of the CTJF concept is that organizations such as NATO should avoid politically grand reforms and pursue change in an evolutionary manner to ensure the greatest chance of success. The importance of gradual change is also highlighted in Gareis' chapter, where he argues that it allows states and societies to become familiar with common armed forces. In an environment of increasing cooperation and collaboration this is an important consideration.

The case studies in this volume also confirm that a new security landscape is emerging, and great power politics is back on the agenda in more absolute terms than since the end of the Cold War. However, the challenges which both China and Russia are posing to the established international order have led some observers to state that Cold War II is emerging and that, in Asia at least, an arms race is taking shape. We do not share these views.

China's rise and military modernization is changing the traditional balance of power in Asia while Europe is in the midst of a significant Russian challenge to the established order. However, as pointed out by Blasko in Chapter 2, China's military modernization and defence budgets are growing in coordination with its economic development. Despite Japan's increased concern about its security environment, the country still spends only 1 per cent of its GDP on defence. New doctrines and technologies may lead to greater instability, but this restraint on spending does not suggest two states gearing up for long-term conflict.

Likewise, the insight which Zysk provided us with in her chapter on Russia shows a military force in dire straits, in particular if we are in for a longer period of decreasing oil prices and continued global sanctions. Thus, Russia is ultimately limited in the existential threat it can pose to Europe.

Long characterizes the US as the 'wary' titan, arguing that US retrenchment, if it happens, will be as a result of choice rather than requirement. In the three largest economies in the world – the US, China and Japan – defence budgets could be increased but for political and domestic economic reasons they are not. The same is the case in Europe. Choice, made on a political level, defines the nature and extent of military change. If a significant or existential threat emerges, major powers have the capacity to ramp up spending, but at the expense of welfare, education and infrastructure, and, in the end, economic growth itself.

Notes

1 Kenneth Waltz, *Theory of International Politics* (New York: McGraw-Hill, 1979), 127.
2 Barry Posen, *The Sources of Military Doctrine: France, Britain, and Germany Between the World Wars* (Ithaca, NY: Cornell University Press 1984); Stephen Peter Rosen, *Winning the Next War: Innovation and the Modern Military* (Ithaca, NY: Cornell University Press, 1991); Morton Halperin, *Bureaucratic Politics and Foreign Policy* (Washington, DC: Brookings Institution, 1974); Jeffrey Isaacson, Christopher Layne and John Arquilla, *Predicting Military Innovation* (Santa Monica, CA: RAND, 1999).
3 Emily Goldman and Thomas Mahnken (eds), *The Information Revolution in Military Affairs in Asia* (New York: Palgrave Macmillan, 2004).
4 Rosen, *Winning the Next War*.
5 Posen, *The Sources of Military Doctrine*.
6 Vincent Davis, *The Politics of Innovation: Patterns in Navy Cases* (Monograph Series in World Affairs, University of Denver, 1967); Bradd Hayes and Douglas Smith (eds), *The Politics of Naval Innovation* (Newport, RI: US Naval War College, 1994);

Owen Cote, *The Politics of Innovative Military Doctrine: The US Navy and Fleet Ballistic Missiles* (PhD Dissertation, MIT, 1996); Rosen, *Winning the Next War.*
7 Dima Adamsky, *The Culture of Military Innovation: The Impact of Cultural Factors on the Revolution in Military Affairs in Russia, the US, and Israel* (Palo Alto, CA: Stanford University Press, 2010), 1; Williamson Murray, 'Does Military Culture Matter?', *Orbis*, vol. 45, no. 1 (1999), 27–57.

Index

A2/AD 253–4, 256, 267, 271
Abe, Shinzo 36, 52n9
ADIZ *see* China, People's Republic of: ADIZ
Afghanistan 81, 115, 128, 140, 150, 158, 178, 197, 204, 207, 209, 211, 212, 235, 242, 247, 249–50, 259
AirSea Battle 253–4, 256, 267, 271
alliance security dilemma 181–2, 186, 191
Andoman and Nicobar Islands 91
Arctic 161, 165, 199, 201, 216–17
arms dynamic 74
arms race 2, 272
austerity 1, 3, 4, 9, 10, 140, 179, 183, 184, 189, 197, 204, 205, 206, 208, 209, 211, 214, 216, 241, 246, 249, 252, 258, 259, 260, 261, 267

Balkans, the 137, 138, 150, 178, 180, 197, 202, 225, 229, 231, 259
Bay of Bengal 91, 94, 103
Briggs Plan 123, 126
Bundeswehr *see* German armed forces

civil–military relations 22–3, 58
China, People's Republic of: ADIZ 19, 44–5, 256; civil–military integration 22–3; Coast Guard 20; defence industry 21–2; defence spending 20–1; maritime ambitions and environment 16; National Security Council 18; Politburo Standing Committee 18; threat perception 15, 16
Chinese armed forces: active defence 17, 29; air force 27–8; amphibious capability 28; army 28–9; cyber warfare 29; exercises 23–4; historic missions 30; international operations 24; joint doctrine 29–30; navy 27; political control 17–19; public relations 26; recruitment 25–6; Second Artillery 28; training 30; two inadequacies 30; view of the US armed forces 16
COIN *see* counter-insurgency
Cold Start Doctrine 104–5
Cold War, the 37, 40, 63, 71, 74, 184, 186, 187, 197, 199, 202, 207, 214, 215, 225, 234, 242, 243, 244, 248, 252, 254, 255, 257, 261, 267, 271
colour revolutions 162, 163, 173
Combined Joint Task Force (NATO): France, role of 227, 234; origin 223–4; structure 230–3; Turkey, role of 229, 234; US, role of 225–6, 228
Combined Security and Defence Policy 135–6, 180–1, 182
Communications Security Establishment 249
conscription 28, 76, 77, 83, 165, 179, 203, 206, 242
counter-insurgency 4, 115–16, 180–1, 182, 271
counter-terrorism 82
Crimea *see* Ukraine crisis
cyber warfare 29, 159, 246

Danish armed forces: air force 199–200; army 200; future force models 212–13; International Brigade 202–3; lack of coherence 211; Lego model 212; mission statement 200; navy 199; quantity v. quality 207; subcontractor model 208, 211, 214; teeth to tail ratio 210–11

Denmark: Afghanistan 204, 210–12; Arctic 216–17; benchmarking 206; Cold War Business Model 201; defence spending 198, 202–4, 214; future challenges 216; Intervention Business Model 202; Post-Cold War Business Model 202
doctrine 5, 6, 7, 8, 178, 267, 272

Europe: defence spending 3, 141–2, 185; globalization, impact of 137, 139; security relevance 180; security structures 143; US defence posture, impact of 140, 180, 181, 191, 228; weaknesses 139, 179
European Defence Agency 181, 189
European Security and Defence Identity 225, 226, 232
European Security and Defence Policy 138
European Security Strategy 139

F-35 Joint Strike Fighter 61, 73, 77, 79
financial crisis (global) 1, 56, 136, 140, 148, 158, 204, 208, 243
Five Eyes 248
France: return to NATO 187–8

Government Communication Headquarters 248–9
German armed forces 179
Germany: defence industry 188–9; defence policy 186; defence spending 179; differences with the UK 182; energy policy 183; Libya *see* Libya crisis; multilateral deployments 187; reliance on US 181
Ghent Framework 180
grand strategy 7, 8, 11n28, 40, 89, 90, 214, 267
Government Communications Security Bureau 249

Hu Jintao 30

India 1965 War 96; 1971 War 93–4, 99–101; Armed Forces Special Powers Act 125, 127; China, relations with 92–3, 98; defence spending 90, 91; Kargil Conflict 102–3; strategic interest 91; strategic rivalry 89; threats to 2
Indian armed forces 2004 Army Doctrine 127; Army 2000 101; counter-insurgency 118; defence goals 90–1; doctrinal adaptation 99, 101–2, 104–5, 116–17, 126; Doctrine for Sub Conventional Operations 119–20; nuclear weapons 102; organizational adaptation 120–1; Strategic Forces Command 92; strategy of annihilation 93–4, 99; strategy of exhaustion 102; two-front war 91; XVII Corps 91
Indonesia 2010 Strategic Defence Plan 56; China, relations with 61–2; civil–military relations 58, 65–6; defence industry 67; defence spending 59–60, 66; EEZ 59; Malaysia, relations with 61; maritime awareness 57, 59, 63, 67, 68; Singapore, relations with 61, 71; strategic importance 55; threat perception 60–1, 62; Total People's Defence 57
Indonesian Armed Forces (TNI): air force 64; army 64; Flash-point Defence 62; force improvement 57–8, 63–5; KODAM 58, 64; military change 55; Minimum Essential Force 62–3, 64; navy 63; peacekeeping operations 65; tasks 59
Iraq 3, 4, 49, 82, 115, 138, 139, 140, 158, 178, 179, 197, 204, 207, 209, 215, 241, 246, 247, 249, 250, 251, 254, 261
ISIS 4

Jammu & Kashmir *see* Kashmir
Japan: Ballistic Missile Defence 37; Basic Policy for National Defence 36, 39–40; budget cap 50; China, relations with 16, 17, 44–5, 47; collective self-defence 38, 52n8, 52n9; National Defence Guidelines 36–7; National Security Council 36; National Security Strategy 36, 40; pacifism 48; Peace Constitution 38; politics and self-defence 47; strategic outlook 47; US Alliance 39, 42, 50
Japanese Armed Force (Self Defence Forces): air force 43; army 43, 46–7; Basic Defence Force Concept 37, 41, 52–3n22; Dynamic Defence Force 45, 48; Dynamic Joint Defence Force 45–6; international operations 49; navy 43, 46–7; power projection 255; role, expansion of 42–3; submarines 46–7, 53n31; use of technology 47–8

276 Index

Jiang Zemin 28
joint operations 9, 16–17, 24, 29–30, 41, 76, 91, 144, 147–8, 164, 201, 205, 270, 271

Kashmir 2, 89, 94, 95, 96, 101, 102, 103, 104, 115, 116, 120, 122, 124, 125
Kargil Conflict *see* India: Kargil Conflict

Libya crisis 3, 24, 138, 140, 158, 178, 182–3, 186, 245, 251, 258
Line of Control 107, 120, 126, 131n28

McKinsey & Company 206, 209, 213
Malacca Straits 55, 62, 92, 256
maritime domain 1, 2, 16, 44, 55, 62, 267, 271
Melian Dialogue 71
military best practice 190
military change 1, 5, 6, 7, 8, 9, 10, 15, 31, 36, 37, 40, 44, 47, 50, 51, 55, 67, 75, 80, 83, 89, 105, 128, 129, 135, 136, 155, 157, 158, 163, 164, 165, 171, 172, 173, 191, 221, 222, 234, 235, 236, 241, 246, 249, 251, 252, 253, 257, 266, 267, 268, 269, 270, 271, 272
military revolution 6, 7, 246
military technical revolution 6
Mizo National Front 122
Moore's Law 75
Multinational military structures 147

National Security Agency 247, 248, 249
NATO 2014 Cardiff Summit 221, 236; Afghanistan 235; Combined Joint Task Force *see* Combined Joint Task Force; New Strategic Concept 142; Partnership for Peace 223; Readiness Action Plan 221; Smart Defence Initiative 180; special operations forces 260; Special Operations Headquarters 258–9; Very High Readiness Joint Task Force 221, 236
NATO Response Force 203, 207, 215, 221, 233, 234, 236
Nehru, Jawaharlal 117–18
Neoclassical Realism 190–1
network centric warfare 70, 76, 156, 168, 246, 247, 249, 250, 251, 252, 254, 271

Operation Grand Slam 96

Pakistan 2, 89, 94, 102, 104, 105, 112n113, 115, 129n3, 251, 257
procurement 2, 58, 61, 139, 180–1, 216, 270

Rashtriya Rifles 121–2, 124
rebalance to Asia *see* US: pivot to Asia
Revolution in Military Affairs 6, 7, 70, 74–5, 76, 77, 81, 82, 83, 128, 178, 190, 192n1
Russia: Arctic, in the 161; China, relations with 161; civil–military relations 157, 170; defence industry 156, 166, 168; defence spending 158; economic weakness 4, 158, 170–1; spheres of privileged interest 159–60; strategic perception 159–62, 171; threat perception 160, 162; Ukraine *see* Ukraine crisis; use of force 160
Russian armed forces: air force modernisation 167; corruption 156, 160, 169, 172; exercises 165; Georgia War 156–7; naval modernisation 167; new view of warfare 163; nuclear capabilities 168; personnel 165–8, 170; problems 156; structural reform 164; tactical nuclear use 161

security landscape 1–4, 7
Senkaku Islands 20
Serdyukov, Anatoli 157, 166
Shoigu, Sergey 164, 166, 168, 169, 173
signals intelligence 247–8
Singapore: China, relations with 73; defence budget 77–8; defence industry 79–80; India, relations with 73; Malaysia, relations with 71; National Security Coordinating Committee 82; perception of vulnerability 72; strategic situation 71–2, 79; UK, withdrawal of 71–2; US, relations with 73, 75
Singapore Armed Forces 3G SAF 70; conscription 76–7; operations other than war 80–3; sustainability 83–4
Singh, J.J. 119–20
Smart Defence Intiative 180
strategic culture 56, 65, 68, 71–2, 139, 146, 147, 186, 189–90, 269
Strategic Defence and Security Review *see* United Kingdom: SDSR
strategy 6, 7, 8, 50, 55, 74, 76, 173, 214, 267, 270

submarines 46–7, 53n31, 63, 168
Suharto 57, 67
Sundarji Doctrine 101
Suter 251

Tea Party 244

UAV 64
Ukraine crisis 1, 136, 162, 163, 172, 215
UKUSA Agreement 248–9
United Kingdom: defence industry 189; euro scepticism 184–5, 188; internal politics 185–6; Libya, role in *see* Libya crisis reduction in forces 179; SDSR 178–9; US, reliance on 181, 184

United States: alliances, roles of 256–7; China, relations with 252–3; defence spending 242–5; Europe, relations with 258–61; India, relations with 257; internal politics 244, 245; Joint Special Operations Command 250; pivot to Asia 136, 180, 252; primacy 242

War on Terror 202, 250, 253
Western European Union 229, 237n16
Widodo, Joko 68
winning hearts and minds 119, 130n23

Xi jinping 17, 18

Yoshida Doctrine 39–40, 50

eBooks
from Taylor & Francis

Helping you to choose the right eBooks for your Library

Add to your library's digital collection today with Taylor & Francis eBooks. We have over 50,000 eBooks in the Humanities, Social Sciences, Behavioural Sciences, Built Environment and Law, from leading imprints, including Routledge, Focal Press and Psychology Press.

Choose from a range of subject packages or create your own!

Benefits for you
- Free MARC records
- COUNTER-compliant usage statistics
- Flexible purchase and pricing options
- All titles DRM-free.

Benefits for your user
- Off-site, anytime access via Athens or referring URL
- Print or copy pages or chapters
- Full content search
- Bookmark, highlight and annotate text
- Access to thousands of pages of quality research at the click of a button.

REQUEST YOUR FREE INSTITUTIONAL TRIAL TODAY

Free Trials Available
We offer free trials to qualifying academic, corporate and government customers.

eCollections

Choose from over 30 subject eCollections, including:

Archaeology	Language Learning
Architecture	Law
Asian Studies	Literature
Business & Management	Media & Communication
Classical Studies	Middle East Studies
Construction	Music
Creative & Media Arts	Philosophy
Criminology & Criminal Justice	Planning
Economics	Politics
Education	Psychology & Mental Health
Energy	Religion
Engineering	Security
English Language & Linguistics	Social Work
Environment & Sustainability	Sociology
Geography	Sport
Health Studies	Theatre & Performance
History	Tourism, Hospitality & Events

For more information, pricing enquiries or to order a free trial, please contact your local sales team:
www.tandfebooks.com/page/sales

www.tandfebooks.com